CAMBRIDGE MONOGRAPHS ON APPLIED AND COMPUTATIONAL MATHEMATICS

40 **Discrete Variational Problems with Interfaces**

The *Cambridge Monographs on Applied and Computational Mathematics* series reflects the crucial role of mathematical and computational techniques in contemporary science. The series publishes expositions on all aspects of applicable and numerical mathematics, with an emphasis on new developments in this fast-moving area of research.

State-of-the-art methods and algorithms as well as modern mathematical descriptions of physical and mechanical ideas are presented in a manner suited to graduate research students and professionals alike. Sound pedagogical presentation is a prerequisite. It is intended that books in the series will serve to inform a new generation of researchers.

A complete list of books in the series can be found at
www.cambridge.org/mathematics.
Recent titles include the following:

24. Learning theory, *Felipe Cucker & Ding Xuan Zhou*
25. Algebraic geometry and statistical learning theory, *Sumio Watanabe*
26. A practical guide to the invariant calculus, *Elizabeth Louise Mansfield*
27. Difference equations by differential equation methods, *Peter E. Hydon*
28. Multiscale methods for Fredholm integral equations, *Zhongying Chen, Charles A. Micchelli & Yuesheng Xu*
29. Partial differential equation methods for image inpainting, *Carola-Bibiane Schönlieb*
30. Volterra integral equations, *Hermann Brunner*
31. Symmetry, phase modulation and nonlinear waves, *Thomas J. Bridges*
32. Multivariate approximation, *V. Temlyakov*
33. Mathematical modelling of the human cardiovascular system, *Alfio Quarteroni, Luca Dede', Andrea Manzoni & Christian Vergara*
34. Numerical bifurcation analysis of maps, *Yuri A. Kuznetsov & Hil G.E. Meijer*
35. Operator-adapted wavelets, fast solvers, and numerical homogenization, *Houman Owhadi & Clint Scovel*
36. Spaces of measures and their applications to structured population models, *Christian Düll, Piotr Gwiazda, Anna Marciniak-Czochra & Jakub Skrzeczkowski*
37. Quasi-interpolation, *Martin Buhmann & Janin Jäger*
38. The Christoffel–Darboux kernel for data analysis, *Jean Bernard Lasserre, Edouard Pauwels & Mihai Putinar*
39. Geometry of the phase retrieval problem, *Alexander H. Barnett, Charles L. Epstein, Leslie Greengard & Jeremy Magland*

Discrete Variational Problems with Interfaces

ROBERTO ALICANDRO
University of Naples

ANDREA BRAIDES
SISSA (International School for Advanced Studies)

MARCO CICALESE
Technical University of Munich

MARGHERITA SOLCI
University of Sassari

CAMBRIDGE
UNIVERSITY PRESS

CAMBRIDGE
UNIVERSITY PRESS

Shaftesbury Road, Cambridge CB2 8EA, United Kingdom

One Liberty Plaza, 20th Floor, New York, NY 10006, USA

477 Williamstown Road, Port Melbourne, VIC 3207, Australia

314–321, 3rd Floor, Plot 3, Splendor Forum, Jasola District Centre, New Delhi – 110025, India

103 Penang Road, #05–06/07, Visioncrest Commercial, Singapore 238467

Cambridge University Press is part of Cambridge University Press & Assessment, a department of the University of Cambridge.

We share the University's mission to contribute to society through the pursuit of education, learning and research at the highest international levels of excellence.

www.cambridge.org
Information on this title: www.cambridge.org/9781009298780

DOI: 10.1017/9781009298766

First published 2024

A catalogue record for this publication is available from the British Library

A Cataloging-in-Publication data record for this book is available from the Library of Congress

ISBN 978-1-009-29878-0 Hardback

Contents

Preface

This is a book for mathematical analysts and applied mathematicians who are interested in understanding how simple discrete interactions may give rise to a variety of macroscopic phenomena as the number of sites under consideration increases. Such types of problems have incredibly diversified origins and implications. This monograph only treats a relatively simple case, when the systems can be parameterized on portions of lattices, are governed by a variational principle, and the relevant energy scaling is such that the overall behavior of the system is driven by some form of interfacial energy, even though the method and results presented go far beyond this setting.

The core of this book is structured following a course given at SISSA in Trieste in 2021 focused on the passage from discrete to continuum for pairwise positive spin interactions on a lattice, comprising Chapters 2, 3, and 8 of the book, the initial section of Chapter 4, and some results on the homogenization on random lattices (Section 5.2). Those parts of the book just by themselves can be used as an independent advanced introductory graduate course. The rest of the book is subdivided into two main themes. The first one is concerned with the treatment of general lattices and interactions and focuses on a novel compactness and integral-representation result (Chapter 4), which allows one to generalize the discrete-to-continuum approach to multiparameter, many-body, long-range interactions with a great freedom in the geometry of the underlying set of nodes. In particular, that set can be a stochastic lattice, for which we have almost certain results (Chapter 5). The second theme is the development and adaptation of the general approach to treat a number of prototypical advanced research topics, each of which is open to a large number and variety of further developments. The approach is flexible enough to allow variations on the form of interfacial energies and additions of nonlocal parts (Chapter 6), and, partly, to allow the treatment of negative spin interaction where frustration plays an important role (Chapter 7).

During the long and eventful history of the writing of this book, a number of scholars and friends have interacted with the authors in different degrees of involvement. We are particularly grateful to Andrey Piatnitski, who contributed a large part of the results and has been a continuous source of inspiration and joyful collaboration, and to Andrea Causin, who followed with intellectual curiosity and creative contributions the development of the book, providing all illustrations and producing literally hundreds of figures. Andrea Braides is indebted to Gianni Dal Maso for many inspiring discussions we had a long time ago, which started his interest in discrete systems. The content of this book has slowly evolved through a number of graduate courses and summer schools, in which little by little the interest in interfaces emerged, within a general approach to discrete-to-continuum problems, from a course at the TIFR, Bangalore, in 2004 and a VIGRE summer school in Salt Lake City (2005), to the summer school on calculus of variations and applications in Ponta Delgada (2006), the winter school calculus of variations in physics and materials science in Würzburg (2012), the "Summer School on Recent Advances in the Theory of Homogenization" in Chicago (2012), to a course at the Thematic Program on Nonlinear Flows at the Erwin Schrödinger Institute in Vienna (2016), to the "Summer School: Multiscale Phenomena" at TU Munich (2019), to the "Winterschool on Analysis and Applied Mathematics" in Münster (2021), to the "Winter School: Analytical Methods in Quantum and Continuum Mechanics" at the Politecnico di Torino (2021). We are grateful to the organizers of these events, which have also allowed continuous improvement, through interaction and discussion, of what has become the material of this book. Among the people who have interacted with the authors on the content of this book or related issues, we want to acknowledge in particular Luigi Ambrosio, John Ball, Xavier Blanc, Antonin Chambolle, Andrey Cherkaev, Valeria Chiadò Piat, Sergio Conti, Irene Fonseca, Gilles Francfort, Gero Friesecke, Maria Stella Gelli, Antoine Gloria, Dick James, Roman Kotecky, Leonard Kreutz, Claude Le Bris, Frederic Legoll, Stefan Luckhaus, Errico Presutti, Mathias Ruf, Sylvia Serfaty, Valery Shmishlyaev, Laura Sigalotti, Takis Souganidis, Florian Theil, Antonio Tribuzio, and Lev Truskinovsky.

Symbols

General

$o(1)_{t\to t_0}$	little-o as $t \to t_0$; that is, any infinitesimal function as $t \to t_0$
$\lfloor t \rfloor$	integer part of $t \in \mathbb{R}$
$\sum\limits_{P(i)} f(i)$	summation of the values $f(i)$ over the set of all i satisfying property $P(i)$ (the function f in general may depend also on other parameters than i)

Sets in $\mathbb{R}^d$

$e_1, \ldots, e_d$	the elements of the canonical orthonormal basis in $\mathbb{R}^d$
$\overline{A}$	closure of the set A
$A \triangle B$	symmetric difference of A and B, defined by $(A \setminus B) \cup (B \setminus A)$
$Q_R^\nu(x_0)$	an open d-dimensional cube with center x_0, a face orthogonal to ν and side length R
$Q_R(x)$	the open coordinate cube centered at x and with side length R
$B_R(x)$	the open ball centered at x with radius R
Π^ν	the hyperplane orthogonal to ν
H^ν	the half space of vectors with nonnegative component in direction ν
$\langle x, y \rangle$	the scalar product between x and y in $\mathbb{R}^d$
$\|x\|$	the Euclidean norm of x
$\|x\|_\infty$	the maximum norm of x, defined by $\max_k |x_k|$
$\|x\|_1$	the 1-norm of x, given by $\sum_k |x_k|$
$\mathrm{dist}(x, A)$	the Euclidean distance of x from A
$\mathrm{dist}_\infty(x, A)$	the distance of x from A in the maximum norm
S^{d-1}	the unit sphere in $\mathbb{R}^d$; that is, the set of points with $\|x\| = 1$
ω_d	the volume of the unit ball in $\mathbb{R}^d$

$[a, b)$ the rectangle defined by the vectors a and b in $\mathbb{R}^d$ (in the notation of stochastic processes)
$\mathrm{Per}(A)$ perimeter of the set A
$\mathrm{Per}(A; \Omega)$ perimeter of the set A in Ω
$\partial^* A$ reduced boundary of the set A

Measures

$|A|$ Lebesgue measure of the set A
$\mu_\varepsilon \overset{*}{\rightharpoonup} \mu$ weak*-convergence of measures
$\frac{d\mu}{d\lambda}$ Radon–Nykodim derivative of the measure μ with respect to the measure λ
$\mu \llcorner A$ the restriction of the measure μ to A; that is, $\mu \llcorner A(B) = \mu(A \cap B)$
δ_x Dirac measure concentrated at x
$\mathcal{H}^{d-1}$ d–one-dimensional Hausdorff measure

Functions

χ_A the characteristic function of the set A
$L^1(\Omega)$ Lebesgue space of integrable functions
$L^\infty(\Omega)$ Lebesgue space of essentially bounded functions
$\|u\|_\infty$ norm of u in $L^\infty(\Omega)$
$BV(\Omega; Y)$ the space of functions of bounded variation in Ω with values in Y or the space of Caccioppoli partitions indexed by Y
$S(u)$ the jump set of u
ν_u, ν measure theoretical normal to $S(u)$ or $\partial^* A$
u^+, u^- traces of u on both sides of $S(u)$

Lattices and Spin Functions

$\mathcal{L}$ a lattice or a multilattice, also intended in a broad sense
$\mathcal{L}_\varepsilon(\Omega)$ or $\mathcal{L}_\varepsilon$, the subset $\Omega \cap \varepsilon\mathcal{L}$ of the elements of $\varepsilon\mathcal{L}$ in Ω
$\mathcal{S}_\eta(U)$ the space of the spin functions defined on $\eta\mathbb{Z}^d \cap U$
$\langle i, j\rangle$ the set of nearest neighbors, in the notation of summation
$\langle\langle i, j\rangle\rangle$ the set of nearest and next-to-nearest neighbors, in the notation of summation
$A_\varepsilon(u)$ the union of ε-cubes associated to the spin function u
$\mathcal{Z}$ the dual lattice of $\mathbb{Z}^2$
$\mathbb{T}$ triangular lattice in $\mathbb{R}^2$

1
Introduction

The object of the analysis presented in this book is discrete systems with a large number of nodes, whose overall behavior is driven by what can be considered a surface energy. Many of these systems have some origin or analogy in models in science and technology. It is not our intention to attempt here an impossible list and description of the many problems leading to such models. We only describe a few paradigmatic examples in order to highlight the width of the field of applications. A first clarifying example is variational models in computer vision such as the one by Blake and Zisserman (1987). There, the unknown, representing the output picture, is a real-valued function defined on pixels, whose ensemble can be regarded as a portion of a lattice. Pairs of pixels on which the difference of the values of this function exceeds some threshold are regarded as separated by an interface. This model has a counterpart in the continuum variational approach to image processing by Mumford and Shah (1989). Among the many other examples, we may single out another key model coming from atomic physics with similar features. This regards atomistic pair potentials such as the Lennard–Jones potential. Even though the overall behavior of systems energetically driven by such potentials is extremely complex, such types of energies can be analyzed close to absolute minima (physically, at zero temperature), showing a phenomenon of *crystallization*; that is, ground states tend to arrange on a regular lattice. Even this expected property of minimizers is a very subtle issue and can be formalized as the stability under finite perturbations of the arrangement of atoms in a lattice. It has been proved to hold only in dimension two and for a limited class of interatomic potentials (Theil, 2006). If crystallization holds, then the energies can again be parameterized on an underlying lattice. A possible continuum approximation gives rise to theories of brittle fracture (Braides et al., 2006; Friedrich and Schmidt, 2014), where the main unknown is the crack site, whose atomistic interpretation can be again given in terms of neighboring nodes with large relative displacements.

In variational theories for data science, data are sometimes regarded as randomly distributed objects and are often labeled by parameters on which interfacial energies are used in order to separate families of similar data (García Trillos and Slepčev, 2016), which is a typical problem of machine learning. Such energies are then thought as defined on some type of random lattices. Also spin systems are a classical issue in statistical mechanics, and often the energies driving their behavior are defined on a regular lattice, with a random dependence that can be interpreted as a property of the interactions. Many of the energies just mentioned can be studied within a multiscale perspective, deriving a number of large-scale behaviors depending on the energy level of the system (Blanc et al., 2005; E and Ming, 2007; Le Bris and Lions, 2005).

Early analyses of variational problems involving nonconvex functionals defined on lattices were mainly carried on from the perspective of numerical analysis in order to implement finite-difference or finite-element schemes. As examples we refer to the discretization of the Ambrosio and Tortorelli (1990) functional in computer vision (Bellettini and Coscia, 1994), or the analysis of the Blake–Zisserman model (Chambolle, 1995). A systematic analysis of classes of discrete energies was performed with different techniques and scopes almost at the same time in a number of papers such as the ones by Blanc et al. (2002), Braides and Gelli (2002), E and Ming (2007), and Friesecke and Theil (2002). The first purely variational analysis of surface energies alone was later carried out by Caffarelli and de la Llave (2005), and shortly after by Alicandro et al. (2006), followed by a number of applications and results. In some cases, as in computer vision theories or brittle fracture, these results involve at the same time an interface energy and some bulk energy; nevertheless, in such analyses the surface part can be studied separately and, conversely, results involving bulk and surface energies can be specialized and refined for purely superficial energies. Moreover, note that in order to describe interfacial energies it is not restrictive to assume that discrete systems are defined only on functions taking a finite number of values; for fracture, for example, this amounts to considering interfaces at a scale where the displacement is approximately constant on the two sides of the crack, which can be justified by a blow-up argument.

In view of these considerations, the object of our analysis is energies whose domain is functions defined on a lattice $\mathcal{L}$ in $\mathbb{R}^d$, or a portion of that lattice, and taking values in a finite set Y. The abundance of techniques and results obtained directly for such types of energies, or for problems where these energies are part of the description, has stimulated the need of a systematization both for a unitary formal structure and in order to highlight in a clear way completely novel directions both applied and theoretical, such as links with discrete dynamical systems and graph theory. The book is intended to be a proposal for the formalization of a common language both within the rich and various subject of variational methods, and toward quite different lines of research, which are essentially discrete. The choice of focusing on discrete interfacial energies is due to the ease of expressing in their terms problems that are intrinsically discrete and not only a discretization of a continuous analog, and at the same time to the generality of the methods and results, which can be exported to problems with other scalings and physical interpretations.

The simplest lattice functions are *spin functions*, where the set of values Y has cardinality two, and is usually taken to be $\{-1, 1\}$. Note that here and after we borrow some terminology from the physical literature – "spin" is one of such terms – but we highlight that we claim no direct application to physical theories, the terminology serving just as a suggestion for the writer and hopefully to the reader. The prototypical energies defined on spin functions depend on *pair interactions*; that is, the overall energy is the sum of the energy between pairs of nodes. Analytically, if we denote by u_i the value of a function u at a node i in $\mathcal{L}$, the energy is a sum of terms depending only on u_i and u_j. A class of energy densities is those minimized when $u_i = u_j$, which are called *ferromagnetic*, here using a terminology borrowed from statistical mechanics. In this case, we may suppose that the energy between two points is proportional to $(u_i - u_j)^2$. A typical spin energy is of the form

$$E(u) = \sum_{i,j} a_{ij}(u_i - u_j)^2. \tag{1.1}$$

Note that in statistical mechanics energies usually take the form $-\sum_{ij} a_{ij}u_iu_j$, which is equivalent to the one just shown since the two expressions only differ by constants independent of u and only depending on the set of nodes that are taken into account. Other analytical expressions can be equivalently used, such as $\sum_{ij} a_{ij}|u_i - u_j|$.

The overall properties of such an energy can be studied using a *discrete-to-continuum* approach. We introduce a small parameter ε, and consider a portion of the scaled lattice $\varepsilon\mathcal{L}$ contained in a fixed bounded Lipschitz open subset Ω of $\mathbb{R}^d$. In this way we allow the number of nodes under analysis to diverge as $\varepsilon \to 0$. Correspondingly, we consider energies

$$E_\varepsilon(u) = \sum_{i,j} \varepsilon^{d-1} a^\varepsilon_{ij}(u_i - u_j)^2, \tag{1.2}$$

where now u is a spin function defined on $\Omega \cap \varepsilon\mathcal{L}$ and $u_i = u(\varepsilon i)$. Accordingly, the sum is taken for i, j such that $\varepsilon i, \varepsilon j \in \Omega \cap \varepsilon\mathcal{L}$. Such functions u can be identified with their piecewise-constant interpolations on the Voronoi cells of the lattice, so that the domain of E_ε can be seen as a subset of $L^\infty(\Omega)$, and the behavior of minimum problems related to E_ε can be stated in terms of a continuum approximation defined in a Lebesgue space. Discrete-to-continuum limits in this spirit have been analyzed in many contexts both in terms of pointwise expansions (e.g. Blanc et al., 2005), numerical approximations (e.g. E and Ming, 2007), and variational limits (e.g. Braides and Gelli, 2002). The scaling ε^{d-1} (*surface scaling*) highlights that we expect the relevant limit contribution as $\varepsilon \to 0$ to be described by a surface energy. This is in accord with the constraint that $u \in \{-1, 1\}$, which allows us to identify u with the set $\{u = 1\}$. Note that in (1.2) we include a dependence of the coefficients a_{ij}^ε on the parameter ε in order to allow for the maximal freedom on the modeling assumptions of our energies.

Under a positiveness assumption on a_{ij}^ε, energies $E_\varepsilon(u^\varepsilon)$ along a sequence of lattice spin functions u^ε can be interpreted as interfacial energies taking into account the interactions through the boundary of the sets $\{u^\varepsilon = 1\}$, after introducing a continuum interpolation of the discrete spin functions. The requirement that such sets converge to a set A defines the *discrete-to-continuum convergence of lattice functions* u^ε to A. The problem of the computation of the Γ-*limit* of energies E_ε can then be set within the framework of such interfacial energies (perimeter energies), which have the form

$$F(A) = \int_{\partial^* A} \varphi(x, \nu)\, d\mathcal{H}^{d-1}, \tag{1.3}$$

where A is a set of finite perimeter representing the continuum counterpart of $\{u = 1\}$; $\partial^* A$ denotes its reduced boundary, whose normal at $\mathcal{H}^{d-1}$-almost every point is denoted by ν; and $\mathcal{H}^{d-1}$ denotes the $d-1$-dimensional (surface) Hausdorff measure. In this way a limit set of finite perimeter and a continuum energy are obtained from a family of discrete energies defined on scaled copies of a lattice, as the scaling parameter tends to 0. The limit energies capture the behavior of the discrete ones in the sense of the convergence of minimum problems: the solutions of minimum problems at a discrete level can be seen as discretizations of an effective continuum problem. Typically, such minimum problems are *minimal-cut problems* for discrete interactions, which are approximated by minimal-perimeter problems on the continuum. In particular, we can consider minimum problems in the whole space, in which case the minimizers are the so-called *Wulff shapes*.

Wulff shapes are connected to classical problems of Statistical Mechanics regarding the collective behavior of microscopic spin systems when the number of configurations diverges, and to the analysis of some crystallization problems involving the asymptotic behavior as N diverges of ensembles of N points in $\mathbb{R}^d$ whose location is such that some energy is minimized involving the distances between points. The minimal configurations tend to arrange as a portion of a lattice whose overall shape is a Wulff shape for some surface energy. Even though closely related to our analysis, we do not examine in detail the connections with such interesting problems.

Back to the description of our discrete-to-continuum approach, once an analytic framework has been established, the focus is on conditions that allow one to prove existence of a Γ-limit and characterize the resulting perimeter functional. A first class of energies that can be analyzed elementarily is *homogeneous ferromagnetic energies* defined on Bravais lattices, of the form

$$E_\varepsilon(u) = \sum_{i,j} \varepsilon^{d-1}\alpha_{i-j}(u_i - u_j)^2; \tag{1.4}$$

that is, when the interaction coefficients are independent of ε and are homogeneous, in the sense that $a^\varepsilon_{ij} = \alpha_{i-j}$. We may assume that the indices are taken in $\mathbb{Z}^d$ since all Bravais lattices can be identified with that lattice, up to a change of variables. We may regroup the interactions as

$$E_\varepsilon(u) = \sum_k \sum_i \varepsilon^{d-1}\alpha_k(u_{i+k} - u_i)^2 \tag{1.5}$$

and study separately the terms $\sum_i \varepsilon^{d-1}\alpha_k(u_{i+k} - u_i)^2$ at fixed k. Usually, the Γ-limit of a sum is not the sum of a Γ-limit, but in this case a *superposition principle* holds due to the fact that a recovery sequence for a planar interface is simply its discretization, independently of k. The outcome is that in this case φ is x-independent and is given by

$$\varphi(\nu) = 4 \sum_{k\in\mathbb{Z}^d} \alpha_k|\langle \nu, k\rangle|, \tag{1.6}$$

the factor 4 coming from the fact that $(u_i - u_j)^2 \in \{0,4\}$. This superposition property can be interpreted as an interfacial version of a *Cauchy–Born rule*, which states that macroscopic energies correspond to a regular arrangement of discrete values. This property is often crucial for computational and modeling reasons and is often analyzed in problems in continuum mechanics (Friesecke and Theil, 2002; E and Ming, 2007; Schmidt, 2008).

In order to obtain the convergence just mentioned, two conditions are necessary:

(i) (*coerciveness on nearest neighbors*) $\alpha_k > 0$ if $\|k\| = 1$;
(ii) (*decay of the coefficients*) $\sum_k \alpha_k \|k\| < +\infty$.

Condition (i) allows one to estimate the perimeter of the interpolated sets $\{u = 1\}$ by the energy $E_\varepsilon(u)$ and thus guarantees compactness of families of functions with equibounded energies. Condition (ii) guarantees that indeed the limit energy is finite only on sets of finite perimeter.

The next level in complexity is periodic systems on $\mathbb{Z}^d$; that is, when the coefficients a_{ij} in (1.2) are still ε-independent, but are periodic of some integer period K; that is,

$$a_{ij} = a_{kl} \text{ if } i = k \text{ and } j = l \text{ modulo } K, \tag{1.7}$$

the case $K = 1$ reducing to that of homogeneous coefficients. This is a first case of *homogenization*, in which the Γ-limit exists and the limit energy density $\varphi_{\rm hom}$ is homogeneous, that is, x-independent. The main issue here is the characterization of $\varphi_{\rm hom}$, which can be achieved in various ways. We present a characterization through an *asymptotic homogenization formula*, which turns out to be flexible to treat systems with other geometrical assumptions. If the range of interactions is R, then this formula is

$$\varphi_{\rm hom}(\nu) = \lim_{T\to+\infty} \frac{1}{T^{d-1}} \min\Big\{ \sum_{i,j\in Q_T^\nu} a_{ij}(u_i - u_j)^2 \colon\ u\colon \mathbb{Z}^d \cap Q_T^\nu \to \{-1,+1\},$$
$$u_i = 1 \text{ if and only if } \langle i,\nu\rangle \geq 0 \text{ in } Q_T^\nu \setminus Q_{T-2R}^\nu \Big\},$$

where Q_t^ν denotes a cube of side length t centered in 0 and one face orthogonal to ν. Note that, while analog formulas are valid for the homogenization of continuum energies, here the nonlocal character of discrete energies must be taken into account in the definition of boundary values, which are imposed in a "cubic annulus" close to the boundary. This is a technical point that is often present when defining boundary values for discrete systems, and is slightly more complex if the range of the interaction is infinite. We present two different techniques to obtain such a formula. The first one is based on the Fonseca and Müller (1992) *blow-up method* adapted to lattice problems, and the second one follows De Giorgi's *localization method*. Both methods are relatively self-contained, up to general measure-theoretical arguments, and do not need further functional notions beside the ones related to perimeter functionals. Other arguments that can be used in this context are Caffarelli and de la Llave's (2005) *plane-like minimizer* arguments, or extensions to convex one-homogeneous functionals as by Ambrosio and Braides (1990b) in the continuum and by Chambolle and Kreutz (2023) for lattice energies, the latter requiring bulk-scale homogenization techniques.

The blow-up technique is useful to provide lower bounds along a discrete-to-continuum converging sequence of functions and is based on the idea of interpreting functionals as measures. The key point is then to describe the relevant density of a limit measure concentrated on the perimeter of the limit set in terms of φ_{hom}. Note that, for discrete problems such as the one just presented, the measures to study are

$$\mu_\varepsilon = \sum_i \Big(\sum_j \varepsilon^{d-1} a_{ij}(u_i - u_j)^2 \Big) \delta_{\varepsilon i}. \tag{1.8}$$

The coefficient of the Dirac delta at εi describes the interaction of such points with the remaining points of the lattice. It is therefore a nonlocal quantity, even though its nonlocality is vanishing with ε. This method allows one to clearly separate the estimate of a lower bound and the construction of recovery sequences for the upper bound, which are directly obtained from the homogenization formula and the density of polyhedral sets. An essential technical point in both computations is the use of *discrete coarea arguments* that are used to match boundary data with asymptotically negligible energetic expense cost.

It is worth noting the flexibility of the blow-up method, which is not limited to Bravais-lattice energies or periodic coefficients. In particular, coupling it with a *projection method*, we may use it to derive homogenization theorems for *aperiodic lattices* obtained by projection from higher-dimensional Bravais lattices on incommensurate lower-dimensional linear spaces, such as *quasicrystals* or *Penrose lattices*. This is a genuinely discrete setting with only a partial counterpart on the continuum. Such lattices are not periodic but they retain some *quasiperiodicity* properties: for a relatively dense set of translations, lattices superpose up to well-separated isolated points whose presence does not invalidate the proof of the asymptotic homogenization formula. A simpler setting, corresponding to the case when the projection is on a commensurate lower-dimensional linear space gives rise to a theory of homogenized surface energies on *thin objects*. Even though this has a counterpart in the continuum, the nonlocal nature of discrete energies gives rise to new phenomenon such as a nonadditive dependence of the thin-film thickness, and is closer to a rigorous treatment of atom deposition theories.

Another issue particularly suited to a lattice formulation is that of a *random dependence* on the interactions for systems of independently distributed coefficients, which again can be studied using the blow-up technique. The existence, and deterministic nature, of the homogenization formula in this case is an almost sure property of the system and is closely connected to *percolation theory* (Kesten, 1982; Grimmett, 1999). For systems with coercive and bounded interactions this can be interpreted as a *first-passage percolation* formula (Boivin, 1990). We also can study some *percolation-threshold variational phenomena* depending on some probability parameter. One such case is obtained by considering a system whose interaction coefficients take a positive finite value with probability $p \in [0, 1]$ and the value $+\infty$ with probability $1 - p$. In the latter case the coefficients define what we call *rigid bonds*; for such a bond, having finite energy amounts to requiring that the corresponding pair u_i and u_j have the same value. The corresponding homogenization formula is linked to asymptotic metric properties on the cluster of points with no rigid bonds. In the two-dimensional setting this cluster is almost surely infinite if and only if $p > 1/2$, which is the case when $\varphi_{\rm hom}$ is finite, and it only depends on p. Its form is related to the so-called *chemical distance* of the system (Garet and Marchand, 2007). Another case in which a variational percolation phenomena occurs is that of the so-called *dilute systems*, whose coefficients mix a positive value with probability p and the value 0 (*weak bonds*) with probability $1-p$. In this case, in two dimensions the cluster of points with weak bonds is infinite for $p > 1/2$, which is the case when $\varphi_{\rm hom}$ is identically 0, while otherwise $\varphi_{\rm hom}$ is given by a first-passage percolation formula only depending on p, which holds also in this case (Wouts, 2009; Cerf and Théret, 2011). In these types of results, percolation techniques are used to describe the geometry of the infinite clusters of coercive interactions, when they exist, proving the existence of "lattice-like" subsets, which are sufficient in order to carry on the discrete-to-continuum process.

De Giorgi's localization method allows one to obtain general compactness and integral-representation results under minimal conditions. In the context of ferromagnetic energies E_ε of the form (1.2) it allows one to prove that, upon extraction of subsequences, the discrete-to-continuum limit of every such family exists and is a possibly inhomogeneous perimeter energy of the form (1.3). The localization method consists in introducing a set variable U and considering localized functionals

$$E_\varepsilon(u,U) = \sum_{\varepsilon i,\varepsilon j\in U\cap\varepsilon\mathcal{L}} \varepsilon^{d-1} a^\varepsilon_{ij}(u_i-u_j)^2, \tag{1.9}$$

and studying their behavior both as functionals of the function u and of the set U. By proving abstract properties on the dependence on U we obtain the Γ-convergence to a measure, which then can be represented as an integral. The extension of this method to lattice energies, which are nonlocal by definition, requires some care, since, for example, the functionals in (1.9) are not subadditive in the set variable, contrary to their continuum counterpart. In order for this procedure to work, besides coerciveness assumptions we have to require a uniform decay condition, without which easy examples show that the Γ-limit may fail to be represented by a surface energy.

The generality and flexibility of the localization method allow one to abandon Bravais lattices, spin functions, and pair interactions and guarantee the extension to very general environments and energies. The domain of the energies can be any infinite discrete set with a minimal distance between its points and no balls with a large diameter in its complement; we call such a set $\mathcal{L}$ an *admissible lattice*. The functions we consider are defined on portions of scaled copies of $\mathcal{L}$ with values in a discrete set Y. If $Y = \{-1, 1\}$, we recover spin systems, but we can also consider, for example, *ternary systems* by choosing $Y = \{-1, 0, 1\}$ or $Y = \{e_1, e_2, e_3\} \subset \mathbb{R}^3$. The energies can take into account *many-body interactions* or also interactions between all possible sites. To that end we may rewrite energies as

$$E_\varepsilon(u) = \sum_{\varepsilon i \in \Omega \cap \varepsilon \mathcal{L}} \varepsilon^{d-1} \phi_i^\varepsilon(\{u_{i+j}\}_j), \tag{1.10}$$

where the function ϕ_i^ε takes into account interactions involving the site i; for example, in the case of ferromagnetic interactions,

$$\phi_i^\varepsilon(\{z_k\}_k) = \sum_k a_{i\,i+k}^\varepsilon (z_k - z_0)^2. \tag{1.11}$$

If the functions ϕ_i^ε satisfy suitable coerciveness and growth conditions, which reduce to the abovementioned conditions on a_{ij}^ε for ferromagnetic interactions, then the Γ-limit exists up to subsequences and can be represented as a functional defined on *partitions of sets of finite perimeter* parameterized by a subset Y_0 of Y, and represented as a sum of integrals on the boundary of the elements of the partition. Note that the notions of nearest-neighbor and of discrete-to-continuum convergence must be suitably modified, which can be done using *Voronoi tessellations*. The compactness result can be applied, for example, to ensure that mixtures of two (or more) types of bonds can be represented on the continuum by a perimeter functional, the property of whose integrand can be then described by computing suitable *energy bounds*. This is a fundamental step in the field of *Optimal Design* of networks.

The range of applications of the compactness theorem makes it necessary to allow for the greatest generality, as the geometry of the lattice and the parameters involved are concerned. This is the case both when dealing with pair interactions in random environments that can be described by admissible lattices and when the assumption of positiveness of ferromagnetic interactions is removed, allowing for a multiplicity of ground states, whose overall behavior can be described by partitions. As for random sets, a notion of *stochastic lattices* can be given, which are almost surely admissible lattices and on which we can define random energies

$$E_\varepsilon^\omega(u) = \sum_{\varepsilon i, \varepsilon j \in U \cap \varepsilon \mathcal{L}^\omega} \varepsilon^{d-1} a_{ij}^\omega (u_i - u_j)^2, \tag{1.12}$$

with coefficients depending on the distance between sites of the lattices. In (1.12) we highlight the dependence on ω, the realization of a suitable random variable. Under conditions of *stationarity* and *ergodicity* the Γ-limit is almost surely deterministic and described by an asymptotic homogenization formula that is the stochastic version of the asymptotic homogenization formula. A key ingredient in the proof of the validity of such a formula is a *subadditivity theorem for discrete stochastic processes*. We note that the hypothesis of admissibility of the lattice can sometimes be removed; for example, for *Poisson random sets*, for which compactness properties are achieved by using Percolation techniques (more precisely, a lemma on *polyominos* covering of Voronoi cells). Note that for the simplest nearest-neighbor energies on scaled Poisson random sets the isotropy of the Poisson process guarantees the isotropy of the limit energy; that is, almost surely we have Γ-convergence to a multiple of the Euclidean perimeter, in contrast with the crystallinity of short-range homogenization in Bravais lattices.

We note that the general compactness theorem sometimes must be integrated with other techniques in order to better describe the limit behavior of the system. One example is that of systems with many parameters Y of which only a subset Y_0 participate in the limit description. In this case the effect of the variables $Y \setminus Y_0$ is minimized out in the computation of the interfacial energy $\varphi_{\rm hom}$ given by the compactness theorem. If we want to better keep track of those parameters, we introduce their *measure of concentration* at the interface. This can be done, for example, for ternary systems giving rise to *surfactants*. Such systems can be parameterized on $\{-1,0,1\}$, and their ground states are only the constant states -1 and 1. The effect of the 0-phase can be described by adding to $\varphi_{\rm hom}$ a dependence on the density of a measure μ describing the limit amount of that phase on the interface

$$\int_{\partial^* A} \varphi_{\rm hom}\Big(\frac{d\mu}{d\mathcal{H}^{d-1} \llcorner \partial^* A}, \nu\Big) d\mathcal{H}^{d-1}. \tag{1.13}$$

This energy density can be explicitly computed, for example, for the *Blume–Emery–Griffiths model* for surfactants. Similarly, a correction to the description by simple interfaces, but more for reasons of a geometric origin, is needed for systems with *high-contrast energies*. Models with such types of energies in the continuum case are used in applications, for example, to the study of the flow in a naturally fractured reservoir (Arbogast et al., 1990). In this case, coerciveness conditions are satisfied on one or more admissible sublattices (*perforated domains*) to each of which the compactness theorem can be applied. The remaining connections may give rise to an additional *interaction term* of bulk type describing the separate effect of the lower-coerciveness interactions between the higher-coerciveness sublattices (*double-porosity energies*). Another type of correction can be necessary if the uniform-decay assumption on the interactions is relaxed, in which case an additional term of *nonlocal type* may appear besides the surface energy of the form

$$\int_{\partial^* A} \varphi_{\hom}(\nu) d\mathcal{H}^{d-1} + \int_{A\times(\Omega\setminus A)} K(x,y)\, d\mu(x,y), \qquad (1.14)$$

such as a *nonlocal perimeter* or an *Ohta–Kawasaky type* functional. Another case, in which the methods of the compactness theorem can be applied after an initial modeling analysis, is that of systems describing (*chiral*) *molecules*. In the simplest case these systems can be described by homogenous ferromagnetic interactions where the sites of the relevant parameter (say, where $u_i = 1$) must be arranged as unions of sets of given shape (molecules). The analysis of the possible minimal configurations of such ensembles gives then a way to obtain a set of parameters that play the role of the set Y_0 in the compactness theorem, and obtain a description in terms of partitions into sets of finite perimeter.

The complexity of the compactness theorem is also justified by its application to *frustrated systems*; that is, to spin systems of pair interactions mixing positive and negative coefficients where there is no ground state that minimizes separately all pair interactions. As a consequence, such systems cannot by reduced to a ferromagnetic system and often possess many periodic ground states. If this is the case, the ground states themselves parameterize the set Y_0 of the compactness theorem, up to a *coarse-graining process* at the level of a period, considering Y as all possible arrays of values in such a period. This allows one to describe the behavior of the system again as an interfacial energy on partitions of sets of finite perimeter, which describe different microscopic *patterns*, or *modulated phases* within the same pattern. Note that the number of the limit parameters can be arbitrarily high even though the system we start with only takes the two values -1 and 1 into account. Note also that not all systems can be asymptotically described in this way: some present some degeneracies due to the possibility of having interfaces with zero energy between variants of ground states, and some others present a *total frustration* with an infinite family of periodic and nonperiodic ground states.

In the terminology of *Graph Theory*, the graphs with vertices and edges corresponding to the systems we have just described are essentially *sparse*; that is, the number of (relevant) edges is much lower than the total number of possible connections. More precisely, the assumptions of the compactness theorem are stated in terms of a decay condition, which implies that the relevant connections are of *equibounded range*. This assumption can be relaxed by a *coarse-graining* approach if the connections give a locally strongly dense graph;

that is, a graph where all connections are considered at a scale much larger than that of the lattice dimensions but still infinitesimal as $\varepsilon \to 0$. If this assumption is relaxed, we may have sparse systems with *diffuse interfaces*; that is, whose behavior is determined by the presence of many interfaces that give an overall bulk energy in the limit. These interfaces also are at the basis of the behavior of *dense graphs*; that is, for which the number of edges is of the same order as the total number of possible connections. In this case the geometry of the set of vertices is irrelevant, and we may parameterize the graph as a discrete subset of $[0,1]$. The theory of *graphons* (Lovász, 2012; Janson, 2013) in Combinatorics allows us then to study the Γ-limit with respect to the weak* L^∞-convergence of the interpolations, which is of the form

$$F(u) = \int_{[0,1]^2} W(x,y)(u(x) - u(y))^2 \, dx \, dy, \tag{1.15}$$

where u now takes values in $[-1,1]$ and W is a symmetric positive function, the limit graphon of the dense graphs. Even though the resulting energy is of bulk type, the parameter u is interpreted as a limit density of sets and W describes the overall effect of the diffuse interfaces of such sets.

For general discrete systems, as remarked at the beginning of this chapter, the surface-energy description must be placed in a proper multiscale framework, together with effects related to other types of scaling. Note that, even when only energetic contributions are taken into account in a static picture described by a Γ-limit process, the same type of functionals can be considered with different scaling depending on the energy level. For the same quadratic energies we may have, for example,

(i) (*bulk scaling*) $\sum_{ij} \varepsilon^d a^\varepsilon_{ij} |u_i - u_j|^2$ giving integral energies $\int f(x, u(x)) \, dx$;
(ii) (*surface scaling*) $\sum_{ij} \varepsilon^{d-1} a^\varepsilon_{ij} |u_i - u_j|^2$ giving surface energies as described in the preceding presentation;
(iii) (*vortex scaling*) $\sum_{ij} \varepsilon^{d-2} |\log \varepsilon|^{-1} a^\varepsilon_{ij} |u_i - u_j|^2$ giving *vortex energies* defined on point singularities;
(iv) (*gradient scaling*) $\sum_{ij} \varepsilon^{d-2} a^\varepsilon_{ij} |u_i - u_j|^2$ giving integral energies depending on gradients $\int f(x, \nabla u(x)) \, dx$, and so on.

Such effects, and others, may be present at the same time. For some of them, methods corresponding to those described for surface energies have been developed and used. In general, the different scalings can be analyzed in a multiscale setting, in which the single scaling is a part of a whole (see e.g. Braides and Truskinovsky, 2008).

The unitary description of a subject with such a complexity and number of methods, results, and applications necessarily requires one to make some choice between the possible approaches. In this book we have chosen one that to us seems open enough to enclose the most standpoints and directions of research.

Bibliographical Notes to the Introduction

The Introduction focuses on problems for lattice systems with an emphasis on the surface energies discussed in this book from the standpoint of the direct approach to the Calculus of Variations. For an account of atomistic-to-continuum methods for Computational Materials Science we refer to the review articles by Blanc et al. (2007) and Le Bris and Lions (2005). Computational problems for which details of interfacial interactions are important to obtain a coupling between continuum discretization procedures and atomistic fine-mesh analysis are quasicontinuum models (see e.g. Tadmor et al., 1996; Blanc et al., 2005; Ortner and Süli, 2008). Multiscale Γ-convergence issues in the passage discrete-to-continuum are also dealt with in Section 11 of the handbook by Braides (2006) and are related to the concepts of Γ-development (see Anzellotti et al., 1994) or Γ-expansion (see Braides and Truskinovsky, 2008).

The discrete-to-continuum description of surface energies is also connected to a problem of crystallization, where now this is interpreted as the analysis of the asymptotic arrangement as N diverges of ensembles of N points in $\mathbb{R}^d$ whose location is such that some energy is minimized involving the distances between points. The points tend to arrange in a configuration close to a portion of a lattice, whose asymptotic shape is driven by the boundary interactions, which is then connected to a perimeter energy on that lattice. Again, the analysis of such asymptotic behavior has been carried out only in various simplified settings (see e.g. Heitmann and Radin, 1980; Radin, 1981; E and Li, 2009; Theil, 2011; Blanc and Lewin, 2015; De Luca and Friesecke, 2017).

Another classical problem in Statistical Mechanics connected to the emergence of macroscopic Wulff shapes is the analysis of the collective behavior of microscopic spin systems when the number of configurations diverges. In that context, a different point of view is usually taken and instead of minimizers, whose role is mainly relevant when, in the terminology the temperature is "close to zero," the object of the analysis is "typical configurations"; that is, sets of configurations with high probability according to a properly defined probability measure (see Dembo and Zeitouni, 1998 and the references therein and Cerf, 2006).

Interfacial problems are connected with problems defined on curves and related to metric properties of graphs. This is evident in dimension 2 where interfaces are one-dimensional objects, and the analogy can be further pushed to higher dimension (see e.g. Braides and Piatnitski, 2013). This analysis can be applied to problems on graphs from the standpoint of Hamilton–Jacobi equations (see e.g. Achdou et al., 2013; Imbert et al., 2013; Lions and Souganidis, 2020; Ishii and Kumagai, 2021), traffic flow (Garavello and Piccoli, 2006), or Aubry–Mather Theory (Siconolfi and Sorrentino, 2021).

Lattice systems can be seen as a particular case of nonlocal energies, and in particular as discretizations of double-integral energies such as in peridynamics (Macek and Silling, 2007; Silling and Lehoucq, 2010). In that perspective the discrete-to-continuum process is connected to what is called the limit of peridynamics “when the horizon goes to zero” (Bellido et al., 2015). We also mention the connection with repulsive-attractive interaction energies (e.g. Carrillo et al., 2014).

Reference texts for theory of graphons are Lovász (2012) and Janson (2013). For complex graphs of a fractal form, a different standpoint could be to consider a multiscale approach, for which we refer to Heida et al. (2020).

Dynamical problems on lattices can be framed in a variational setting using modern techniques of gradient-flow type as in the book of Ambrosio et al. (2008). Their analysis is related to issues in the study of motion in heterogenous media, which is a very wide and largely unexplored territory. For some results on evolutions of interfaces in planar lattices we refer to the recent book by Braides and Solci (2021).

2

Preliminaries

This chapter is devoted to the introduction of a prototypical example, that of nearest-neighbor ferromagnetic energies defined on spin functions, and to the necessary tools for its asymptotic description. More general parameters than spin function will be dealt with in Chapter 4 when we consider functions with values in an arbitrary finite set.

The relevant scalings for ferromagnetic interactions are introduced in Section 2.1, leading to the study of families of energies depending on a small parameter ε that can be interpreted as functionals on sets of finite perimeter. The necessary related Γ-convergence notions are recalled in Section 2.2, while in Section 2.3 we summarize the relevant properties of sets of finite perimeter. In Section 2.4 we illustrate the discrete-to-continuum approach and finally apply it to conclude the asymptotic description of the prototypical example in Section 2.5.

2.1 Prototypical Nearest-Neighbor Ferromagnetic Energies

Our model problem is that of finding a subdivision of a set of nodes in $\mathbb{Z}^d$ into two subsets of given cardinality so as to minimize the number of connections between them. In the simplest case, we can consider as connections the set of pairs $(i, j) \in \mathbb{Z}^d \times \mathbb{Z}^d$ at unit distance (*nearest neighbors*) not belonging to the same subset.

This problem can be described as an energy minimization of the functional depending on subsets A of a given sets of indices $\mathcal{L} \subset \mathbb{Z}^d$ defined by

$$E(A) = \#\{(i,j) \in \mathcal{L} \times \mathcal{L} \colon \|i - j\| = 1, \quad i \in A, j \notin A\}. \tag{2.1}$$

Equivalently, we may rewrite this problem as defined on *spin functions*; that is, functions $u \colon \mathcal{L} \to \{-1, 1\}$. In this case, with an abuse of notation, the energy can be written as

$$E(u) = \frac{1}{8} \sum_{\|i-j\|=1} (u_i - u_j)^2. \tag{2.2}$$

Energies defined on spin functions are often referred to as *Ising systems*.

The value $E(u)$ in (2.2) corresponds to the value $E(A)$ in (2.1), where $A = \{i \in \mathcal{L} : u(i) = 1\}$ and $u_i = u(i)$; that is, $u_i = 1$ if $i \in A$ and $u_i = -1$ if $i \notin A$. The factor 8 is due to the fact that $(u_i - u_j)^2 \in \{0, 4\}$ and to the double counting of pairs of indices. Note that the choice of the square is arbitrary since only the values $u_i \in \{-1, 1\}$ are considered. Equivalently we could use

$$E(u) = \frac{1}{4} \sum_{\|i-j\|=1} |u_i - u_j|.$$

The way of writing E is due to traditional choices and to the fact that developing squares is easier. Note that $(u_i - u_j)^2 = 2(1 - u_i u_j)$ so that if $\mathcal{L}$ is finite, up to an additive constant the energy can be rewritten as

$$E(u) = -\frac{1}{4} \sum_{\|i-j\|=1} u_i u_j,$$

which is the usual way of writing Ising systems.

The choice of $\{-1, 1\}$ as parameters for u comes from statistical mechanics and is coherent with a more general notation involving vector spin functions. Equivalently (and sometimes more handily) we might consider $u : \mathcal{L} \to \{0, 1\}$, making the connection between functions and (characteristic functions of) sets more evident.

We are interested in describing approximately the *minimal-cut problem* just described as the number of nodes diverges; that is, $\#\mathcal{L} \gg 1$. This will be done by constructing approximate continuum problems by a Γ-convergence approach.

In order to describe the limit of energies as $\#\mathcal{L} \gg 1$, we use a scaling parameter. Let $\Omega \subset \mathbb{R}^d$ be a Lipschitz open set.

(i) *Scaling of the lattice $\mathcal{L}$*. We introduce a space scale $\varepsilon > 0$ (where ε is a "small" parameter) and consider the scaled lattice

$$\mathcal{L}_\varepsilon = \mathcal{L}_\varepsilon(\Omega) = \Omega \cap \varepsilon\mathbb{Z}^d,$$

whose cardinality is of the order $\frac{1}{\varepsilon^d}$.

(ii) *Scaling of the energies*. We set

$$\mathcal{E}_\varepsilon = \{(i, j) : \varepsilon i, \varepsilon j \in \mathcal{L}_\varepsilon, \|i - j\| = 1\}. \tag{2.3}$$

We use the notation $\langle i, j \rangle$ to indicate pairs in $\mathbb{Z}^d$ such that $(i, j) \in \mathcal{E}_\varepsilon$. The energies are then scaled according to a *surface scaling* factor; namely,

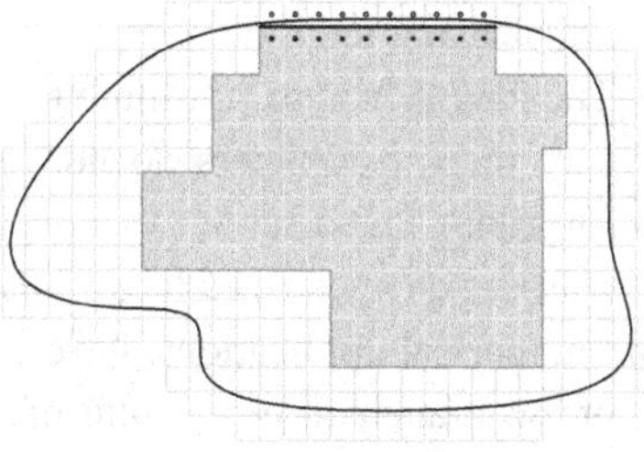

Figure 2.1 Error due to missing interactions at the boundary.

$$E_\varepsilon(u) = \frac{1}{8} \sum_{\langle i,j \rangle} \varepsilon^{d-1} (u_i - u_j)^2, \tag{2.4}$$

where $u\colon \mathcal{L}_\varepsilon \to \{-1, 1\}$ and $u_i = u(\varepsilon i)$. Note (again) the factor $\frac{1}{8}$.

Up to an error close to the boundary of Ω (see Fig. 2.1 for a pictorial representation) the value of $E_\varepsilon(u)$ can be interpreted as the perimeter of the set

$$A_\varepsilon(u) = \bigcup_{\substack{\varepsilon i \in \mathcal{L}_\varepsilon \\ u_i = 1}} \left(\left[-\frac{\varepsilon}{2}, \frac{\varepsilon}{2} \right]^d + \varepsilon i \right), \tag{2.5}$$

which is identified with the spin function u.

The surface scaling guarantees the equicoerciveness of the energies, in the sense that if u^ε is a family of functions such that $E_\varepsilon(u^\varepsilon) \le C$, then for all Ω' regular open subsets of $\mathbb{R}^d$ with $\Omega' \subset\subset \Omega$, for ε small enough, the perimeter of $\Omega' \cap A_\varepsilon(u^\varepsilon)$ is equibounded. This gives precompactness of such sets thanks to the properties of sets of equibounded perimeter, which will be made precise in Section 2.3 and will allow us to define the convergence $u^\varepsilon \to A$ as the convergence of $A_\varepsilon(u^\varepsilon) \to A$ locally in Ω (see Section 2.4). Our asymptotic problem is then translated in the approximation of minimal-cut problems for E_ε by a suitable minimal-perimeter problem for a continuum energy defined on sets of finite perimeter in Ω. This will be done by using the notion of Γ-convergence recalled in Section 2.2 and computing the Γ-limit of E_ε with respect to the convergence $u^\varepsilon \to A$. The limit behavior of the nearest-neighbor energies (2.4) will be described in the last section of the chapter.

2.2 Γ-Convergence

The theory of Γ-convergence has been developed to analyze the behavior of variational problems, such as homogenization and phase changes, where there appear small parameters that make the treatment of the problem complex or numerically expensive. The idea then is to substitute for these problems a new "effective" problem, where these parameters do not appear, or appear in a simplified manner. It is surprising that discrete systems, in which problems are defined, for example, on lattices characterized by a vanishing lattice size, with apparent theoretical and numerical applications, have come rather late to the attention of the experts of this branch of the calculus of variations.

2.2.1 Definition and Properties

The idea of Γ-convergence is to approximate minimum problems

$$\min\{F_\varepsilon(x) \colon x \in X\}$$

by a problem

$$\min\{F(x) \colon x \in X\}$$

independent of the small parameter ε. This is done via a notion of convergence of the energies given in local terms, that is, through the characterization of the behavior of the energies of convergent sequences of points, and is compatible with continuous perturbations. Here and later we will call a *sequence* a family parameterized by either $\varepsilon > 0$, for example, $\{x_\varepsilon\}_\varepsilon$, or by $j \in \mathbb{N}$, for example, $\{x_j\}_j$. If there is no ambiguity, we will simply write $\{x_\varepsilon\}$ and $\{x_j\}$, respectively.

Definition 2.1 (Γ-convergence) Let (X, d) be a metric space and for all $\varepsilon > 0$ let $F_\varepsilon \colon X \to [-\infty, +\infty]$ be a functional defined on X. We say that the sequence $\{F_\varepsilon\}$ *Γ-converges* to $F \colon X \to [-\infty, +\infty]$, or that such F is the *Γ-limit* of $\{F_\varepsilon\}$, as $\varepsilon \to 0$ (with respect to the metric d), and we will write

$$F = \Gamma\text{-}\lim_{\varepsilon \to 0} F_\varepsilon \tag{2.6}$$

if, for all $x \in X$, the two following conditions hold:

(i) (liminf inequality) for all $x_\varepsilon \to x$ we have $F(x) \le \liminf_{\varepsilon \to 0} F_\varepsilon(x_\varepsilon)$;
(ii) (limsup inequality) there exist $x_\varepsilon \to x$ such that $F(x) \ge \limsup_{\varepsilon \to 0} F_\varepsilon(x_\varepsilon)$.

Taking (i) into account, condition (ii) can be rephrased as the existence of a *recovery sequence*; that is, $x_\varepsilon \to x$ such that $F(x) = \lim_{\varepsilon \to 0} F_\varepsilon(x_\varepsilon)$.

Remark 2.2 (Energies on varying domains) In general, we consider $F_\varepsilon \colon X_\varepsilon \to [-\infty, +\infty]$ where X_ε is (identified with) a subset of X. In this case the preceding definition is understood to hold for $x_\varepsilon \in X_\varepsilon$ and, if necessary, we set $F(x) = +\infty$ if no such $x_\varepsilon \to x$ exists.

The first condition in Definition 2.1 requires that F provide an estimate from below of the limit (of the minima) of the energies F_ε, while the second ensures that this limit be reached. The definition is designed so that the following theorem applies.

Theorem 2.3 (Fundamental theorem of Γ-convergence) *Let F be the Γ-limit of the sequence $\{F_\varepsilon\}$. If a compact $K \subset X$ exists such that $\inf_X F_\varepsilon = \inf_K F_\varepsilon$, then there exists $\min_X F$ and we have*

$$\min_X F = \lim_{\varepsilon\to 0} \inf_X F_\varepsilon.$$

Moreover, if $\{x_\varepsilon\}$ is a precompact sequence such that $F_\varepsilon(x_\varepsilon) = \inf_X F_\varepsilon + o(1)_{\varepsilon\to 0}$ with a subsequence converging to $\overline{x} \in X$, then $F(\overline{x}) = \min_X F$.

Proof It suffices to prove the second claim. We can then consider a sequence $\{x_{\varepsilon_j}\}$ such that at the same time we have $\lim_{j\to+\infty} F_{\varepsilon_j}(x_{\varepsilon_j}) = \liminf_{\varepsilon\to 0} \inf_X F_\varepsilon$ and $x_{\varepsilon_j} \to \overline{x}$. We then have

$$F(\overline{x}) \le \liminf_{\varepsilon\to 0} F_\varepsilon(x_\varepsilon) \le \lim_{j\to+\infty} F_{\varepsilon_j}(x_{\varepsilon_j}) = \liminf_{\varepsilon\to 0} \inf_X F_\varepsilon,$$

thanks to (i). From (ii) in contrast for all $x \in X$ there exists a sequence $\{\overline{x}_\varepsilon\}$ such that

$$F(x) \ge \limsup_{\varepsilon\to 0} F_\varepsilon(x_\varepsilon) \ge \limsup_{\varepsilon\to 0} \inf_X F_\varepsilon,$$

which concludes the proof by taking the infimum in x and comparing with the previous inequality. □

2.2.2 Computation of Γ-Limits

In this subsection we make some observations that are useful in the actual computation of a Γ-limit.

Proposition 2.4 (Compactness of Γ-convergence) *If (X,d) is a separable metric space, then from every sequence $\{F_j\}$ with $F_j : X \to [-\infty, +\infty]$ we can extract a Γ-converging subsequence.*

Proposition 2.5 (Stability under subsequences) *A sequence $\{F_\varepsilon\}$ Γ-converges to F if and only if from every subsequence $\{F_{\varepsilon_j}\}$ we can extract a subsequence Γ-converging to F.*

These results are useful when first we extract converging subsequences by compactness, and then we characterize the Γ-limit, showing it does not depend on the subsequence.

Proposition 2.6 (Stability under continuous perturbations) *If $\{F_\varepsilon\}$ Γ-converges to F and G is continuous, then $\{F_\varepsilon + G\}$ Γ-converges to $F + G$.*

The use of this proposition is twofold. On one hand it can be applied to the study of sequences $\{H_\varepsilon\}$ whose elements can be written as $H_\varepsilon = F_\varepsilon + G$ with G continuous and $\{F_\varepsilon\}$ having a known Γ-limit. On the other hand, by writing $F_\varepsilon = (F_\varepsilon - G) + G$, one can sometimes choose G such that the computation of the Γ-limit of $\{F_\varepsilon - G\}$ is more handy or meaningful, and then add G to the result.

2.2.3 Upper and Lower Bounds

The computation of a Γ-limit can be reduced to two separate estimates in the same way as the computation of an ordinary limit can be seen as the equality of an upper and a lower limit.

Definition 2.7 (Γ-lim inf and Γ-lim sup) If $F_\varepsilon : X \to [-\infty, +\infty]$, we define the Γ-*lower limit* and the Γ-*upper limit* of the sequence $\{F_\varepsilon\}$ as

$$\Gamma\text{-}\liminf_{\varepsilon\to 0} F_\varepsilon(x) = \inf\Big\{\liminf_{\varepsilon\to 0} F_\varepsilon(x_\varepsilon) : x_\varepsilon \to x\Big\} \tag{2.7}$$

and

$$\Gamma\text{-}\limsup_{\varepsilon\to 0} F_\varepsilon(x) = \inf\Big\{\limsup_{\varepsilon\to 0} F_\varepsilon(x_\varepsilon) : x_\varepsilon \to x\Big\}, \tag{2.8}$$

respectively.

Note the asymmetry of the definition. Note, moreover, that the quantities (2.7) and (2.8) always exist.

The two inequalities characterizing the Γ-limit can be rewritten as a lower and an upper estimate for the Γ-lim inf and the Γ-lim sup

$$F(x) \le \Gamma\text{-}\liminf_{\varepsilon\to 0} F_\varepsilon(x), \qquad \Gamma\text{-}\limsup_{\varepsilon\to 0} F_\varepsilon(x) \le F(x), \tag{2.9}$$

which can be treated separately. As for the latter, a trivial upper bound is obtained by choosing $x_\varepsilon = x$ where possible (for lattice energies taking x_ε as the discretization of x). Two useful observations for the computation of Γ-limits are contained in the following proposition.

Proposition 2.8 (Lower semicontinuity and Γ-limits) *The following properties hold.*

(i) *The functionals F' and F'' defined by*

$$F'(x) = \Gamma\text{-}\liminf_{\varepsilon\to 0} F_\varepsilon(x), \qquad F''(x) = \Gamma\text{-}\limsup_{\varepsilon\to 0} F_\varepsilon(x) \tag{2.10}$$

are lower semicontinuous on X.

(ii) *If* $\overline{F}_\varepsilon$ *denotes the* lower-semicontinuous envelope *of* F_ε, *that is,*

$$\overline{F}_\varepsilon(x) = \sup\{G(x) : G \leq F_\varepsilon,\ G \text{ lower semicontinuous}\}, \tag{2.11}$$

then we also have

$$F'(x) = \Gamma\text{-}\liminf_{\varepsilon\to 0} \overline{F}_\varepsilon(x), \qquad F''(x) = \Gamma\text{-}\limsup_{\varepsilon\to 0} \overline{F}_\varepsilon(x). \tag{2.12}$$

The first observation allows us to limit our choice of candidate Γ-limits to lower-semicontinuous functionals. The second one allows us to sometimes simplify the form of F_ε.

Remark 2.9 (Lower bound by supremum of measures) The way one often proves a lower bound is by optimizing a family of inequalities; that is, by finding a family of lower-semicontinuous functionals G_λ such that

$$F'(x) = \Gamma\text{-}\liminf_{\varepsilon\to 0} F_\varepsilon(x) \geq G_\lambda(x). \tag{2.13}$$

For discrete-to-continuous processes, in general those will be functionals on the continuum for which lower semicontinuity translates into simplifying convexity conditions. Condition (2.13) is satisfied in particular if the inequality $F_\varepsilon \geq G_\lambda$ holds on the domain of F_ε. From (2.13) we then deduce that

$$\Gamma\text{-}\liminf_{\varepsilon\to 0} F_\varepsilon \geq \sup_\lambda G_\lambda \tag{2.14}$$

(we recall that the supremum of a family of lower semicontinuous functionals is lower semicontinuous).

This inequality can be further sharpened whenever we have a localized version of $\Gamma\text{-}\liminf_{\varepsilon\to 0} F_\varepsilon$ by using the following lemma with $\mu(A)$ the Γ-lim inf localized on A. Note that, for integral functionals, localizing on a set A means simply to integrate only on A, and the corresponding Γ-limits are computed by only requiring convergence on A itself.

Lemma 2.10 (Supremum of measures) *Let* μ *be a function defined on the open bounded subsets of* $\mathbb{R}^d$ *such that* $\mu(A \cup B) \geq \mu(A) + \mu(B)$ *if* $\overline{A} \cap \overline{B} = \emptyset$, *and let* σ *be a Borel measure and* $\{f_\lambda\}$ *a countable family of Borel functions. If* $\mu(A) \geq \int_A f_\lambda d\sigma$, *for all* λ *and for all* A *bounded open set, then* $\mu(A) \geq \int_A \sup_\lambda f_\lambda d\sigma$.

Remark 2.11 (Upper bound by relaxation) The upper bound depends on a construction or *ansatz* on recovery sequences. By the lower-semicontinuity properties of the Γ-limsup in general it is not necessary to construct such sequences for all $x \in X$, but we may proceed as follows.

(i) We choose a subset $D \subset X$, and for all $x \in D$ we construct $\overline{x}_\varepsilon \to x$ so that

$$F''(x) = \limsup_{\varepsilon \to 0} F_\varepsilon(\overline{x}_\varepsilon) \le G(x) \quad \text{for all } x \in D \tag{2.15}$$

for some functional G.

(ii) From this inequality we deduce that

$$F''(x) \le \overline{G}_D(x) \quad \text{for all } x \in X, \tag{2.16}$$

where

$$G_D(x) = \begin{cases} G(x) & \text{if } x \in D, \\ +\infty & \text{otherwise,} \end{cases}$$

and $\overline{G}_D$ is its lower-semicontinuous envelope.

Remark 2.12 (Upper bound by density) In the particular case when in the previous remark $\overline{G}_D = G$ – that is, when D is dense in X, G is lower semicontinuous and for all $x \in X$ there exists $\{x_j\} \subset D$, $x_j \to x$, such that $G(x_j) \to G(x)$ – we deduce that

$$F''(x) \le G(x) \quad \text{for all } x \in X. \tag{2.17}$$

Hence, we obtain an upper bound thanks to a construction on a dense set D and the computation of a lower-semicontinuous envelope. Particular choices of D will be regular or piecewise simple functions or sets, which are dense in a stronger topology for which G is continuous.

2.3 Surface Energies

As we have seen, functions taking two values; that is, spin functions, can be identified with characteristic functions and then with sets. We then introduce the notion of sets of finite perimeter, which is particularly useful when dealing with energies that can be interpreted as surface energies on sets and provide compactness properties.

2.3.1 Sets of Finite Perimeter

A good notion of the perimeter of a set from the standpoint of the calculus of variations is the maximal extension of the usual notion of perimeter for polytopes (polyhedral sets in d dimensions) lower semicontinuous with respect to the convergence in measure.

Let $\Omega \subset \mathbb{R}^d$. If $\{A_n\}$ is a sequence of measurable sets in $\mathbb{R}^d$ and $A \subset \mathbb{R}^d$, we say that A_n converges to A (in Ω) if

$$|(A_n \triangle A) \cap \Omega| \to 0 \quad \text{as } n \to +\infty. \tag{2.18}$$

In this case, we write $A_n \to A$. Note that this is equivalent to requiring that $\chi_{A_n} \to \chi_A$ in $L^1(\Omega)$, so that sometimes we will also say that $A_n \to A$ in $L^1(\Omega)$. Likewise, we define the local convergence of A_n to A in $L^1_{\mathrm{loc}}(\Omega)$; that is, we say A_n locally converges to A (in Ω) if, for any $\Omega' \subset\subset \Omega$,

$$|(A_n \triangle A) \cap \Omega'| \to 0 \quad \text{as } n \to +\infty. \tag{2.19}$$

If A is a polytope in $\mathbb{R}^d$, then the perimeter of A, denoted by $\mathrm{Per}(A)$, is elementarily defined. We can now give the definition of perimeter.

Definition 2.13 (Perimeter) Let $A \subset \mathbb{R}^d$; the *perimeter* of A is defined by

$$\mathrm{Per}(A) = \inf\left\{\liminf_{n\to+\infty} \mathrm{Per}(A_n) : A_n \to A, A_n \text{ polytope}\right\}.$$

The set A is a *set of finite perimeter* if $\mathrm{Per}(A) < +\infty$.

In the sequel, we will use the notation $\langle x, y\rangle$ for the scalar product in $\mathbb{R}^d$. Note that this is the same symbol used for the summation of nearest neighbors; this abuse of notation will not affect comprehension, as will be clear from the context or explicitly specified.

Remark 2.14 (Distributional definition of the perimeter)

(i) An alternative way to define the perimeter of A is in a distributional way as

$$\mathrm{Per}(A) = \sup\left\{\int_A \operatorname{div}\varphi\, dx : \varphi \in C_0^\infty(\mathbb{R}^d;\mathbb{R}^d), \|\varphi\|_\infty \le 1\right\},$$

which states that the characteristic function of A is a function of bounded variation. This characterization allows us to prove functional-analytic properties of the space of sets of finite perimeter.

(ii) If A is a sufficiently regular subset of $\mathbb{R}^d$ whose surface element is denoted by $d\Sigma$, then $\int_{\partial A}\langle\varphi, \nu_A\rangle\, d\Sigma = -\int_A \operatorname{div}\varphi\, dx$ for any $\varphi \in C_0^\infty(\mathbb{R}^d;\mathbb{R}^d)$. Taking φ equal to (an approximation of) $-\nu$ on ∂A, we obtain its surface area; hence, the definition of the perimeter is an extension of the usual definition.

Remark 2.15 (Perimeter in Ω) The notion of perimeter can be localized on open subsets Ω of $\mathbb{R}^d$, setting

$$\mathrm{Per}(A;\Omega) = \sup\left\{\int_A \operatorname{div}\varphi\, dx : \varphi \in C_0^\infty(\Omega;\mathbb{R}^d), \|\varphi\|_\infty \le 1\right\}.$$

Sets A such that $\mathrm{Per}(A;\Omega) < +\infty$ will be called *sets of finite perimeter in* Ω. If Ω is regular, then sets of finite perimeter can be locally approximated by polytopes in Ω.

The following theorem summarizes the main structure properties of sets of finite perimeter. We denote by $\mathcal{H}^{d-1}$ the $d-1$-*dimensional Hausdorff measure*, which generalizes the notion of surface area.

Theorem 2.16 (Reduced boundary and inner normal) *If* $A \subset \mathbb{R}^d$ *has finite perimeter in* Ω, *then there exists a set* $\partial^* A \subseteq \partial A$, *called the* reduced boundary *of* A, *such that*

(i) *the total variation of the distributional derivative of the characteristic function* χ_A *is given by* $\mathcal{H}^{d-1} \llcorner \partial^* A$; *in particular,*

$$\mathrm{Per}(A) = \mathcal{H}^{d-1}(\partial^* A);$$

(ii) $\partial^* A$ *is rectifiable; that is, there exists* $N \subset \mathbb{R}^d$ *with* $\mathcal{H}^{d-1}(N) = 0$ *such that*

$$\partial^* A \subset \bigcup_{h \in \mathbb{N}} \Gamma_h \cup N$$

where $\{\Gamma_h\}$ *is a sequence of compact subsets of* C^1*-hypersurfaces.*

Moreover (by the implicit-function theorem) there exists ν, *the* (inner) normal *to* $\partial^* A$, *defined* $\mathcal{H}^{d-1}$*-almost everywhere on* $\partial^* A$. *This normal corresponds to the (common) normal to the hypersurfaces* Γ_h.

Remark 2.17 (Blow up at the reduced boundary) From the preceding result we deduce that there exists a function $\nu_A \colon \partial^* A \to S^{d-1}$, the *inner normal* to A (coinciding almost everywhere with ν defined earlier), such that

$$\lim_{\varrho \to 0} \frac{|B^+_\varrho(x, \nu_A(x)) \setminus A|}{\varrho^d} = 0 \quad \text{for all } x \in \partial^* A,$$

where $B^+_\varrho(x,\nu) = \{y \in \mathbb{R}^d \colon \|y - x\| < \varrho,\ \langle y - x, \nu\rangle > 0\}$.

As a consequence, if A is a set of finite perimeter, then for $\mathcal{H}^{d-1}$-almost all $x_0 \in \partial^* A$ we have the convergence of the blown-up sets

$$\frac{1}{\varrho}(A - x_0) \to \{x \in \mathbb{R}^d \colon \langle x, \nu\rangle \geq 0\} \text{ in } L^1_{\mathrm{loc}}(\mathbb{R}^d) \text{ as } \varrho \to 0.$$

The introduction of the sets of finite perimeter is motivated by the following compactness theorem.

Theorem 2.18 (Compactness) *Let* $\{A_n\}$ *be a sequence of sets of finite perimeter such that* $\sup_n \mathrm{Per}(A_n;\Omega) < +\infty$. *Then there exists a subsequence* $\{A_{n_k}\}$ *(locally) converging in* Ω *to a set* A *of finite perimeter.*

Note that if Ω is bounded, then the convergence is in $L^1(\Omega)$.

2.3.2 Perimeter Functionals

Theorem 2.16 allows one to define perimeter functionals as integrals on the reduced boundary. They will be obtained as limits of energies E_ε under general conditions.

Definition 2.19 (Perimeter functionals) Let Ω be an open subset of $\mathbb{R}^d$ and let $X = \{A\colon A \text{ of finite perimeter in } \Omega\}$. A functional $F\colon X \to \mathbb{R}$ is a (homogeneous) *perimeter functional* if it is given by

$$F(A) = \int_{\Omega\cap\partial^* A} \varphi(\nu(x))\, d\mathcal{H}^{d-1}(x), \tag{2.20}$$

where $\nu(x)$ is the inner normal to $\partial^* A$ at x and $\varphi\colon S^{d-1} \to [0, +\infty)$ is a continuous function. Note that the usual perimeter is obtained by taking the constant 1 as φ in (2.20).

Since Γ-limits are lower semicontinuous with respect to the convergence in which they are computed, we may restrict our analysis to lower-semicontinuous functionals, characterized in the following theorem.

Theorem 2.20 (Semicontinuity of perimeter functionals) *The (homogeneous) perimeter functional F defined in* (2.20) *is lower semicontinuous with respect to the $L^1_{\rm loc}$ convergence if and only if φ defines a norm; that is, (the one-homogeneous extension of) φ is convex; that is, the function (again denoted by φ) defined in $\mathbb{R}^d$ by*

$$\varphi(z) = \begin{cases} \|z\|\,\varphi\Big(\dfrac{z}{\|z\|}\Big) & \text{if } z \neq 0,\\ 0 & \text{if } z = 0 \end{cases}$$

is convex. Furthermore, if Ω is a Lipschitz set, then for all A there exists a sequence of polytopes A_n converging to A such that $F(A_n) \to F(A)$.

Remark 2.21 (Representation of perimeter functionals; Wulff shapes) Let $\varphi\colon S^{d-1} \to \mathbb{R}$ be such that $\varphi \geq c > 0$, extended to $\mathbb{R}^d$ by one-homogeneity.

(i) We can characterize the perimeter functional F given by (2.20) by the set

$$B_\varphi = \{z \in \mathbb{R}^d : \varphi(z) \leq 1\},$$

which is a convex set if F is lower semicontinuous. Identifying F with B_φ is a handy way to describe the limit energies pictorially.

(ii) Another equivalent characterization of F is the identification with a *Wulff shape* of φ, that is, a (convex) set W_φ maximizing

$$\max\{|A|: F(A) \le 1\}.$$

Equivalently, we may define Wulff shapes as

$$W_\varphi = \frac{B}{F(B)^{\frac{1}{d-1}}},$$

where B is a solution of $\min\{F(B): |B| \ge 1\}$. Conversely, minimizers of this last problem are given by

$$\frac{W_\varphi}{|W_\varphi|^{\frac{1}{d}}}.$$

The equivalence of the two problems is obtained by using the fact that $\mathcal{H}^{d-1}$ is a positively $d-1$-homogeneous set function.

The sets W_φ can be obtained from B_φ by a duality argument. Note that minimizers W_φ satisfy $F(A) = 1$ by the homogeneity of F. If φ is even, that is, $\varphi(\nu) = \varphi(-\nu)$, then we can choose uniquely W_φ by centering it at 0.

2.4 Discrete-to-Continuous Convergence of Lattice Functions

Keeping the prototypical example in Section 2.1 in mind, our energies will be defined on spaces of functions whose domain is a portion of a lattice. We can take as a model the cubic lattice $\mathbb{Z}^d$, but we can also think of different lattices (for example, a triangular lattice in $\mathbb{R}^2$) and not necessarily of Bravais lattices (for example, the hexagonal lattice in $\mathbb{R}^2$). In this context, we briefly recall the notions introduced in Section 2.1. Our variables are functions

$$u: \varepsilon\mathbb{Z}^d \cap \Omega \to \{-1, 1\}, \tag{2.21}$$

where $\Omega \subset \mathbb{R}^d$ is a reference open set, and ε is the characteristic lattice size. Note that in the sequel we will consider, more generally, functions u with values in a finite set Y; they will be identified as a subset of some $\mathbb{R}^m$ (see Chapter 4). For brevity of notation we set

$$\mathcal{L}_\varepsilon(\Omega) = \varepsilon\mathbb{Z}^d \cap \Omega, \qquad u_i = u(\varepsilon i) \quad \text{for } \varepsilon i \in \mathcal{L}_\varepsilon(\Omega). \tag{2.22}$$

We associate to any function $u: \mathcal{L}_\varepsilon(\Omega) \to \{-1, 1\}$ the corresponding set

$$A_\varepsilon(u) = \bigcup_{\substack{\varepsilon i \in \mathcal{L}_\varepsilon(\Omega)\\ u_i = 1}} \left(\left[-\frac{\varepsilon}{2}, \frac{\varepsilon}{2}\right]^d + \varepsilon i\right). \tag{2.23}$$

Now we can give a definition of convergence for spin functions.

Definition 2.22 (Convergence of spin functions and of discrete sets) For all $\varepsilon > 0$ let u^ε be a spin function defined on $\mathcal{L}_\varepsilon(\Omega)$ and let A be a subset of Ω. Then we define the *convergence* $u^\varepsilon \to A$ by setting

$$u^\varepsilon \to A \text{ if } A_\varepsilon(u^\varepsilon) \to A \text{ in } L^1_{\mathrm{loc}}(\Omega) \tag{2.24}$$

with $A_\varepsilon(u^\varepsilon)$ as in (2.23). Correspondingly, we say that a sequence of discrete sets $A_\varepsilon \subset \mathcal{L}_\varepsilon(\Omega)$ converges to A if $u^\varepsilon = 2\chi_{A_\varepsilon} - 1 \to A$.

Remark 2.23 (Convergence on Bravais lattices) An analogous definition can be given for spin functions on arbitrary Bravais lattices. Indeed, if $\mathcal{L}$ in $\mathbb{R}^d$ is a Bravais lattice generated by independent vectors $v_1, \ldots, v_d$, we can give the corresponding identification of spin functions u defined on $\varepsilon\mathcal{L}$ with sets $A_\varepsilon(u)$ upon taking the elementary cell

$$C = \left\{ \sum_{j=1}^{d} t_j v_j : |t_j| \le \frac{1}{2} \right\} \tag{2.25}$$

in the place of $[-\frac{1}{2}, \frac{1}{2}]^d$.

We can now specialize the definition of Γ-convergence to the discrete setting as follows. We will take all Γ-limits with respect to this convergence, or with respect to some variants that will be clear from the context and often considered as understood not to overburden the notation. Note that this convergence and all its variants can be seen as convergences derived from the topology of a separable metric space, so that standard Γ-convergence arguments can be applied.

Definition 2.24 (Discrete-to-continuum Γ-convergence) For any $\varepsilon > 0$, let E_ε be a functional defined on the set $\{u \colon \mathcal{L}_\varepsilon(\Omega) \to \{-1; 1\}\}$, and F be a functional defined on the family of sets of finite perimeter in Ω. The sequence E_ε *Γ-converges to F at A* with respect to the convergence defined in (2.24) if and only if

(i) (liminf inequality) for all u^ε converging to A we have $F(A) \le \liminf_{\varepsilon\to 0} E_\varepsilon(u^\varepsilon)$;
(ii) (existence of a recovery sequence) there exist u^ε converging to A such that $F(A) = \lim_{\varepsilon\to 0} E_\varepsilon(u^\varepsilon)$.

If this holds for any A of finite perimeter, we say that F is the *Γ-limit* of E_ε and write $F = \Gamma\text{-}\lim_{\varepsilon\to 0} E_\varepsilon$.

Note that if (i) holds, in order to prove (ii) it suffices to show that for all $\eta > 0$ there exists u^η_ε converging to A such that

$$F(A) \geq \limsup_{\varepsilon\to 0} E_\varepsilon(u^\eta_\varepsilon) - \eta$$

(*approximate limsup inequality*). Indeed, by a diagonal argument, we then obtain a sequence $\{u^\varepsilon\}$ converging to A that satisfies $F(A) \geq \limsup_{\varepsilon\to 0} E_\varepsilon(u^\varepsilon)$, which is a recovery sequence by (i).

Remark 2.25 (One-dimensional limits) Note that for $d = 1$, sets of finite perimeter are simply (sets equivalent to) finite unions of intervals, whose end-points are the (reduced) boundary. In this case the dependence on the normal is trivial, which makes it more convenient to use functions as parameters and the corresponding notation for the convergence. Given a sequence of spin functions $u^\varepsilon : \varepsilon\mathbb{Z} \to \{-1, +1\}$ converging to a set of finite perimeter A in the sense of (2.24), we say that u^ε converges to $u = 2\chi_A - 1$, and instead of $\partial^* A$ we use the notation $S(u)$; that is, $S(u)$ is the *set of discontinuity points* of the piecewise-constant function u.

2.5 Γ-Limit of Nearest-Neighbor Ferromagnetic Energies

We now conclude the analysis of the model example introduced in Section 2.1 by computing the Γ-limit of the discrete energies

$$E_\varepsilon(u) = \frac{1}{8}\sum_{\langle i,j\rangle} \varepsilon^{d-1}(u_i - u_j)^2.$$

The compactness theorem for sets of finite perimeter allows us to deduce a compactness property for E_ε as follows.

Remark 2.26 (Compactness of E_ε) Let $\{u^\varepsilon\}$ be a family of spin functions defined on $\mathcal{L}_\varepsilon(\Omega)$ such that $E_\varepsilon(u^\varepsilon) \leq C$. Then, for any $\Omega' \subset\subset \Omega$,

$$\mathcal{H}^{d-1}(\Omega' \cap A_\varepsilon(u^\varepsilon)) \leq E_\varepsilon(u^\varepsilon) \leq C \tag{2.26}$$

for ε small enough. By Theorem 2.18 we obtain that there exist a subsequence (again denoted by $\{u^\varepsilon\}$) and a set of finite perimeter A such that $u^\varepsilon \to A$ in the sense of the convergence (2.24).

Now we compute the Γ-limit of the sequence $\{E_\varepsilon\}$ with respect to convergence (2.24).

Optimization of the Lower Bound Let $\{u^\varepsilon\}$ be such that $E_\varepsilon(u^\varepsilon)$ is equibounded. Remark 2.26 ensures that (up to subsequences) $u^\varepsilon \to A$, where A is a set of finite perimeter. We may improve the lower estimate (2.26) by observing that the inner normal to $\partial A_\varepsilon(u^\varepsilon)$, denoted by ν_ε, may only take the values $-e_k$ and e_k for $k \in \{1,\dots,d\}$, and then, for all $\Omega' \subset\subset \Omega$, we have

$$E_\varepsilon(u^\varepsilon) \geq \int_{\Omega' \cap \partial A_\varepsilon(u^\varepsilon)} \varphi(\nu_\varepsilon)\, d\mathcal{H}^{d-1} \tag{2.27}$$

for every norm φ such that $\varphi(e_k) \leq 1$ for all $k \in \{1,\dots,d\}$. The largest such norm is the 1-norm given by

$$\|\nu\|_1 = \sum_{k=1}^{d} |\nu_k|, \quad \nu = (\nu_1,\dots,\nu_d).$$

Hence, thanks to (2.27) and the lower semicontinuity of the perimeter functionals, we have

$$\liminf_{\varepsilon\to 0} E_\varepsilon(u^\varepsilon) \geq \int_{\Omega' \cap \partial^* A} \|\nu\|_1\, d\mathcal{H}^{d-1}.$$

The arbitrariness of $\Omega' \subset\subset \Omega$ finally gives

$$\liminf_{\varepsilon\to 0} E_\varepsilon(u^\varepsilon) \geq \int_{\Omega \cap \partial^* A} \|\nu\|_1\, d\mathcal{H}^{d-1}.$$

Upper Bound by Density As for the Γ-lim sup inequality, it is sufficient to construct a recovery sequence for a polytope A; indeed, by Theorem 2.20, they are (strongly) dense in the space of characteristic functions of sets of finite perimeter, and we can use a diagonal argument.

We now construct a recovery sequence for a polytope A. The sequence $\{u^\varepsilon\}$ defined by $u_i^\varepsilon = 2\chi_A(\varepsilon i) - 1$ can be chosen as a recovery sequence, since

$$\lim_{\varepsilon\to 0} E_\varepsilon(u^\varepsilon) = \int_{\Omega \cap \partial A} \|\nu\|_1\, d\mathcal{H}^{d-1}.$$

Characterization of the Γ-Limit By gathering the two preceding estimates, we obtain

$$\Gamma\text{-}\lim_{\varepsilon\to 0} E_\varepsilon(A) = \int_{\Omega \cap \partial^* A} \|\nu\|_1\, d\mathcal{H}^{d-1}. \tag{2.28}$$

The functional

$$F(A) = \int_{\Omega \cap \partial^* A} \|\nu\|_1\, d\mathcal{H}^{d-1}$$

is the so-called 1-*crystalline perimeter* of A. The Wulff shape of the 1-crystalline norm is the d-dimensional coordinate cube with measure 1.

Application to Minimal-Cut Problems We can apply this result to the asymptotic description of minimal-cut problems. We fix a sequence $\{N_\varepsilon\}$ of integer numbers such that $N_\varepsilon \varepsilon^d \to C$, with $0 < C < |\Omega|$. Note that if a set A is such that $|A| = C$, then we may construct a recovery sequence $\{u^\varepsilon\}$ satisfying the constraint $\#\{i : u_i^\varepsilon = 1\} = N_\varepsilon$. This implies that the constrained functionals

$$E_\varepsilon^C(u) = \begin{cases} E_\varepsilon(u) & \text{if } \#\{i : u_i = 1\} = N_\varepsilon, \\ +\infty & \text{otherwise} \end{cases}$$

converge to the functional

$$F^C(A) = \begin{cases} F(A) & \text{if } |A| = C, \\ +\infty & \text{otherwise.} \end{cases}$$

Theorem 2.3 ensures that the discrete minimal-cut problems

$$\min\{E_\varepsilon(u) \colon \#\{i : u_i = 1\} = N_\varepsilon\}$$

can be approximated by the continuum least-perimeter problem

$$\min\{F(A) : |A| = C\}$$

in the sense that the minimizers of the latter are exactly the cluster points of the minimizers of the former.

Bibliographical Notes to Chapter 2

The notion of Γ-convergence was introduced by De Giorgi (see De Giorgi and Franzoni, 1975). An in-depth illustration of the general properties of Γ-convergence is contained in the monograph by Dal Maso (1993). An introductory presentation of Γ-convergence from the standpoint of applications is contained in Braides (2002) and further developed in Braides (2006). The treatment of the paradigmatic example in this chapter follows Alicandro et al. (2006).

The theory of sets of finite perimeter is dealt with in detail in Maggi (2012). A general introduction to BV function can be found in the book by Ambrosio et al. (2000). A brief account of both subjects can be found in the lecture notes by Braides (1998). The characterization of Wulff shapes can be found in Morgan (1998).

Perimeter functionals are a special case of functionals on partitions into sets of finite perimeter, of which an in-depth treatment has been carried out in Ambrosio and Braides (1990a,b). Integral-representation results for such functionals are studied by Bouchitté et al. (2002) and also by Braides and Chiadò Piat (1995).

3
Homogenization of Pairwise Systems with Positive Coefficients

In this chapter, we consider a family a_{ij} of nonnegative coefficients describing the set of the connections between nodes of $\mathbb{Z}^d$, each considered with a possible weight. With the model case of nearest-neighbor interactions treated in the previous chapter in mind, we study the asymptotic behavior of functionals of the form

$$E_\varepsilon(u) = \sum_{\varepsilon i, \varepsilon j \in \mathcal{L}_\varepsilon(\Omega)} \varepsilon^{d-1}\, a_{ij}(u_i - u_j)^2, \tag{3.1}$$

where Ω is an open subset of $\mathbb{R}^d$ with Lipschitz boundary. We use the notation

$$\mathcal{L}_\varepsilon = \mathcal{L}_\varepsilon(\Omega) = \varepsilon\mathbb{Z}^d \cap \Omega$$

as in (2.22), and for $u\colon \mathcal{L}_\varepsilon(\Omega) \to \{-1, 1\}$ we set $u_i = u(\varepsilon i)$. Such systems with positive coefficients are commonly denoted as *ferromagnetic systems*.

In particular, we consider the following *homogenization* problem: show that, under suitable conditions on a_{ij}, the Γ-limit is a homogeneous perimeter functional; that is, a functional of the form

$$F(A) = \int_{\Omega \cap \partial^* A} \varphi(\nu(x))\, d\mathcal{H}^{d-1}(x).$$

Various problems and techniques involving oscillating energies are labeled with the term homogenization, which sometimes is a synonym of averaging or weak convergence. In this context we will talk about homogenization whenever in the discrete-to-continuum process the limit energy is described by a homogeneous energy and the whole sequence E_ε converges.

3.1 Homogeneous Systems

This section is devoted to homogeneous, that is, translation-invariant systems of coefficients. As an example, we can consider the set of connections in $\mathbb{Z}^2$ given

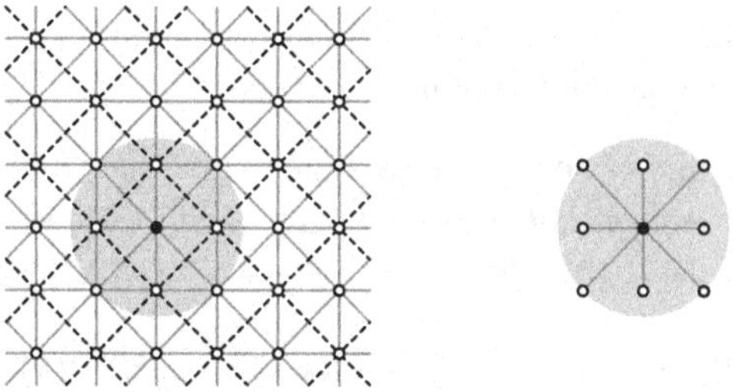

Figure 3.1 Nearest and next-to-nearest-neighbor interactions (a) and connections of a given point (b).

by the nearest and next-to-nearest neighbors $\{(i,j)\colon \|i-j\| \le \sqrt{2}\}$. The set of such connections is indicated by $\langle\langle i,j\rangle\rangle$, meaning that summation labeled with this symbol indicates summation on such pairs of indices. This corresponds in (3.1) to the choice $a_{ij} = 1$ if $\|i-j\| \le \sqrt{2}$, and $a_{ij} = 0$ otherwise. The network of interactions is represented in Fig. 3.1, where, in part (b), we have highlighted the sites interacting with a given site. Note that a ferromagnetic energy with connections beyond nearest neighbors can also be expressed as depending on subsets A as in (2.1), but its expression in terms of the spin function u is simpler and handier.

Our assumptions, which comprise both nearest and next-to-nearest-neighbor interactions, are that the coefficients a_{ij} in (3.1) satisfy the following hypotheses:

(i) (*ferromagnetic energies*) $a_{ij} \ge 0$ for any i, j; this implies that the ground states are the uniform states; that is, u is identically 1 or -1;

(ii) (*coerciveness of nearest-neighbor interactions*) $a_{ij} \ge c > 0$ if $\|i-j\| = 1$;

(iii) (*homogeneity*) $a_{ij} = \alpha_{i-j} = \alpha_{j-i}$ for any i, j. Note that the symmetry can be assumed without loss of generality by possibly replacing both coefficients a_{ij} and a_{ji} with $\frac{a_{ij}+a_{ji}}{2}$ and noting that this change does not influence the value of the energy.

Moreover, we suppose as a *working hypothesis* that the coefficients α_k are of finite range; that is,

(iv) (*finite range) there exists $R > 0$ such that $\alpha_k = 0$ if $\|k\| > R$.*

This hypothesis can be relaxed and will be eventually substituted by a summability hypothesis on the family $\{\alpha_k\}$.

This section will be devoted to the proof of the following result and to the illustration of a superposition argument.

Theorem 3.1 (Limits of homogeneous systems) *Under hypotheses* (i)–(iv), *the* Γ*-limit of the sequence of energies* E_ε *defined by* (3.1) *with respect to the convergence* $u^\varepsilon \to A$ *is given by*

$$F(A) = \int_{\Omega\cap\partial^* A} 4 \sum_{k\in\mathbb{Z}^d} \alpha_k |\langle k, \nu(x)\rangle| \, d\mathcal{H}^{d-1}(x)$$

for all A sets of finite perimeter.

Remark 3.2 (Coerciveness and domain of the Γ-limit) If a sequence $\{u^\varepsilon\}$ is such that $E_\varepsilon(u^\varepsilon) \le C$, then the coerciveness of the nearest-neighbor interactions ensures that the corresponding sets $A_\varepsilon(u^\varepsilon)$ have equibounded perimeter, so that they are precompact thanks to Theorem 2.18. Up to subsequences, we can assume that $u^\varepsilon \to A$ for a set A of finite perimeter. Since the coefficients are of finite range, by choosing $u^\varepsilon = 2\chi_A - 1$ on $\varepsilon\mathbb{Z}^d$ we obtain $\lim_{\varepsilon\to 0} E_\varepsilon(u^\varepsilon) \le C_R\mathcal{H}^{d-1}(\Omega\cap\partial^* A)$, where the positive constant C_R depends only on $\max_k \alpha_k$ and the range R in (iv). Hence, the domain of the Γ-limit of E_ε is the family of sets with finite perimeter.

Remark 3.3 (A generalization of the coerciveness hypothesis) We can replace the coerciveness hypothesis on nearest neighbors with the following condition:

> there exists $R > 0$ such that, for any $i \in \mathbb{Z}^d$, the set $\{k \in \mathbb{Z}^d \cap B_R(0),$
> $a_{i\,i+k} \ge c > 0\}$ generates $\mathbb{Z}^d$ (with coefficients in $\mathbb{Z}$),

since in this case the strictly positive contribution to the energy due to nearest neighbors with u of changing sign can be recovered using a cycle in the set just mentioned. In Fig. 3.2 some examples of this condition are pictured. The solid lines with arrows represent strictly positive interactions.

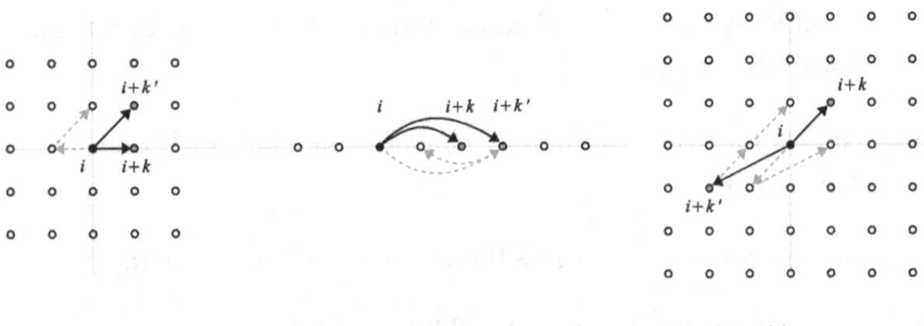

Figure 3.2 Examples of coercive systems not satisfying nearest-neighbor coerciveness.

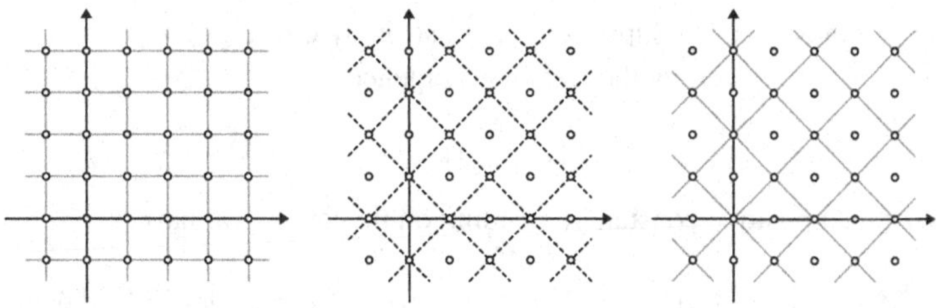

Figure 3.3 Decomposition of a square lattice with next-to-nearest-neighbor interactions.

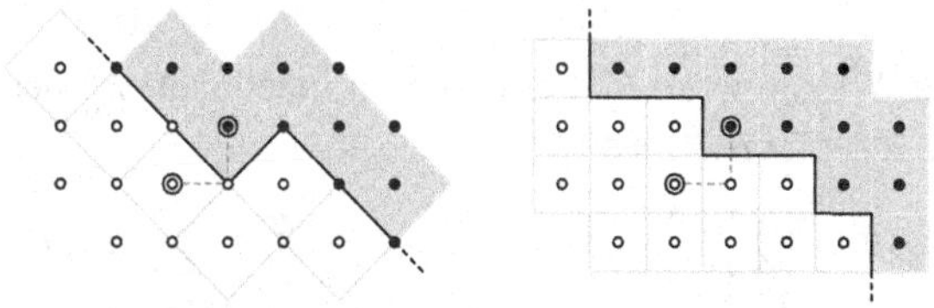

Figure 3.4 Coerciveness of next-to-neighbor interactions.

Computation of the Γ-Limit by Superposition We start with an example to describe the superposition of lattices, with the aid of a pictorial description. Again, we consider the lattice $\mathbb{Z}^2$ with the set of connections given by the nearest and next-to-nearest interactions. This corresponds to assuming that the range of a_{ij} is $R = \sqrt{2}$. We can decompose the lattice into three lattices, as pictured in Fig. 3.3, and separately estimate the contribution of the energy on the different sublattices of the decomposition. To that end, we need to demonstrate that sequences that converge on the whole lattice still converge to the same limit when considered on each such sublattice. This will be done in a general framework in the following remark.

Remark 3.4 (Coerciveness and convergence on sublattices) Let $\{u^\varepsilon\}$ be an equibounded sequence. We can assume that $A_\varepsilon = A_\varepsilon(u^\varepsilon) \to A$. Note that the convergence is strong in L^1 and that the perimeters of A_ε are equibounded. Let $\tilde{\mathcal{L}}$ be a sublattice of $\mathbb{Z}^d$ and define $\tilde{u}^\varepsilon$ as the restriction of u^ε to $\Omega \cap \varepsilon\tilde{\mathcal{L}}$. Let C be the elementary cell of $\tilde{\mathcal{L}}$, and let $\tilde{A}_\varepsilon$ denote the set corresponding to $\tilde{u}^\varepsilon$ by interpolation in $\Omega \cap \varepsilon\tilde{\mathcal{L}}$ (see Definition (2.5) with C in the place of $[-\frac{1}{2}, \frac{1}{2}]^d$). Then we still have $\tilde{A}_\varepsilon \to A$ strongly in L^1. Indeed, $\mathcal{H}^{d-1}(\Omega \cap \partial\tilde{A}_\varepsilon) \leq C\mathcal{H}^{d-1}(\Omega \cap \partial A_\varepsilon)$, since to a change of sign of $\tilde{u}^\varepsilon$ between i and j there corresponds (at least) a change of sign of u^ε along a path joining i and j (see Fig. 3.4).

By compactness we can suppose that $\tilde{A}_\varepsilon$ strongly converges to a set of finite perimeter $\tilde{A}$. Moreover, by the weak convergence

$$\chi_{\varepsilon(\tilde{\mathcal{L}}+C\cap[-\frac{1}{2},\frac{1}{2}]^d)} \rightharpoonup c_{\tilde{\mathcal{L}}},$$

where $c_{\tilde{\mathcal{L}}}$ is a positive constant depending on the lattice, we have

$$\chi_{\tilde{A}_\varepsilon\cap\varepsilon(\tilde{\mathcal{L}}+C\cap[-\frac{1}{2},\frac{1}{2}]^d)} \rightharpoonup c_{\tilde{\mathcal{L}}}\chi_{\tilde{A}} \text{ and } \chi_{A_\varepsilon\cap\varepsilon(\tilde{\mathcal{L}}+C\cap[-\frac{1}{2},\frac{1}{2}]^d)} \rightharpoonup c_{\tilde{\mathcal{L}}}\chi_{A}.$$

Since $\tilde{A}_\varepsilon \cap \varepsilon\tilde{\mathcal{L}} = A_\varepsilon \cap \varepsilon\tilde{\mathcal{L}}$, we deduce that $\tilde{A} = A$.

Now we can compute the Γ-limit of the sequence $\{E_\varepsilon\}$.

Lower Bound We can write E_ε as the sum of functionals E_ε^k depending on a parameter $k \in \mathbb{Z}^d$; that is,

$$E_\varepsilon(u) = \sum_{k\in\mathbb{Z}^d}\sum_{\varepsilon i\in\mathcal{L}_\varepsilon} \varepsilon^{d-1}\alpha_k(u_{i+k}-u_i)^2 = \sum_{k\in\mathbb{Z}^d} E_\varepsilon^k(u).$$

For a fixed $k \in \mathbb{Z}^d$ we consider a lattice $\mathcal{L}^k$ obtained by using k and a basis of the orthogonal space (B denotes the corresponding cell in this $d-1$ subspace, as pictured in Fig. 3.5), and define the functional

$$E_\varepsilon^{\mathcal{L}^k}(u) = \sum_{i\in\mathcal{L}^k\cap\frac{1}{\varepsilon}\Omega} \varepsilon^{d-1}\alpha_k(u_{i+k}-u_i)^2,$$

which is a nearest-neighbor energy in the lattice $\mathcal{L}^k$, where all the interactions in the directions orthogonal to k have coefficient equal to 0. Recalling Remark 3.4, we have that, if $u^\varepsilon \to A$, then the restriction of u^ε to the sublattice $\varepsilon\mathcal{L}^k \cap \Omega$ (again denoted by u^ε) converges strongly to the same A. Denoting the

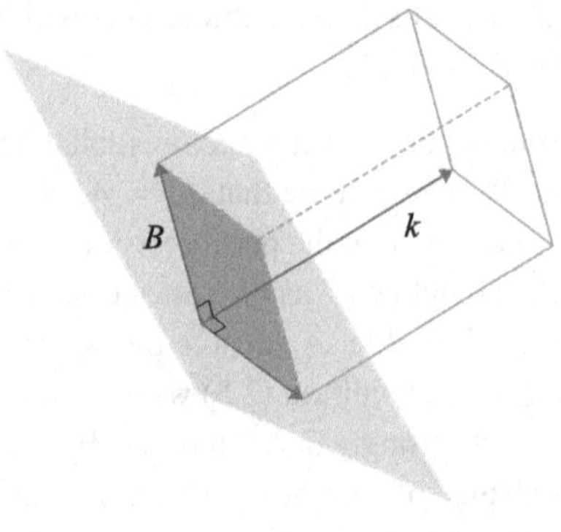

Figure 3.5 The elementary cell of $\mathcal{L}^k$.

interpolation set of u^ε on the sublattice $\varepsilon\mathcal{L}^k \cap \Omega$ by $A^k_\varepsilon(u^\varepsilon)$, we can write, for any open set $\Omega' \subset\subset \Omega$ and for ε small enough,

$$E^{\mathcal{L}^k}_\varepsilon(u^\varepsilon) \geq \frac{4}{\mathcal{H}^{d-1}(B)} \int_{\Omega'\cap\partial A^k_\varepsilon(u^\varepsilon)} \alpha_k \Big|\Big\langle \frac{k}{\|k\|}, \nu\Big\rangle\Big| \, d\mathcal{H}^{d-1}$$
$$= \frac{4}{|C^k|} \int_{\Omega'\cap\partial A^k_\varepsilon(u^\varepsilon)} \alpha_k |\langle k, \nu\rangle| \, d\mathcal{H}^{d-1},$$

where C^k is the fundamental cell of $\mathcal{L}^k$ as in (2.25). By lower semicontinuity we then obtain

$$\liminf_{\varepsilon\to 0} E^{\mathcal{L}^k}_\varepsilon(u^\varepsilon) \geq \frac{4}{|C^k|} \int_{\Omega'\cap\partial^* A} \alpha_k |\langle k, \nu\rangle| \, d\mathcal{H}^{d-1}.$$

Taking the supremum over all sets Ω' compactly contained in Ω, we get

$$\liminf_{\varepsilon\to 0} E^{\mathcal{L}^k}_\varepsilon(u^\varepsilon) \geq F^k(A) = \frac{4}{|C^k|} \int_{\Omega\cap\partial^* A} \alpha_k |\langle k, \nu\rangle| \, d\mathcal{H}^{d-1}.$$

Now, we have to count how many different (translated) lattices we have with one side of the elementary cell equal to k; we note that $\mathbb{Z}^d$ contains $|C^k|$ disjoint copies of the lattice $\mathcal{L}^k$; hence, going back to the functionals E^k_ε,

$$\liminf_{\varepsilon\to 0} E^k_\varepsilon(u^\varepsilon) \geq 4 \int_{\Omega\cap\partial^* A} \alpha_k |\langle k, \nu\rangle| \, d\mathcal{H}^{d-1},$$

since the sum of the lim inf is lower than the lim inf of the sum. We conclude that

$$\liminf_{\varepsilon\to 0} E_\varepsilon(u^\varepsilon) = \liminf_{\varepsilon\to 0} \sum_{k\in\mathbb{Z}^d} E^k_\varepsilon(u^\varepsilon) \geq 4 \sum_{k\in\mathbb{Z}^d} \alpha_k \int_{\Omega\cap\partial^* A} |\langle k, \nu\rangle| \, d\mathcal{H}^{d-1}.$$

Note that the factor 4 (instead of 8) is due to the fact that the vectors k and $-k$ are both accounted for.

The candidate limit energy density is then

$$\varphi(\nu) = 4 \sum_{k\in\mathbb{Z}^d} \alpha_k |\langle k, \nu\rangle|. \tag{3.2}$$

Upper Bound Note that, in general, the Γ-limit of a sum does not coincide with the sum of the Γ-limits, since the recovery sequences can be different. In this case, we obtain the upper inequality by noting that the same recovery sequence can be used for all k. As in the case of nearest-neighbor interactions, we prove the lim sup inequality for a polytope A and conclude by using a density argument.

If A is a polytope, then locally it is a portion of a half-space, with boundary orthogonal to a direction ν. Then, by choosing $u^\varepsilon = 2\chi_{A\cap\varepsilon\mathbb{Z}^d} - 1$, we obtain that the restriction to each $\varepsilon\mathcal{L}^k$ is a recovery sequence for $|C^k|F^k(A)$. Indeed, if we consider these restrictions, that is, we limit the interactions to the sublattice $\varepsilon\mathcal{L}^k$, we have that the measure of the boundary of the set $A^k_\varepsilon(u^\varepsilon)$ is the measure of the projection on the hyperplane orthogonal to ν, and then proportional to $|\langle k,\nu\rangle|$. This implies that $\{u^\varepsilon\}$ is a recovery sequence for any sequence of functionals E^k_ε of the decomposition, and we have an upper bound with the same φ.

This concludes the proof of Theorem 3.1.

Crystalline Norms If the range of α_k is finite, then $B_\varphi = \{z\colon \varphi(z) \le 1\}$ is a polytope, since by (3.2) it is the intersection of half-spaces and the vectors k for which $\alpha_k \ne 0$ span $\mathbb{Z}^d$. Since the coefficients are symmetric, B_φ is symmetric with respect to the origin. Then, the Wulff shape W_φ also is a polytope; more precisely, due to its symmetry properties, it is a *zonotope*, which is a strict subclass of polytopes symmetric with respect to the origin. A norm φ such that W_φ is a polytope is called a *crystalline norm.*

Example 3.5 (Next-to-nearest neighbors in $\mathbb{Z}^2$) We apply Theorem 3.1 to the case of nearest and next-to-nearest interactions in $\mathbb{Z}^2$ with weighted coefficients; that is, we fix

$$\alpha_k = \begin{cases} \alpha & \text{if } k \in \{-e_1, e_1, -e_2, e_2\},\\ \beta & \text{if } k \in \{-(e_1+e_2), (e_1+e_2), -(e_1-e_2), (e_1-e_2)\},\\ 0 & \text{otherwise,}\end{cases}$$

with $\alpha, \beta > 0$. The energy density is

$$\begin{aligned}\varphi(\nu) &= 8\Big(\alpha(|\langle e_1,\nu\rangle| + |\langle e_2,\nu\rangle|) + \beta(|\langle e_1+e_2,\nu\rangle| + \langle e_1-e_2,\nu\rangle)\Big)\\ &= 8(\alpha\|\nu\|_1 + 2\beta\|\nu\|_\infty).\end{aligned}$$

The Wulff shape W_φ is an octagon, as shown in Fig. 3.6.

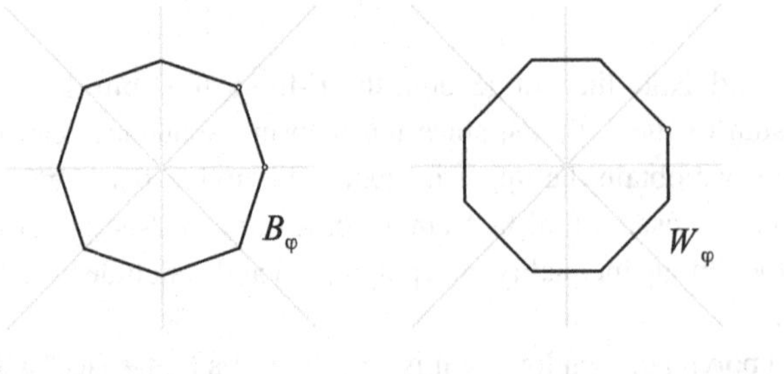

Figure 3.6 B_φ and W_φ for next-to-nearest-neighbor systems.

Interactions on General Bravais Lattices The Γ-convergence result of Theorem 3.1 can be generalized by considering a general (d-dimensional) Bravais lattice $\mathcal{L}$ in $\mathbb{R}^d$. In this case, the limit energy density depends on the volume of the cell of the lattice; we have

$$\varphi(v) = 4c_{\mathcal{L}} \sum_{k \in \mathbb{Z}^d} \alpha_k |\langle k, v\rangle|, \tag{3.3}$$

where $c_{\mathcal{L}} = |C|^{-1}$ and C is the fundamental cell of the lattice $\mathcal{L}$ as in (2.25).

Example 3.6 (Nearest-neighbor interactions on the triangular lattice) As an example we consider a triangular lattice in two dimensions. Let $v_n = \big(\cos((n-1)\frac{\pi}{3}), \sin((n-1)\frac{\pi}{3})\big)$ for $n \in \{1,2,3\}$, and let $\mathcal{L} = \mathbb{T}$ be the Bravais lattice given by

$$\mathbb{T} = \mathbb{Z}v_1 + \mathbb{Z}v_2. \tag{3.4}$$

We consider the nearest-neighbor interactions on $\mathbb{T}$; that is, the coefficients a_{ij} are given by

$$a_{ij} = \begin{cases} 1 & \text{if } i - j \in \{-v_n, v_n\}_{n=1}^3, \\ 0 & \text{otherwise.} \end{cases}$$

Since $c_{\mathbb{T}} = \frac{2}{\sqrt{3}}$, by (3.3) we get

$$\varphi(v) = \frac{8}{\sqrt{3}} \sum_{n=1}^{3} |\langle v, v_n\rangle|.$$

The corresponding Wulff shape is a hexagon as shown in Fig. 3.7.

Systems with Infinite Range of Interaction Following the proof of Theorem 3.1, we note that we can drop the hypothesis of finite range. If the range of the coefficients is not finite but

$$\sum_{k \in \mathbb{Z}^d} \alpha_k \|k\| < +\infty, \tag{3.5}$$

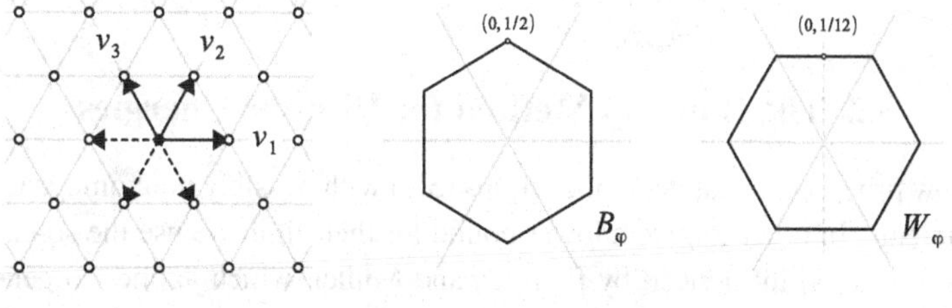

Figure 3.7 B_φ and W_φ for the triangular lattice.

the same proof can be repeated (almost) word for word. Indeed, we can estimate $E_\varepsilon(u^\varepsilon)$ from below by the same energy limited to the interactions with range less than R; that is, $\|i-j\| \le R$, indicated by $E_\varepsilon^{(R)}(u^\varepsilon)$. Since $E_\varepsilon^{(R)}$ is of finite range, the sequence Γ-converges to $F^{(R)}(A) = \int_{\Omega\cap\partial^* A} \varphi_R(\nu)\, d\mathcal{H}^{d-1}$, where

$$\varphi_R(\nu) = 4 \sum_{\|k\|\le R} \alpha_k |\langle k, \nu\rangle|.$$

Noting that $\varphi_R(\nu)$ increasingly converges to $\varphi(\nu)$, the lower bound can be optimized by taking the supremum over $R > 0$, obtaining

$$\liminf_{\varepsilon\to 0} E_\varepsilon(u^\varepsilon) \ge \sup_R \liminf_{\varepsilon\to 0} E_\varepsilon^{(R)}(u^\varepsilon)$$
$$\ge \sup_R \int_{\Omega\cap\partial^* A} \varphi_R(\nu)\, d\mathcal{H}^{d-1} = \int_{\Omega\cap\partial^* A} \varphi(\nu)\, d\mathcal{H}^{d-1}.$$

As for the upper bound, the same recovery sequence $\{u_\varepsilon\}$ as in the finite-range case can still be used, since the infinite sum $\sum_{k\in\mathbb{Z}^d} E_\varepsilon^k(u^\varepsilon)$ is uniformly convergent by hypothesis (for A a portion of a half-space, it is bounded by the perimeter of the boundary of A times $4\sum_{k\in\mathbb{Z}^d} \alpha_k\|k\|$).

Note that if the range is not finite, we can have energy densities φ that are not crystalline but have a Wulff shape that is a limit of zonotopes. Such a set is called a *zonoid*. Indeed, all φ positive, symmetric, convex, and positively homogeneous of degree one whose Wulff shape is a zonoid can be obtained as limits of functions as in (3.2) for a suitable family α_k satisfying (3.5) (see also Section 4.10). In particular, since a ball is a zonoid, functions φ in (3.2) may be arbitrarily close to a constant on S^{d-1}; that is, the limit perimeter may be very close to the Euclidean one. Note that in dimension $d = 2$ all symmetric convex sets are zonoids, while this is not true for $d \ge 3$.

An interesting observation in the theory of zonoids is that formula (3.2) can be interpreted as an integration on the unit sphere with respect to a sum of Dirac deltas. By the density of such sums of Dirac deltas zonotopes are identifies as symmetric positive measures on the sphere, and Wulff shapes of Ising systems as converging sums of Dirac deltas concentrated on rational directions.

3.2 The Blow-up Method for Discrete Energies

We now proceed to treat general energies (3.1) with possibly nonhomogeneous coefficients. In order to give a lower bound for their limit we use the so-called *blow-up method* introduced by Fonseca and Müller, which applies to general classes of functionals. This method is used to prove a lower estimate for energies F_ε, showing that, if $u_\varepsilon \to u$, then

$$\liminf_{\varepsilon\to 0} F_\varepsilon(u_\varepsilon) \geq F(u) = \int_\Omega f(x)\,d\lambda,$$

for some given measure λ and for some f that can be characterized in terms of local quantities depending on u only (e.g. $u(x)$ or $\nabla u(x)$). With a fixed sequence $\{u_\varepsilon\}$, the idea is to interpret $F_\varepsilon(u_\varepsilon)$ as a sequence of (total variations of) equibounded positive measures; that is, $F_\varepsilon(u_\varepsilon) = \lambda_\varepsilon(\Omega)$. Since $\lambda_\varepsilon(\Omega) \leq C$, we get that, up to subsequences, λ_ε weak*-converges to a measure μ. Hence, to prove the lower estimate it is sufficient to show that

$$\frac{d\mu}{d\lambda}(x) \geq f(x) \text{ for } \lambda\text{-almost all } x \in \Omega$$

and then integrate with respect to the measure λ.

In the following, we specialize the blow-up method to the case of discrete energies

$$E_\varepsilon(u) = \sum_{\varepsilon i, \varepsilon j \in \mathcal{L}_\varepsilon(\Omega)} \varepsilon^{d-1} a_{ij}(u_i - u_j)^2.$$

We assume that the coefficients a_{ij} satisfy the following hypotheses:

(i) (*ferromagnetic energies*) $a_{ij} \geq 0$ for any i, j;
(ii) (*coerciveness of nearest-neighbor interactions*) $a_{ij} \geq c > 0$ if $\|i - j\| = 1$;
(iii) (*finite range*) there exists $R > 0$ such that $a_{ij} = 0$ if $\|j - i\| > R$.

With these assumptions, the sequence $\{E_\varepsilon\}$ is equicoercive with respect to the convergence $u^\varepsilon \to A$. In order to have a lower bound in terms of a perimeter functional, the target measure λ is the restriction of $\mathcal{H}^{d-1}$ to the reduced boundary of the limit A of u^ε, and we will exhibit $f(x)$ that can be written as a function of the normal $\nu(x)$.

Let $\{u^\varepsilon\}$ be a sequence with equibounded energy, and $u^\varepsilon \to A$ as $\varepsilon \to 0$. We define the sequence of measures $\{\mu_\varepsilon\}$ given by

$$\mu_\varepsilon = \sum_{\varepsilon i \in \Omega} \Big(\sum_{\varepsilon j \in \Omega} \varepsilon^{d-1} a_{ij}(u_i^\varepsilon - u_j^\varepsilon)^2 \Big)\, \delta_{\varepsilon i},$$

where δ_x is the Dirac measure concentrated at x; that is, for a Borel set B,

$$\mu_\varepsilon(B) = \sum_{\varepsilon i \in B} \sum_{\varepsilon j \in \Omega} \varepsilon^{d-1} a_{ij}(u_i^\varepsilon - u_j^\varepsilon)^2.$$

The value $\mu_\varepsilon(B)$ takes into account interactions between nodes in B and nodes in the whole Ω, but, due to the finite-range hypothesis, indeed the latter can be limited to an εR-neighborhood of B.

Since $\mu_\varepsilon(\Omega) = E_\varepsilon(u^\varepsilon)$, the measures are equibounded and up to subsequences we can assume

$$\mu_\varepsilon \overset{*}{\rightharpoonup} \mu.$$

We want to show that there exists a function φ such that

$$\frac{d\mu}{d\mathcal{H}^{d-1} \llcorner \partial^* A}(x_0) \geq \varphi(\nu(x_0)) \quad \text{for } \mathcal{H}^{d-1}\text{-almost every } x_0 \in \partial^* A. \tag{3.6}$$

Indeed, if (3.6) holds for some φ (independent of the subsequence and of Ω), we obtain the lower bound for the Γ-limit since

$$\begin{aligned}\liminf_{\varepsilon\to 0} E_\varepsilon(u^\varepsilon) = \liminf_{\varepsilon\to 0} \mu_\varepsilon(\Omega) \geq \mu(\Omega) \geq \int_{\partial^* A\cap\Omega} \frac{d\mu}{d\mathcal{H}^{d-1} \llcorner \partial^* A}(x_0)\, d\mathcal{H}^{d-1} \\ \geq \int_{\partial^* A\cap\Omega} \varphi(\nu(x_0))\, d\mathcal{H}^{d-1}.\end{aligned}$$

Recalling Remark 2.17, we have that, for almost all $x_0 \in \partial^* A$, the blown-up sets converge to the half space $H^\nu = \{x \in \mathbb{R}^d : \langle x, \nu\rangle \geq 0\}$; that is,

$$\frac{1}{\varrho}(A - x_0) \to H^\nu \quad \text{for } \mathcal{H}^{d-1}\text{-almost every } x_0 \in \partial^* A. \tag{3.7}$$

We can then prove (3.6) for points $x_0 \in \partial^* A$ such that the measure-theoretical derivative of the limit measure μ with respect to $\mathcal{H}^{d-1} \llcorner \partial^* A$ exists, and (3.7) holds.

To obtain a lower estimate of the derivative of the measure μ, we note that we can write

$$\frac{d\mu}{d\mathcal{H}^{d-1} \llcorner \partial^* A}(x_0) = \lim_{\varrho\to 0} \frac{\mu(Q^\nu_\varrho(x_0))}{\varrho^{d-1}},$$

where $Q^\nu_\varrho(x_0)$ is a d-dimensional cube with center x_0, a face orthogonal to ν and side length ϱ. For any $\varrho > 0$ such that $\mu(\partial Q^\nu_\varrho(x_0)) = 0$ (hence, for all but a countable number of ϱ) we have that $\mu_\varepsilon(Q^\nu_\varrho(x_0)) \to \mu(Q^\nu_\varrho(x_0))$, which implies

$$\frac{d\mu}{d\mathcal{H}^{d-1} \llcorner \partial^* A}(x_0) = \lim_{\varrho\to 0} \lim_{\varepsilon\to 0} \frac{\mu_\varepsilon(Q^\nu_\varrho(x_0))}{\varrho^{d-1}}.$$

Hence, we can choose a sequence $\varrho_\varepsilon \to 0$ as $\varepsilon \to 0$ such that $\varrho_\varepsilon \gg \varepsilon$, and

$$\frac{d\mu}{d\mathcal{H}^{d-1} \llcorner \partial^* A}(x_0) = \lim_{\varrho\to 0} \frac{\mu(Q^\nu_\varrho(x_0))}{\varrho^{d-1}} = \lim_{\varepsilon\to 0} \frac{\mu_\varepsilon(Q^\nu_{\varrho_\varepsilon}(x_0))}{\varrho_\varepsilon^{d-1}}.$$

The problem is now to estimate $\frac{\mu_\varepsilon(Q^\nu_{\varrho_\varepsilon}(x_0))}{\varrho_\varepsilon^{d-1}}$ by using the definition of E_ε. (Up to this point, we have only used the fact that the measures μ_ε are equibounded.) The idea is to obtain a lower bound by minimizing the effect of the sequence $\{u^\varepsilon\}$ with the given condition that the related sets $A_\varepsilon(u^\varepsilon)$ are close

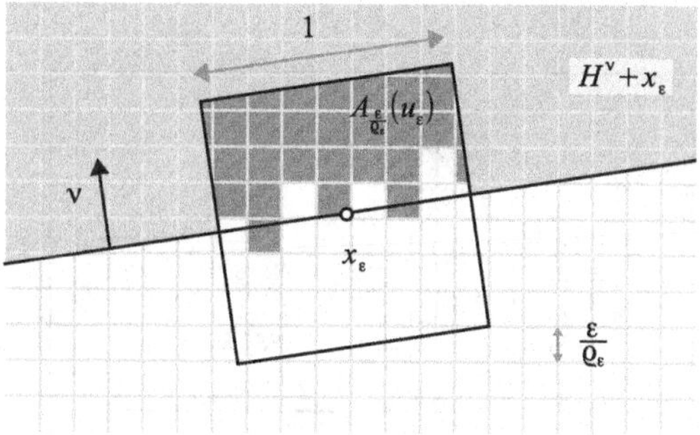

Figure 3.8 Scaling to a cube of side length 1.

to a hyperplane orthogonal to ν and through x_0. We start by rescaling the cube $Q^\nu_{\varrho_\varepsilon}(x_0)$ to a cube with side length 1 (Fig. 3.8). Since the coefficients a_{ij} are positive, for the lower bound we can consider only the interactions between points inside the cube, and we obtain

$$\frac{\mu_\varepsilon(Q^\nu_{\varrho_\varepsilon}(x_0))}{\varrho_\varepsilon^{d-1}} \geq \sum_{\frac{\varepsilon}{\varrho_\varepsilon}i,\frac{\varepsilon}{\varrho_\varepsilon}j\in Q^\nu_1(x_\varepsilon)} \frac{\varepsilon^{d-1}}{\varrho_\varepsilon^{d-1}} a_{ij}(u_i^\varepsilon - u_j^\varepsilon)^2,$$

where $x_\varepsilon = \frac{x_0}{\varrho_\varepsilon}$. Since giving a boundary condition is easier to handle than that of being close to $H^\nu(x_\varepsilon) = H^\nu + x_\varepsilon$, we show that we can modify the sequence $\{u^\varepsilon\}$ close to the boundary of the cube $Q^\nu_1(x_\varepsilon)$ without essentially modifying the energies.

Lemma 3.7 (Variation of boundary data) *For any fixed $\delta \in (0, \frac{1}{4})$, there exists $\tilde{u}^\varepsilon : Q^\nu_1(x_\varepsilon) \cap \frac{\varepsilon}{\varrho_\varepsilon}\mathbb{Z}^d \to \{-1, 1\}$ such that*

$$A_{\frac{\varepsilon}{\varrho_\varepsilon}}(\tilde{u}^\varepsilon) \cap \frac{\varepsilon}{\varrho_\varepsilon}\mathbb{Z}^d = H^\nu(x_\varepsilon) \cap \frac{\varepsilon}{\varrho_\varepsilon}\mathbb{Z}^d \quad \text{in a } \delta\text{-neighborhood of } \partial Q^\nu_1(x_\varepsilon),$$

and

$$\sum_{\frac{\varepsilon}{\varrho_\varepsilon}i,\frac{\varepsilon}{\varrho_\varepsilon}j\in Q^\nu_1(x_\varepsilon)} \frac{\varepsilon^{d-1}}{\varrho_\varepsilon^{d-1}} a_{ij}(u_i^\varepsilon - u_j^\varepsilon)^2$$
$$\geq \sum_{\frac{\varepsilon}{\varrho_\varepsilon}i,\frac{\varepsilon}{\varrho_\varepsilon}j\in Q^\nu_1(x_\varepsilon)} \frac{\varepsilon^{d-1}}{\varrho_\varepsilon^{d-1}} a_{ij}(\tilde{u}_i^\varepsilon - \tilde{u}_j^\varepsilon)^2 + o(1)_{\varepsilon\to 0} - c\delta.$$

Remark 3.8 (A discrete coarea argument) The proof of Lemma 3.7 relies on a discrete coarea argument, which we briefly describe in a simple case in dimension $d = 2$. Let $R \subset \mathbb{R}^2$ be a coordinate rectangle and b denote the length

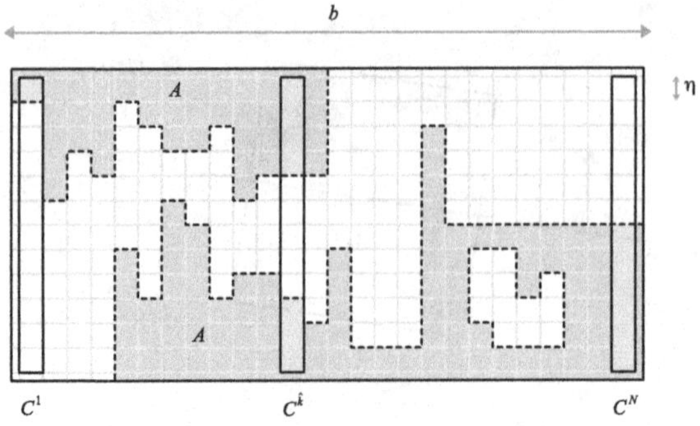

Figure 3.9 An illustration of Remark 3.8.

of the basis. Let A be a subset of R given by the union of coordinate squares with side length η and centered at points of $\eta\mathbb{Z}^2$. We consider the columns of squares in A such that the centers have the same first coordinate, and indicate them by $C^1,\dots,C^N$ with $N = \lfloor \frac{b}{\eta} \rfloor - 1$ (see Fig. 3.9).

For any $k \in \{1,\dots,N\}$ we have that

$$\mathcal{H}^1(\partial C^k) \leq \#(\eta\mathbb{Z}^2 \cap C^k)4\eta \leq \frac{4|C^k|}{\eta}.$$

Since $\sum_{k=1}^N |C^k| \leq |A|$, there exists an index $\hat{k}$ such that $|C^{\hat{k}}| \leq \frac{|A|}{N}$, hence

$$\mathcal{H}^1(\partial C^{\hat{k}}) \leq C\frac{|A|}{b}.$$

Proof of Lemma 3.7 We suppose that $x_0 = 0$ and $\nu = e_d$, in order to avoid the translations of the centers of the cubes since $x_\varepsilon = 0$. We set $Q^\nu_\varrho = Q^\nu_\varrho(0)$. This results in a simplification of the notation without changing the main features of the proof. We make another simplifying hypothesis, supposing that we have only nearest-neighbor interactions; that is, $a_{ij} = 0$ if $\|i - j\| > 1$. This assumption allows one to represent the interactions more easily, which for nearest neighbor correspond to interfaces on the reference lattice. Note that we can generalize the proof to the case of finite-range interactions with some more technical details.

We set $\eta_\varepsilon = \frac{\varepsilon}{\varrho_\varepsilon}$. Let $A_\varepsilon = A_{\eta_\varepsilon}(u_\varepsilon)$, and H^ν_ε denote the set given by the union of cubes $Q^\nu_{\eta_\varepsilon}(x)$ with $x \in \eta_\varepsilon\mathbb{Z}^d \cap H^\nu$. We define the set $\tilde{A}_\varepsilon$ (and the corresponding discrete function $\tilde{u}^\varepsilon$) by setting

$$\tilde{A}_\varepsilon = \begin{cases} A_\varepsilon & \text{in } Q^\nu_{(2k+1)\eta_\varepsilon}, \\ H^\nu_\varepsilon & \text{otherwise in } Q^\nu_1, \end{cases}$$

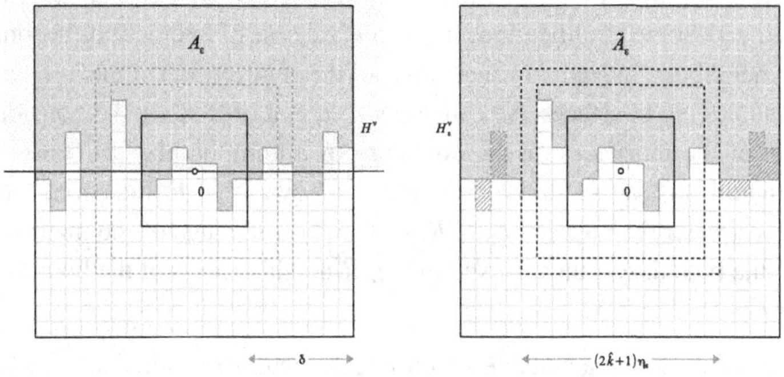

Figure 3.10 Modification of A_ε close to the boundary of Q_1^ν.

where $k \in \mathbb{N}$ is such that $\frac{1-2\delta}{2\eta_\varepsilon} - \frac{1}{2} < k < \frac{1}{2\eta_\varepsilon} - \frac{1}{2}$ and $\delta \in (0, \frac{1}{4})$. We estimate the energy $E_\varepsilon(\tilde{u}^\varepsilon)$ inside the cube Q_1^ν, obtaining

$$\sum_{\eta_\varepsilon i, \eta_\varepsilon j \in Q_1^\nu} \eta_\varepsilon^{d-1} a_{ij}(\tilde{u}_i^\varepsilon - \tilde{u}_j^\varepsilon)^2 \leq \sum_{\eta_\varepsilon i, \eta_\varepsilon j \in Q_1^\nu} \eta_\varepsilon^{d-1} a_{ij}(u_i^\varepsilon - u_j^\varepsilon)^2 + C\delta + r(\varepsilon, k),$$

where $C\delta$ estimates the contribution of the boundary of H^ν in $Q_1^\nu \setminus Q_{1-2\delta}^\nu$ and $r(\varepsilon, k)$ is the contribution of the additional boundary of $\tilde{A}_\varepsilon$ in $\partial Q_{(2k+1)\eta_\varepsilon}^\nu$ (see Fig. 3.10). To prove that we can choose k such that this term is infinitesimal as $\varepsilon \to 0$, we use a discrete coarea argument as described in Remark 3.8 with $d = 2$. Setting

$$C^k = (Q_{(2k+1)\eta_\varepsilon}^\nu \setminus Q_{(2k-1)\eta_\varepsilon}^\nu) \cap A_\varepsilon$$

for all admissible k, which are less than $\frac{\delta}{\eta_\varepsilon}$, we have that there exists $\hat{k}$ such that

$$|C^{\hat{k}}| \leq \frac{\eta_\varepsilon}{\delta} |A_\varepsilon \triangle H_\varepsilon^\nu|.$$

As in Remark 3.8 applied to $A_\varepsilon \triangle H_\varepsilon^\nu$, we deduce that

$$\mathcal{H}^{d-1}(A_\varepsilon \cap \partial Q_{(2\hat{k}+1)\eta_\varepsilon}^\nu) \leq C\#(\eta_\varepsilon \mathbb{Z}^d \cap C^{\hat{k}})\eta_\varepsilon^{d-1} \leq C\frac{|C^{\hat{k}}|}{\eta_\varepsilon} \leq \frac{C}{\delta}|A_\varepsilon \triangle H_\varepsilon^\nu|,$$

where C is a positive constant depending on d. Hence, by choosing $k = \hat{k}$, we obtain

$$\sum_{\eta_\varepsilon i, \eta_\varepsilon j \in Q_1^\nu} \eta_\varepsilon^{d-1} a_{ij}(\tilde{u}_i^\varepsilon - \tilde{u}_j^\varepsilon)^2 \leq \sum_{\eta_\varepsilon i, \eta_\varepsilon j \in Q_1^\nu} \eta_\varepsilon^{d-1} a_{ij}(u_i^\varepsilon - u_j^\varepsilon)^2 + C\delta + \frac{C}{\delta}|A_\varepsilon \triangle H_\varepsilon^\nu|$$

concluding the proof. □

Remark 3.9 (Finite-range interactions) The proof of Lemma 3.7 can be repeated with some modifications in the general case of interactions beyond the nearest neighbors. Indeed, in the estimate of the contribution of the interactions due to additional interfaces, we can apply a variant of Remark 3.8 considering the sum of the interfaces on the boundary of a finite number of consecutive cubic annuli, instead of only one. Note, moreover, that in the proof we can choose $\hat{k}$ greater than a fixed $k_0 > R$, where R is the range of the interactions. Hence, the boundary data are fixed in a $\eta_\varepsilon R$-neighborhood of the boundary of the cube Q_1^ν.

We can now define a suitable function φ for which (3.6) holds. By Lemma 3.7, since $|A_\varepsilon \cap (H_\varepsilon^\nu + x_\varepsilon)| \to 0$ as $\varepsilon \to 0$ and δ is arbitrarily small, we get

$$\liminf_{\varepsilon\to 0} \frac{\mu_\varepsilon(Q^\nu_{\varrho_\varepsilon}(x_0))}{\varrho_\varepsilon^{d-1}} \geq \liminf_{\varepsilon\to 0} \sum_{\frac{\varepsilon}{\varrho_\varepsilon}i,\frac{\varepsilon}{\varrho_\varepsilon}j\in Q_1^\nu(x_\varepsilon)} \Big(\frac{\varepsilon}{\varrho_\varepsilon}\Big)^{d-1} a_{ij}(\tilde{u}_i^\varepsilon - \tilde{u}_j^\varepsilon)^2. \quad (3.8)$$

To shorten the notation, we let

$$\mathcal{S}_\eta(U) = \{u\colon \mathcal{L}_\eta(U) \to \{-1,1\}\} \quad (3.9)$$

where $\mathcal{L}_\eta(U) = \eta\mathbb{Z}^d \cap U$ as in (2.22). If $\eta = 1$, we simply write $\mathcal{S}(U)$. Thanks to (3.8) and to Remark 3.9, we can estimate, up to an infinitesimal term as $\varepsilon \to 0$,

$$\frac{\mu_\varepsilon(Q^\nu_{\varrho_\varepsilon}(x_0))}{\varrho_\varepsilon^{d-1}} \geq \min\Big\{ \sum_{\frac{\varepsilon}{\varrho_\varepsilon}i,\frac{\varepsilon}{\varrho_\varepsilon}j\in Q_1^\nu(x_\varepsilon)} \Big(\frac{\varepsilon}{\varrho_\varepsilon}\Big)^{d-1} a_{ij}(u_i - u_j)^2 \colon\ u \in \mathcal{S}_{\frac{\varepsilon}{\varrho_\varepsilon}}(Q_1^\nu(x_\varepsilon)),$$
$$u = 2\chi_{H^\nu(x_\varepsilon)} - 1 \text{ in } Q_1^\nu(x_\varepsilon)\setminus Q^\nu_{1-\frac{2\varepsilon R}{\varrho_\varepsilon}}(x_\varepsilon)\Big\},$$

where R is the range of the interactions. We now set

$$\varphi(\nu) = \liminf_{\eta\to 0} \min\Big\{ \sum_{\eta i,\eta j\in Q_1^\nu(y)} \eta^{d-1} a_{ij}(u_i - u_j)^2 \ :\ u \in \mathcal{S}_\eta(Q_1^\nu(y)),$$
$$u = 2\chi_{H^\nu(y)} - 1 \text{ in } Q_1^\nu(y)\setminus Q^\nu_{1-2\eta R}(y),\ y \in \mathbb{R}^d\Big\}$$
$$= \liminf_{T\to+\infty} \frac{1}{T^{d-1}} \min\Big\{ \sum_{i,j\in Q_T^\nu(y)} a_{ij}(u_i - u_j)^2 \ :\ u \in \mathcal{S}(Q_T^\nu(y)),$$
$$u = 2\chi_{H^\nu(y)} - 1 \text{ in } Q_T^\nu(y)\setminus Q^\nu_{T-2R}(y),\ y \in \mathbb{R}^d\Big\},$$

where we have rescaled by setting $T = \frac{1}{\eta}$ and minimized over all possible centers of the cubes. The definition of φ, corresponding to a minimization over all oscillations close to a flat interface orthogonal to ν, does not depend on the

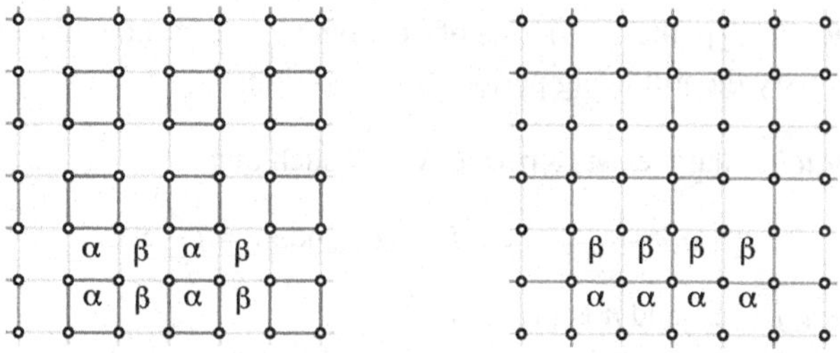

Figure 3.11 Two nonhomogeneous systems in $\mathbb{Z}^2$. The Greek letters refer to the the link above them.

subsequence and on the choice of the center of the cube, and depends only on the normal ν.

We then have the lower bound for the energies E_ε

$$\liminf_{\varepsilon\to 0} E_\varepsilon(u_\varepsilon) \geq \int_{\Omega\cap\partial^* A} \varphi(\nu(x))\, d\mathcal{H}^{d-1} \tag{3.10}$$

for φ defined as earlier.

3.3 Periodic Homogenization

In this section we consider periodic arrangements of a_{ij}. If K denotes the period of the system, then periodicity gives an invariance by εK translations of the energies at a discrete level and eventually ensures the homogeneity of the perimeter functionals in the limit.

The homogenization process is already interesting when we only have nearest-neighbor interactions. In Fig. 3.11 we picture two two-dimensional nearest-neighbor systems with a_{ij} taking only the values α and β. On the left-hand side picture the connections are located in series, on the right-hand side picture in parallel. While the connections are in equal number, their geometrical properties are different; in particular, in the system on the left-hand side connections with strength β form isolated loops, If $\beta > \alpha$, such loops are sometimes called *hard inclusions*. Both systems have the property of being periodic of period 2 in each direction.

We now formalize the hypotheses on the system of coefficients. Let $\Omega \subset \mathbb{R}^d$ be an open set, and $\mathcal{L}_\varepsilon = \Omega \cap \varepsilon\mathbb{Z}^d$. We again consider the family of energies given by (3.1).

In addition to hypotheses (i)–(iii) of Section 3.2, we assume that the coefficients a_{ij} satisfy the following periodicity condition:

(iv) (*periodicity*) there exists a period $K \in \mathbb{N}$ such that

$$a_{ij} = a_{lm}, \text{ if } l = i + Ke_n \text{ and } m = j + Ke_n$$

for any $i, j \in \mathbb{Z}^d$ and $n \in \{1, \dots, d\}$.

This hypothesis generalizes the homogeneity assumption in Section 3.1, which holds if $K = 1$.

The coefficients satisfy the hypotheses of the previous sections, which guarantee that the domain of the Γ-limit (which exists up to subsequences) is given by sets of finite perimeter. Moreover, the Γ-limit is bounded from above and from below by (multiples of) the perimeter, so that we can conjecture that it is concentrated on the reduced boundary of A. Now we prove that there exists a homogeneous limit energy density $\varphi_{\hom}$ such that

$$\Gamma\text{-}\lim_{\varepsilon\to 0} E_\varepsilon(A) = \int_{\Omega\cap\partial^* A} \varphi_{\hom}(\nu)\, d\mathcal{H}^{d-1}.$$

The superposition arguments used in the proof of the Γ-convergence result for homogeneous networks do not work for nonhomogeneous energies since recovery sequences depend on the structure of the interactions. This can be checked in the systems of interactions represented in Fig. 3.11, where recovery sequences follow least-energy paths. While in the example pictured in the right-hand side the Γ-limit is a 1-crystalline perimeter with coefficient $\frac{\alpha+\beta}{2}$, corresponding to the trivial recovery sequences obtained by discretizing the target set, the hard-inclusion system on the left-hand side leads to a 1-crystalline perimeter with coefficient α (if $\alpha < \beta$), with recovery sequences avoiding the stronger β-connections (see Fig. 3.12).

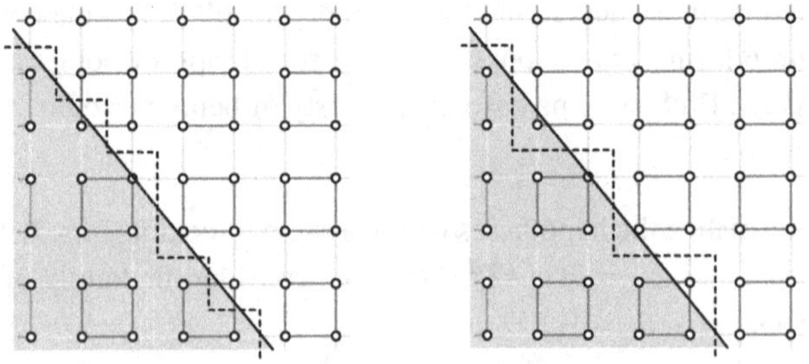

Figure 3.12 Discretization of a half-space and a corresponding recovery sequence.

3.3.1 The Asymptotic Homogenization Formula

By the blow-up method we have a lower bound with the function φ defined as in Section 3.2. We note that by the periodicity of a_{ij}, the minimum problems defining φ are invariant by translations in $K\mathbb{Z}^d$. Hence, we can consider only cubes centered in 0, obtaining

$$\varphi(\nu) = \liminf_{T\to+\infty} \frac{1}{T^{d-1}} \min\Big\{ \sum_{i,j\in Q_T^\nu} a_{ij}(u_i - u_j)^2 : u \in \mathcal{S}(Q_T^\nu), \quad u = 2\chi_{H^\nu} - 1 \text{ in } Q_T^\nu \setminus Q_{T-2R}^\nu \Big\}. \quad (3.11)$$

We now prove that the lim inf in this definition of φ is indeed a limit; that is, the function

$$\varphi_{\mathrm{hom}}(\nu) = \lim_{T\to+\infty} \frac{1}{T^{d-1}} \min\Big\{ \sum_{i,j\in Q_T^\nu} a_{ij}(u_i - u_j)^2 : u \in \mathcal{S}(Q_T^\nu), \quad u = 2\chi_{H^\nu} - 1 \text{ in } Q_T^\nu \setminus Q_{T-2R}^\nu \Big\} \quad (3.12)$$

is well defined. To that end, we define

$$g(T,\nu) = \min\Big\{ \sum_{i,j\in Q_T^\nu} a_{ij}(u_i - u_j)^2 : u \in \mathcal{S}(Q_T^\nu), u = 2\chi_{H^\nu} - 1 \text{ in } Q_T^\nu \setminus Q_{T-2R}^\nu \Big\}$$

and show that there exists

$$\lim_{T\to+\infty} \frac{g(T,\nu)}{T^{d-1}}.$$

To prove the existence of the limit, we use a *subadditivity argument*. Upon identifying spin functions with sets, this can be described as follows. Given an optimal set A_T (corresponding to a minimizer u^T) in the cube Q_T^ν, we want to construct a test set in a larger cube Q_S^ν, with $S \gg T$, providing an estimate showing that $S^{d-1}g(S,\nu)$ is not larger than $T^{d-1}g(T,\nu)$ up to negligible terms as $T \to +\infty$. This is done by constructing a competitor A_S patching up translated copies of A_T; this is a usual argument in *nonlinear homogenization* theories.

Note that we have translational invariance of the optimal set if we move the cube Q_T^ν by any vector Kw, where K is the period and $w \in \mathbb{Z}^d$. The elements of $K\mathbb{Z}^d$ are referred to as *admissible translations*. The translation moving Q_T^ν to an adjacent cube sharing a $d-1$ face and centered on the hyperplane orthogonal to ν is not in general an admissible translation for arbitrary ν and T. However, we can find an admissible translation such that the distance of the center of the translated cube from the hyperplane ∂H^ν is of order K and the distance between the centers of the cubes is of order $T + K$, as pictured (for $d = 2$) in Fig. 3.13. Then the distance between the cubes is uniformly bounded by a

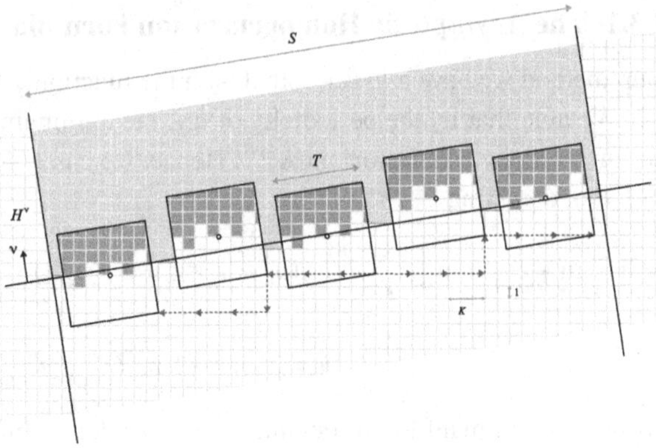

Figure 3.13 A pictorial proof of the homogenization formula.

constant C, and the same holds for the distance of the centers from ∂H^ν. We repeat this construction by invading $\partial H^\nu \cap Q^\nu_S$ with non overlapping cubes until the distance of the translated cubes from the boundary of Q^ν_S is less than $2T$, and define a set given by the translation of A_T in each of these cubes and by the discretization of the half-space H^ν otherwise in Q^ν_S. Hence, since the number of the cubes is less than $(\frac{S}{T})^{d-1}$, we have

$$\frac{g(S,\nu)}{S^{d-1}} \leq \frac{1}{S^{d-1}}\Big(\frac{S^{d-1}}{T^{d-1}}g(T,\nu) + \frac{S^{d-1}}{T^{d-1}}CT^{d-2} + (S^{d-1} - (S-2T)^{d-1})\Big),$$

where the term CT^{d-2} estimates the additional interface close to each translated cube with side length T, and the last term in the sum estimates the additional interface close to the boundary of Q^ν_S. By taking the lim sup as $S \to +\infty$ we obtain

$$\limsup_{S\to+\infty} \frac{g(S,\nu)}{S^{d-1}} \leq \frac{g(T,\nu)}{T^{d-1}} + \frac{C}{T}$$

since $S^{d-1} - (S-2T)^{d-1} = o(S^{d-1})_{S\to+\infty}$. Now, taking the lim inf as $T \to +\infty$, it follows that

$$\limsup_{S\to+\infty} \frac{g(S,\nu)}{S^{d-1}} \leq \liminf_{T\to+\infty} \frac{g(T,\nu)}{T^{d-1}},$$

which implies that the lim inf in the definition of φ is in fact a limit.

3.3.2 Upper Bound

We note that the construction of the test set in Q^ν_S used to prove the existence of the limit is in fact the construction of a recovery sequence for the Γ-limit of E_ε for a half-space. Indeed, let w^T denote a minimizer of the energy in the cube Q^ν_T with $w^T = 2\chi_{H^\nu} - 1$ close to the boundary of Q^ν_T; that is,

$$g(T,\nu) = \sum_{i,j\in Q^T_\nu} a_{ij}(w^T_i - w^T_j)^2.$$

We indicate by $\tilde{w}^T$ the function constructed by translating the cube Q^ν_T, as in the proof of the existence of the limit, to "almost invade" the whole hyperplane ∂H^ν, which corresponds to letting S go to $+\infty$. If A is a polytope, we localize the construction of the recovery sequence by considering A as a portion of a half-space with boundary orthogonal to ν. We leave to the reader the details of the construction for a general polytope, noting that a small variation of this construction is necessary close to the boundary of each face where the recovery sequence can be simply taken as the discretization of the polytope itself. We define u^ε by setting $u^\varepsilon_i = \tilde{w}^T_i$. Then

$$\begin{aligned}\limsup_{\varepsilon\to 0} E_\varepsilon(u^\varepsilon) &= \limsup_{\varepsilon\to 0} \sum_{\varepsilon i,\varepsilon j\in\mathcal{L}_\varepsilon(\Omega)} \varepsilon^{d-1} a_{ij}(u^\varepsilon_i - u^\varepsilon_j)^2 \\ &\le \limsup_{\varepsilon\to 0} \varepsilon^{d-1}\frac{\mathcal{H}^{d-1}(\Omega\cap\partial A)}{\varepsilon^{d-1}T^{d-1}}(g(T,\nu) + CT^{d-2}) \\ &\le \mathcal{H}^{d-1}(\Omega\cap\partial A)\,\varphi_{\rm hom}(\nu) + o(1)_{T\to+\infty},\end{aligned}$$

which shows that $\{u^\varepsilon\}$ is a recovery sequence up to a term arbitrarily small. As we noticed in the definition of Γ-convergence, this is sufficient to give an upper bound for the Γ-limit. We can use again the density of polyhedral sets to conclude that for any A with finite perimeter,

$$\Gamma\text{-}\lim_{\varepsilon\to 0} E_\varepsilon(A) \le \int_{\Omega\cap\partial^* A} \varphi_{\rm hom}(\nu)\, d\mathcal{H}^{d-1}.$$

This concludes the proof of the following result.

Theorem 3.10 (Homogenization of periodic networks) *Let a_{ij} be a periodic system of ferromagnetic interactions satisfying hypotheses* (i)–(iv). *Then the sequence of functionals E_ε defined in* (3.1) *Γ-converges with respect to the convergence $u^\varepsilon \to A$ to the functional F defined on sets of finite perimeter by*

$$F(A) = \int_{\Omega\cap\partial^* A} \varphi_{\rm hom}(\nu)\, d\mathcal{H}^{d-1},$$

where $\varphi_{\rm hom}$ satisfies the asymptotic formula (3.12).

Definition 3.11 (Homogenized energy density of the system a_{ij}) Given a periodic system a_{ij} of ferromagnetic interactions satisfying hypotheses (i)–(iv), the corresponding function $\varphi_{\rm hom}$ given by the asymptotic formula (3.12) is the *homogenized energy density of the system* a_{ij}, or the *surface tension of the system* a_{ij}.

Remark 3.12 (Crystallinity of $\varphi_{\rm hom}$ and a dual formula) The hypothesis that a_{ij} is of finite range guarantees that $\varphi_{\rm hom}$ is crystalline as in the homogeneous case. The proof of this fact can be achieved by showing that an alternative formula for $\varphi_{\rm hom}$ can be obtained by minimization on functions $w\colon \mathbb{Z}^d \to \mathbb{R}$ satisfying a periodicity assumption; namely,

$$\varphi_{\rm hom}(\nu) = \frac{4}{K^d} \min\Big\{ \sum_{\substack{i\in\{1,\dots,K\}^d\\ j\in\mathbb{Z}^d}} a_{ij}|w_i - w_j| : w_i - \langle \nu, i\rangle \ K\text{-periodic}\Big\}, \quad (3.13)$$

where K is the period of the coefficients. Such a formula is achieved by remarking that the homogenization of periodic lattice energies defined on spin functions is equivalent to the homogenization of their one-homogeneous extensions defined on functions with values in $\mathbb{R}$, in analogy with a similar result in the continuum case. Since these energies are convex, their homogenized energy density is defined by a cell-problem formula, which is the one just shown. We do not include the proof of this result, since it would require a treatment of discrete energies at the bulk scaling outside the scope of this text and with little relation to the rest of the book.

Note that in the case of homogeneous coefficients $a_{ij} = \alpha_{j-i}$ formula (3.13) agrees with (3.2). Also, note that in Theorem 3.10 the finite-range hypothesis can be relaxed (in which case $\varphi_{\rm hom}$ may not be crystalline). This will be done in Section 4.8.

3.3.3 Plane-Like Minimizers

The homogenization of ferromagnetic spin systems can be deduced from a stronger property; namely, the existence of *plane-like minimizers*.

Definition 3.13 (Plane-like minimizer) Let a_{ij} be a family of nonnegative coefficients and let $\nu \in S^{d-1}$. A function $u\colon \mathbb{Z}^d \to \{-1,1\}$ is a *plane-like minimizer* for an energy

$$E(u) = \sum_{i,j\in\mathbb{Z}^d} a_{ij}(u_i - u_j)^2$$

in the direction ν if

(i) the interface of u is contained in a strip orthogonal to ν; that is, there exists $R > 0$ such that $u_i = 1$ if $\langle i, \nu\rangle \geq R$ and $u_i = -1$ if $\langle i, \nu\rangle \leq -R$;

(ii) u is minimal with respect to compactly supported perturbations; that is,

$$\sum_{i,j\in\mathbb{Z}^d} a_{ij}\Big((u_i - u_j)^2 - (v_i - v_j)^2\Big) \le 0$$

if $u_i = v_i$ except for a finite number of indices i.

The existence of plane-like minimizers simplifies the computation of the homogenized energy density.

Proposition 3.14 *Let a_{ij} satisfy the hypotheses of Theorem* 3.10 *and suppose that u^ν is a plane-like minimizer for E in the direction ν. If $\varphi_{\rm hom}$ is defined as in* (3.12), *then*

$$\varphi_{\rm hom}(\nu) = \lim_{T\to+\infty} \frac{1}{T^{d-1}} \sum_{i,j\in Q_T^\nu} a_{ij}(u_i^\nu - u_j^\nu)^2. \tag{3.14}$$

Proof It suffices to note that in the homogenization formula for $\varphi_{\rm hom}$ we can take test functions u, which satisfy $u = u^\nu$ in a neighborhood of the boundary of Q_T^ν. For such problems u^ν is a minimizer by definition, and formula (3.14) immediately follows. □

The following result holds for periodic systems. We do not include its proof, which is due to Caffarelli and de la Llave.

Theorem 3.15 (Existence of plane-like minimizers) *Let a_{ij} be a periodic system of coefficients as in Theorem* 3.10. *Then for every $\nu \in S^{d-1}$ there exists a plane-like minimizer for the energy E in direction ν.*

3.3.4 Multi-Bravais Lattices

We can generalize the homogenization results of the previous sections to the case of a multi-Bravais lattice. Here we only consider a simplified homogeneous model on a multi-Bravais lattice given by the union of a Bravais lattice and a single translated copy. The case of more complex structures can be analyzed in complete analogy. For periodic coefficients, the blow-up argument can be extended almost word for word.

Given a basis $\{v_1, v_2, \dots, v_d\}$ of $\mathbb{R}^d$, let $\mathcal{L}^0$ be the associated d-dimensional Bravais lattice. For $\tau \in \mathbb{R}^d \setminus \mathcal{L}^0$ we set $\mathcal{L}^\tau = \mathcal{L}^0 + \tau$, the translation of $\mathcal{L}^0$ by τ. We define the *multi-Bravais lattice* $\mathcal{L} = \mathcal{L}^0 \cup \mathcal{L}^\tau$.

Example 3.16 (The honeycomb lattice) Let $\mathcal{L}^0 = \mathbb{T}$ be the triangular lattice in $\mathbb{R}^2$ generated by the vectors $v_1 = (1,0)$ and $v_2 = (\frac{1}{2}, \frac{\sqrt{3}}{2})$, and let $\tau = (0, \frac{\sqrt{3}}{3})$. The *honeycomb lattice* is the lattice $\mathcal{L} = \mathcal{L}^0 \cup \mathcal{L}^\tau$ (see Fig. 3.14).

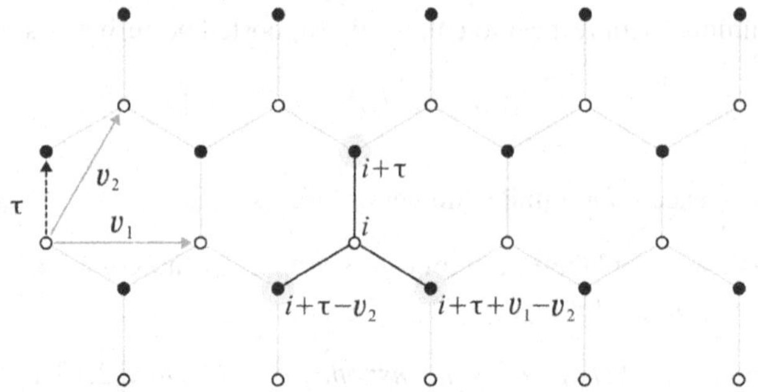

Figure 3.14 The honeycomb lattice.

As in the previous sections, for a given lattice $\widetilde{\mathcal{L}}$ and an open set Ω we set $\widetilde{\mathcal{L}}_\varepsilon(\Omega) = \varepsilon\widetilde{\mathcal{L}} \cap \Omega$. We consider the energies

$$\begin{aligned} E_\varepsilon(u) = &\sum_{k\in\mathcal{L}^0} \alpha_k^0 \sum_{\varepsilon i,\varepsilon(i+k)\in\mathcal{L}_\varepsilon^0(\Omega)} \varepsilon^{d-1}(u_i - u_{i+k})^2 \\ &+ \sum_{k\in\mathcal{L}^0} \alpha_k^\tau \sum_{\varepsilon i,\varepsilon(i+k)\in\mathcal{L}_\varepsilon^\tau(\Omega)} \varepsilon^{d-1}(u_i - u_{i+k})^2 \\ &+ \sum_{k\in\mathcal{L}^0} \alpha_k \sum_{\varepsilon i\in\mathcal{L}_\varepsilon(\Omega),\varepsilon(i+k+\tau)\in\mathcal{L}_\varepsilon^\tau(\Omega)} \varepsilon^{d-1}(u_i - u_{i+k+\tau})^2 \end{aligned}$$

defined for $u\colon \mathcal{L}_\varepsilon(\Omega) \to \{-1,1\}$, where for $k \in \mathcal{L}^0$, α_k^0, α_k^τ, and α_k are non-negative coefficients satisfying the finite-range assumption. Then, the following result holds.

Theorem 3.17 (Limits of homogeneous systems on multi-Bravais lattices) *The Γ-limit of the sequence of energies E_ε defined previously with respect to the convergence $u^\varepsilon \to A$ is given by*

$$F(A) = \int_{\Omega\cap\partial^* A} 4c_{\mathcal{L}^0}\Big(\sum_{k\in\mathcal{L}^0}(\alpha_k^0+\alpha_k^\tau)|\langle\nu(x),k\rangle| + \sum_{k\in\mathcal{L}^0}\alpha_k|\langle\nu(x),k+\tau\rangle|\Big)\, d\mathcal{H}^{d-1}(x)$$

for all A sets of finite perimeter, where $c_{\mathcal{L}^0} = |C|^{-1}$ and C is the unitary cell of $\mathcal{L}^0$ defined as in (2.25).

Example 3.18 (Nearest neighbors on the honeycomb lattice) On the honeycomb lattice defined earlier we consider nearest-neighbor interactions; that is, we take $\alpha_k^0 = \alpha_k^\tau = 0$ and

$$\alpha_k = \begin{cases} 1 & \text{if } k \in \{0, -v_2, v_1 - v_2\}, \\ 0 & \text{otherwise.} \end{cases}$$

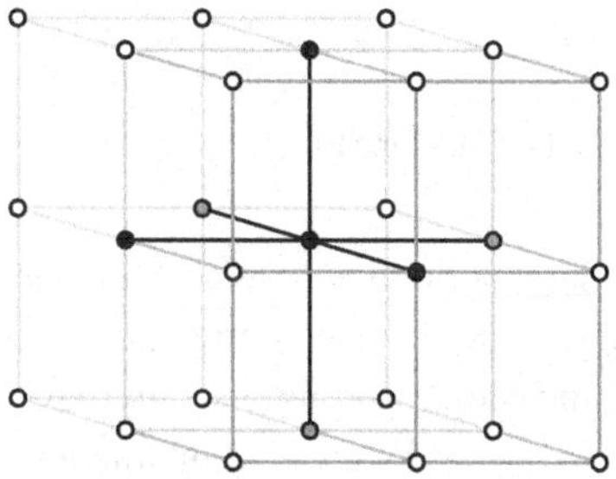

Figure 3.15 Connected perforated domain with connected complement.

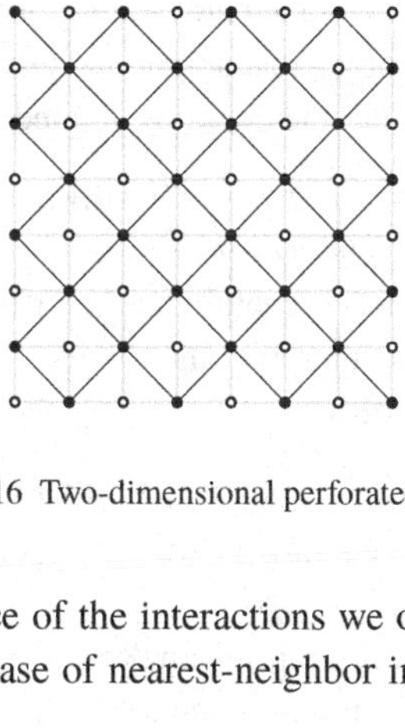

Figure 3.16 Two-dimensional perforated domain.

Note that for such a choice of the interactions we obtain the same interfacial energy density as in the case of nearest-neighbor interactions on a triangular lattice.

3.3.5 Perforated Domains

The scope of this section is to show that the homogenization theorem holds also in some cases when the coerciveness conditions, that is, the positiveness of coefficients, are not satisfied for all nearest-neighbor connections. We have in mind what in the continuum counterpart are called *(periodically) perforated domains*. Such domains can be viewed as a coercive medium to which a periodic array of bounded regions has been removed, or, more in general, as a periodic connected coercive medium.

In the discrete setting with nearest-neighbor interactions a periodicity cell of such a domain in dimension three is pictured in Fig. 3.15. A next-to-nearest-neighbor perforated domain with inclusions in dimension two is represented in Fig. 3.16, where the black and gray sites represent the elements in the medium (the black ones being those in the periodicity cell, and the white sites those in the complement).

Figure 3.17 One-dimensional perforated domain.

Note that for nearest-neighbor interactions both the medium and the complement may be connected only in dimension three or higher, as in the continuum case, and in the one-dimensional setting we only have the trivial case with empty complement. If we allow long-range interactions (e.g. next-to-nearest interactions), then we may have connected perforated domains also in one dimension, as pictured in Fig. 3.17.

We will model perforations in a general setting, using possibly vanishing interaction coefficients, even for nearest neighbors. Given Ω a bounded Lipschitz open subset of $\mathbb{R}^d$, we consider the energies defined by

$$E_\varepsilon(u) = \sum_{\varepsilon i, \varepsilon j \in \mathcal{L}_\varepsilon(\Omega)} \varepsilon^{d-1} a_{ij}(u_i - u_j)^2,$$

with the nonnegative symmetric coefficients a_{ij} periodic with period K and $u \colon \mathcal{L}_\varepsilon(\Omega) \to \{-1, 1\}$. We assume that the range of a_{ij} is finite, but we allow coefficients to be 0, so that the set

$$\mathcal{B} = \{i \in \mathbb{Z}^d : a_{ij} = 0 \text{ for all } j \in \mathbb{Z}^d\} \tag{3.15}$$

may be not empty, and the generalized coerciveness condition of Remark 3.3 does not hold.

The missing coerciveness assumption on $\mathcal{B}$ requires a new argument to prove compactness of sequences with equibounded energy in a suitable sense. The idea, as mentioned earlier, is to consider the missing sites in $\mathcal{B}$ as "holes" inside the domain, and (with some connectedness hypothesis on the complement of $\mathcal{B}$), given a sequence $\{u^\varepsilon\}$ with equibounded energy, to extend u^ε in the scaled holes $\varepsilon\mathcal{B}$ by functions Lu^ε for which we can prove compactness.

Definition 3.19 (Connectedness of discrete sets) Let $\mathcal{E}$ be a symmetric subset of $\mathbb{Z}^d \times \mathbb{Z}^d$. Given a set $\mathcal{A} \subset \mathbb{Z}^d$, we say that it is *$\mathcal{E}$-connected* if for all $i, j \in \mathcal{A}$, there exists a path in $\mathcal{A}$ joining i and j with connections in $\mathcal{E}$; that is, there exists a subset $\{i_0, i_1, \ldots, i_n\} \subset \mathcal{A}$ such that $i_0 = i$, $i_n = j$, and $(i_{k-1}, i_k) \in \mathcal{E}$ for all $k \in \{1, \ldots, n\}$. If $\mathcal{E}$ is the set of nearest neighbors, we simply say that $\mathcal{A}$ is *connected in $\mathbb{Z}^d$*.

We define the set of *coercive sites* for the system a_{ij} as

$$C = \{i \in \mathbb{Z}^d : \text{there exists } j \in \mathbb{Z}^d \text{ such that } a_{ij} > 0\}, \tag{3.16}$$

and the set of connections

$$\mathcal{E} = \{(k, l) \in \mathbb{Z}^d \times \mathbb{Z}^d : a_{kl} > 0\}. \tag{3.17}$$

Proposition 3.20 (Extension and compactness) *Let the set C, defined in (3.16), be $\mathcal{E}$-connected and not empty. Let $\{u^\varepsilon\}$ be such that $E_\varepsilon(u^\varepsilon)$ is equibounded. Then there exist a sequence $\{Lu^\varepsilon\}$ and a set A of finite perimeter in Ω such that $Lu^\varepsilon \to A$ as $\varepsilon \to 0$ and $Lu^\varepsilon = u^\varepsilon$ in $\Omega \cap \varepsilon C$.*

Proof Let $\{u^\varepsilon\}$ be such that $E_\varepsilon(u^\varepsilon) \leq S < +\infty$. For any $\varepsilon > 0$ we define the two sets of indices $\mathcal{I}_\varepsilon^+$ and $\mathcal{I}_\varepsilon^-$ given by

$$\mathcal{I}_\varepsilon^\pm = \{j \in \mathbb{Z}^d : u^\varepsilon = \pm 1 \text{ on } \varepsilon(C \cap K(j + [0,1)^d))\}.$$

We let $M \in \mathbb{N}$ be such that for all $i, j \in [0,K)^d \cap C$, there exists a path connected in $\mathcal{E}$ contained in $[-MK, K + MK)^d$. We fix $\Omega' \subset\subset \Omega$ and ε small enough so that $\varepsilon(K(j + [-M, M + 1)^d)) \subset \Omega$ for all j such that $\varepsilon K j \in \Omega'$.

Now, let $j \in \mathcal{I}_\varepsilon^+$ be such that there exists $j' \notin \mathcal{I}_\varepsilon^+$ with $\|j - j'\| = 1$. Then there exist i and i' in $Kj + C \cap [-MK, K + MK]^d$ such that $(i, i') \in \mathcal{E}$ and $u_i^\varepsilon \neq u_{i'}^\varepsilon$. Hence, for each such j the contribution to the energy is at least $C\varepsilon^{d-1}$, where the constant C takes into account that $\min\{a_{ij} : (i,j) \in \mathcal{E}\} > 0$ and that each (i, i') as earlier may be shared by at most $(1+2M)^d$ neighboring periodicity cubes. The same holds for $\mathcal{I}_\varepsilon^-$, and we obtain $\#\mathcal{I}_\varepsilon^\pm \leq \frac{S}{C\varepsilon^{d-1}}$. Then, the sets

$$A_\varepsilon^\pm = \bigcup_{j \in \mathcal{I}_\varepsilon^\pm} \varepsilon(Kj + [0,K]^d)$$

have equibounded perimeter in Ω'. Setting

$$\mathcal{I}_\varepsilon^0(\Omega') = \left(\mathbb{Z}^d \cap \frac{1}{\varepsilon}\Omega'\right) \setminus (\mathcal{I}_\varepsilon^+ \cup \mathcal{I}_\varepsilon^-),$$

we note that for any $j \in \mathcal{I}_\varepsilon^0(\Omega')$, there exists a pair $(i, i') \in \mathcal{E}$ as earlier. Hence, again we have $\#\mathcal{I}_\varepsilon^0(\Omega') \leq \frac{S}{C\varepsilon^{d-1}}$. Moreover,

$$\left| \bigcup_{j \in \mathcal{I}_\varepsilon^0(\Omega')} \varepsilon(Kj + [0,K]^d) \right| \leq \frac{S}{C}\varepsilon.$$

Now, we can define the *extension* Lu^ε by setting

$$Lu_i^\varepsilon = \begin{cases} u_i^\varepsilon & \text{if } \varepsilon i \in \varepsilon C \cap \Omega, \\ 1 & \text{if } \varepsilon i \in \varepsilon(Kj + \mathbb{Z}^d \cap [0,K)^d) \text{ for } j \in \mathcal{I}_\varepsilon^+, \\ -1 & \text{if } \varepsilon i \in \varepsilon(Kj + \mathbb{Z}^d \cap [0,K)^d) \text{ for } j \in \mathcal{I}_\varepsilon^-, \\ 1 & \text{otherwise,} \end{cases}$$

the last value being arbitrary. By the previous considerations, we deduce that $\mathcal{H}^{d-1}(\Omega' \cap A_\varepsilon(Lu^\varepsilon))$ is equibounded and we obtain the claim. □

Remark 3.21 (Independence from the extension) The limit set A does not depend on the choice of the extended sequence $\{Lu^\varepsilon\}$. Indeed, let w be the weak limit of the piecewise-constant interpolation w^ε of the function $\tilde{u}^\varepsilon$ given by u^ε in $\Omega \cap \varepsilon C$ and 0 on $\varepsilon(\mathbb{Z}^d \setminus C)$. Since $Lu^\varepsilon = u^\varepsilon$ on $\Omega \cap \varepsilon C$ we have that w equals the limit of $\chi_{\varepsilon C} Lu^\varepsilon$, which weakly converge to $c\chi_{A\cap\Omega}$, c being the constant weak limit of $\chi_{\varepsilon C}$, so that $\chi_{A\cap\Omega} = \frac{w}{c}$.

Note that we cannot in general have an L^1 convergence on the whole Ω since $\Omega \cap \varepsilon C$ may have a "disconnected" part close to the boundary of Ω, on which we cannot have any control.

The proposition and Remark 3.21 ensure the coerciveness of energies E_ε with respect to the convergence $u^\varepsilon \to A$ understood as the convergence of $Lu^\varepsilon \to A$. This point is the only one where the coerciveness on nearest neighbors has been used in the proof of Theorem 3.10, which then holds unchanged, with the limit energy density $\varphi_{\rm hom}$ still characterized by the homogenization formula.

3.3.6 Almost-Periodic Homogenization

The homogenization theorem can be extended to almost-periodic coefficients. The set of coefficients a_{ij} is *almost periodic* if for all $\eta > 0$ there exists a *relatively dense set* $\mathcal{T}_\eta \subset \mathbb{Z}^d$; that is, there exists $L_\eta > 0$ such that $\mathcal{T}_\eta + [0, L_\eta]^d = \mathbb{R}^d$, such that for all $\tau \in \mathcal{T}_\eta$, we have

$$|a_{i+\tau\, j+\tau} - a_{ij}| \leq \eta$$

for all i and j. The elements of $\mathcal{T}_\eta$ are called *η-almost periods* of a_{ij}.

An example of such coefficients is obtained by taking $a_{ij} = f(i + j)$, where f is a continuous periodic function in $\mathbb{R}^d$ not of integer period; for example, in one dimension $f(t) = \sin t$.

Theorem 3.22 (Almost-periodic homogenization) *Let a_{ij} be a set of almost-periodic coefficients satisfying the hypotheses of Section* 3.2. *Then the conclusions of the homogenization Theorem* 3.10 *hold unchanged.*

Proof The proof is essentially the same as in the periodic case (and analogous to the corresponding result for the homogenization of perimeter energies in the continuum), using $\mathcal{T}_\eta$ in the place of $K\mathbb{Z}^d$ and the arbitrariness of $\eta > 0$. Indeed, even though energies are not invariant by translations of $\mathcal{T}_\eta$, they give an error which is negligible as $\eta \to 0$ and we may restrict the computation of a lower bound to cubes centered in 0 as in (3.11). Similarly, we may use translations in $\mathcal{T}_\eta$ to prove the existence of the limit in the homogenization formula, and to construct an upper bound, up to an error of order η. □

It may be interesting to note that, while the homogenization theorem is essentially unchanged from the periodic to the almost-periodic setting, in the latter there may not exist plane-like minimizers.

Example 3.23 (Nonexistence of plane-like minimizers) We consider nested sets of nearest neighbors. For $n \geq 1$ we define

$$B_n = \Big\{(i,j) \in \mathbb{Z}^2 \times \mathbb{Z}^2 : \|i-j\| = 1 \text{ and } \frac{i_1 + j_1}{2} \text{ or } \frac{i_2 + j_2}{2} \in \frac{1}{2} + 2{\cdot}3^n + 4{\cdot}3^n\mathbb{Z}\Big\}.$$

Since $2\cdot 3^{n+1} + 4\cdot 3^{n+1}\mathbb{Z} \subset 2\cdot 3^n + 4\cdot 3^n\mathbb{Z}$ we have $B_{n+1} \subset B_n$. We set $B_0 = \{(i,j) \in \mathbb{Z}^2 \times \mathbb{Z}^2 \colon \|i-j\| = 1\}$. For all $i, j \in \mathbb{Z}^2$ with $\|i-j\| = 1$, we define

$$a_{ij} = 2^{-n} \quad \text{if } (i,j) \in B_n \setminus B_{n+1}.$$

Note that $B_n \cap ([-3^n, 3^n]^2 \times [-3^n, 3^n]^2) = \emptyset$. Hence, we deduce that $\bigcap_n B_n = \emptyset$ and the coefficients a_{ij} are well defined whenever $\|i-j\| = 1$. Note, moreover, that we can write

$$a_{ij} = 1 - \sum_{n=1}^{+\infty} 2^{-n} \chi_{B_n}(i,j),$$

where χ_{B_n} is the characteristic function of B_n. This observation implies that minimal connections will use paths in B_n with the largest possible n and hence cannot be confined to stripes around a linear interface of a fixed width.

Note that the system of coefficients a_{ij} is the uniform limit of a family of periodic coefficients of increasing periods. This is perhaps the strongest notion of almost periodicity, implying all other types of almost periodicity.

The coefficients in the previous example are not *quasiperiodic*; that is, they cannot be written in terms of sections of periodic functions with noninteger period.

Conjecture 3.24 (Existence of plane-like minimizers for quasiperiodic systems) *We conjecture the existence of plane-like minimizers for quasiperiodic coefficients, for example, if $a_{ij} = f(i,j)$, where f is continuous and periodic of noninteger period in each of its entries.*

3.4 Aperiodic Lattices

In this section we consider energies defined on sets that can be interpreted as projections of sections of higher-dimensional lattices on lower-dimensional subspaces. We first treat the case of noncommensurate sections; a simpler formulation can be given for rational sections and will be briefly dealt with in Section 3.5.

3.4.1 Quasicrystals

Aperiodic objects as quasicrystals are often viewed as projections of a portion of a regular lattice on lower-dimensional (irrational) sections. They are not periodic, but possess some properties of quasiperiodicity that are sufficient to carry out a homogenization process. For the sake of brevity we treat the case of irrational *thin films*; that is, of limit theories parameterized on $\mathbb{R}^{d-1}$, and of energies defined on the cubic lattice $\mathbb{Z}^d$. The case of other lower-dimensional thin objects will be illustrated in a remark at the end of the section.

We consider $\zeta \in S^{d-1}$ such that $\langle \zeta, e_d \rangle \neq 0$. We define the *thin film* with normal ζ and thickness $N \in \mathbb{R}$ as

$$S_N = \{x \in \mathbb{Z}^d : 0 \leq \langle \zeta, x \rangle \leq N\}. \tag{3.18}$$

We suppose that

(i) ζ is an *irrational direction*; that is, having set $\Pi^\zeta = \{x \in \mathbb{R}^d : \langle \zeta, x \rangle = 0\}$,

$$\mathbb{Z}^d \cap \Pi^\zeta = \{0\}. \tag{3.19}$$

(ii) The set S_N is connected in $\mathbb{Z}^d$ (see Definition 3.19); that is, for all $i, j \in S_N$, there exists a subset $\{k_0, \ldots, k_M\} \subset S_N$ such that $k_0 = i$, $k_M = j$, and $\|k_l - k_{l-1}\| = 1$ for any l. This hypothesis is satisfied for large enough N.

Let ω be an open subset of $\mathbb{R}^{d-1}$. The nearest-neighbor energy of the thin film on ω is given by

$$E_\varepsilon(u) = \sum_{\substack{\|i-j\|=1 \\ \varepsilon i, \varepsilon j \in \Omega_\varepsilon}} \varepsilon^{d-2}(u_i - u_j)^2, \tag{3.20}$$

where $\Omega_\varepsilon = \varepsilon S_N \cap (\omega \times \mathbb{R})$ and $u : \Omega_\varepsilon \to \{-1, 1\}$.

In what follows $P : \mathbb{Z}^d \to \mathbb{Z}^{d-1}$ will be the orthogonal projection defined by $P(i_1, \ldots, i_d) = (i_1, \ldots, i_{d-1})$. Using the projection P, to each spin function $v : \varepsilon S_N \to \{-1, 1\}$, we can associate a projected function as in the following definition (see Fig. 3.18).

Definition 3.25 (Dimension reduction) Let ω be an open subset of $\mathbb{R}^{d-1}$ and let $\Omega_\varepsilon = (\omega \times \mathbb{R}) \cap \varepsilon S_N$. Given $v : \Omega_\varepsilon \to \{-1, 1\}$, we define the function

$$\mathcal{P}v : \varepsilon\mathbb{Z}^{d-1} \cap \omega \to \{-1, 0, 1\}$$

by setting

$$\mathcal{P}v_\alpha = \mathcal{P}v(\varepsilon\alpha) = \begin{cases} -1 & \text{if } v(\varepsilon i) = -1 \text{ for all } i \in P^{-1}(\alpha) \cap S_N, \\ 1 & \text{if } v(\varepsilon i) = 1 \text{ for all } i \in P^{-1}(\alpha) \cap S_N, \\ 0 & \text{otherwise;} \end{cases} \tag{3.21}$$

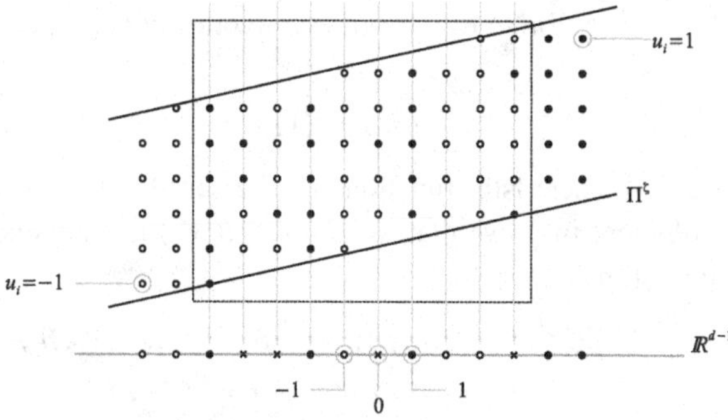

Figure 3.18 Projection of a discrete thin film on a coordinate hyperplane.

that is, $\mathcal{P}v(\varepsilon\alpha) = v(\varepsilon i)$ if $\alpha = P(i)$ and v is constant on the preimage $P^{-1}(\alpha)$ in S_N, while $\mathcal{P}v(\varepsilon\alpha) = 0$ otherwise. Note that, since $\langle\zeta, e^d\rangle \neq 0$, $\mathcal{P}v_\alpha$ is well defined for any α such that $\varepsilon\alpha \in \varepsilon\mathbb{Z}^{d-1} \cap \omega$.

Now we give a definition of convergence to a lower-dimensional object.

Definition 3.26 (Convergence to a dimensionally reduced parameter) Let $u^\varepsilon : \Omega_\varepsilon \to \{-1, 1\}$. We say that u^ε converge to a set of finite perimeter $A \subset \omega$ if $\mathcal{P}u^\varepsilon$ converge to A as in Definition 2.22; that is, $\mathcal{P}u^\varepsilon$ converge to $u = 2\chi_A - 1$ in $L^1_{\rm loc}(\omega)$, where $\mathcal{P}u^\varepsilon$ is identified with the piecewise-constant function defined by

$$\mathcal{P}u^\varepsilon(x) = \mathcal{P}u^\varepsilon_\alpha \ \text{ if } \ x \in \varepsilon\alpha + \Big(-\frac{\varepsilon}{2}, \frac{\varepsilon}{2}\Big)^{d-1}.$$

Proposition 3.27 (Compactness) *Let ω be a bounded open subset of $\mathbb{R}^{d-1}$ with Lipschitz boundary, and let $u^\varepsilon : \Omega_\varepsilon \to \{-1, 1\}$ be such that $E_\varepsilon(u^\varepsilon)$ is equibounded. Then, up to a subsequence, there exists a set of finite perimeter $A \subset \omega$ such that u^ε converge to A in the sense of Definition* 3.26.

Proof The result follows by estimating the perimeter of the sets $\{\mathcal{P}u^\varepsilon = 1\}$ and $\{\mathcal{P}u^\varepsilon = -1\}$, and showing that the sets $\{\mathcal{P}u^\varepsilon = 0\}$ are asymptotically negligible.

First, note that, for all $\alpha \in \mathbb{Z}^{d-1}$ such that $\mathcal{P}u^\varepsilon_\alpha = 0$, there exist nearest neighbors $i, j \in P^{-1}(\alpha) \cap S_N$ such that $u^\varepsilon_i \neq u^\varepsilon_j$; hence

$$4\varepsilon^{d-2}\#\{\alpha : \mathcal{P}u^\varepsilon_\alpha = 0\} \le E_\varepsilon(u^\varepsilon) \le c. \tag{3.22}$$

Moreover, for any $\omega' \subset\subset \omega$, we have

$$|\{x \in \omega' : \mathcal{P}u^\varepsilon(x) = 0\}| \le \varepsilon^{d-1}\#\{\alpha : \mathcal{P}u^\varepsilon_\alpha = 0\} \le c\varepsilon,$$

hence the sets $\{x \in \omega\colon \mathcal{P}u^\varepsilon(x) = 0\}$ are asymptotically negligible. Now, we set

$$V_\varepsilon = \{x \in \omega\colon \mathcal{P}u^\varepsilon(x) = 1\}.$$

To show that the characteristic functions of the sets V_ε are precompact in $L^1_{\mathrm{loc}}(\omega)$, it is sufficient to prove that $\mathcal{H}^{d-2}(\omega' \cap \partial^* V_\varepsilon)$ is equibounded. This follows from the estimate

$$\begin{aligned}\mathcal{H}^{d-2}(\omega' \cap \partial^* V_\varepsilon) &\leq \varepsilon^{d-2}\#\{(\alpha,\beta)\colon \|\alpha-\beta\| = 1,\ \alpha \in V_\varepsilon, \beta \notin V_\varepsilon\}\\ &\leq \varepsilon^{d-2}\#\{\alpha\colon (\mathcal{P}u^\varepsilon)_\alpha = 0\}\\ &\quad + \varepsilon^{d-2}\#\{(i,j)\colon \|i-j\| = 1, u_i^\varepsilon = 1, u_j^\varepsilon = -1\}\\ &= \varepsilon^{d-2}\#\{\alpha\colon (\mathcal{P}u^\varepsilon)_\alpha = 0\} + \frac{1}{8}E_\varepsilon(u^\varepsilon),\end{aligned}$$

which is equibounded by (3.22). This proves the compactness of the sets $\{\mathcal{P}u^\varepsilon = 1\}$ and hence also of the sets $\{\mathcal{P}u^\varepsilon = -1\}$ by complementarity. □

Now, we describe the asymptotic behavior of the energies E_ε through the computation of their Γ-limit with respect to the convergence in Definition 3.26. In order to define the limit surface tension, we introduce some notation in lower dimension, in accordance with the one already used.

For any $\nu \in S^{d-2}$, H^ν denotes the half-space $\{x \in \mathbb{R}^{d-1}\colon \langle x,\nu\rangle \geq 0\}$ and $Q_T^{\nu,d-1} = Q_T^\nu$ denotes a $d-1$-dimensional open cube with center at 0, side length T and one face orthogonal to ν. Moreover, for any $U \subset \omega$ we define the nonscaled localized energy

$$E(u;U) = \sum_{\substack{\|i-j\|=1\\ i,j \in S_N \cap (U\times\mathbb{R})}} (u_i - u_j)^2, \tag{3.23}$$

where $u\colon S_N \cap (\omega \times \mathbb{R}) \to \{-1,1\}$.

The proof of the homogenization theorem differs from the periodic case in that we cannot use a periodicity argument to prove the existence of the limit defining the homogenized energy density. To that end we have to use some quasiperiodicity property of the set S_N itself. We start by proving an almost-periodicity property due to the irrationality of the $d-1$-dimensional linear subspace Π^ζ.

Lemma 3.28 (Almost-periodic properties of S_N) *For all $\eta > 0$ we define*

$$\widehat{T}_\eta = \{\hat{\tau} \in S_N : \mathrm{dist}\,(\hat{\tau}, \Pi^\zeta) < \eta\}.$$

Then

(i) *the set $T_\eta = P(\widehat{T}_\eta)$ is a* relatively dense set; *that is, there exists $L_\eta > 0$ such that $T_\eta + [0, L_\eta]^{d-1} = \mathbb{R}^{d-1}$;*

(ii) *setting*

$$R_\eta = \inf\{\|\hat{\tau} - \hat{\tau}'\| : \hat{\tau}, \hat{\tau}' \in \widehat{T}_\eta,\ \hat{\tau} \neq \hat{\tau}'\}, \tag{3.24}$$

the sequence $\{R_\eta\}$ *diverges as* $\eta \to 0$.

Proof (i) The function $f(x) = \mathrm{dist}(x, \mathbb{Z}^d)$ is periodic and continuous; hence, its restriction to Π^ζ is quasiperiodic and the set $\Lambda_\eta = \{x \in \Pi^\zeta : f(x) < \eta\}$ is relatively dense. For each $x \in \Lambda_\eta$ there exists $\hat{\tau} \in \widehat{T}_\eta$ at a distance less than η, and correspondingly we find $\tau = P(\widehat{\tau}) \in T_\eta$. This shows that T_η is relatively dense.

(ii) If $\{R_\eta\}$ does not diverge, then there exist $\hat{\tau}_\eta, \hat{\tau}'_\eta$ such that the sequence $(\hat{\tau}_\eta - \hat{\tau}'_\eta)$ converges to $w \in \Pi^\zeta \cap \mathbb{Z}^d$. Since $w \neq 0$, this contradicts (3.19). □

We prove the existence of the surface tension for $d > 2$. We do not treat the case $d = 2$, which can be dealt with separately. In the proof that follows, we highlight the arguments where we use the hypothesis $d > 2$.

Proposition 3.29 (Existence of dimensionally reduced surface tension) *Let* $d > 2$. *For all* $\nu \in S^{d-2}$, *there exists the limit*

$$\varphi_{\mathrm{hom}}(\nu) = \lim_{T \to +\infty} \frac{1}{T^{d-2}} g_T(\nu), \tag{3.25}$$

where

$$g_T(\nu) = \min\left\{E(u; Q_T^\nu) \colon\ u \colon S_N \to \{-1, 1\},\ \mathcal{P}(u) = 2\chi_{H^\nu} - 1 \text{ on } \mathbb{Z}^{d-1} \backslash Q_{T-2}^\nu\right\}.$$

The core of the proof of Proposition 3.29 is the construction of test functions for $E(u; \tau + Q_T^\nu)$ starting from minimizers of $E(u; Q_T^\nu)$ for suitable integer translations τ in order to construct test functions for $g_S(\nu)$ with $S \gg T$ and apply a subadditive argument. The set of admissible translations will be the set T_η defined in Lemma 3.28. Note that, if u is an admissible test function for $E(u; Q_T^\nu)$, we cannot simply set $u_i^\tau = u_{i-\hat{\tau}}$ (with $P(\hat{\tau}) = \tau$) since in general,

$$\big((\hat{\tau} + (Q_T^\nu \times \mathbb{R})) \cap S_N\big) \setminus \big(\hat{\tau} + ((Q_T^\nu \times \mathbb{R}) \cap S_N)\big) \neq \emptyset;$$

that is, not all points where we consider the values of u^τ in the computation of the energy are $\hat{\tau}$-translations of points where we control the values of u. The idea of the construction of u^τ is that we can define it by translation except close to some set of exceptional points. In the following proposition, which is the key point of the proof of Proposition 3.29, we will show that this exceptional set is not negligible, but that the function u must be constant on some sets enclosing most of these points, so that u^τ can be defined as these constant values in the translations of the "enclosing" sets.

Proposition 3.30 (Test functions for "almost translated" domains) *Fixed $\eta > 0$, let R_η be defined in (3.24). Let $T > 2R_\eta$ and $u^T : S_N \to \{-1,1\}$ be such that $\mathcal{P}(u^T) = 2\chi_{H^\nu} - 1$ on $\mathbb{Z}^{d-1} \setminus Q^\nu_{T-2}$ and $E(u^T; Q^\nu_T) \leq CT^{d-2}$. Then for all $\tau \in T_\eta$ there exists $u^{T,\tau} : S_N \to \{-1,1\}$ such that $\mathcal{P}(u^{T,\tau})(\alpha) = 2\chi_{H^\nu}(\alpha - \tau) - 1$ for any $\alpha \in \mathbb{Z}^{d-1} \setminus (\tau + Q^\nu_{T-2})$, and*

$$E(u^{T,\tau}; \tau + Q^\nu_T) \leq E(u^T; Q^\nu_T) + \frac{c}{R_\eta} T^{d-2}, \tag{3.26}$$

where the constant c depends only on d, S_N, and C.

Proof Let $\tau \in T_\eta$ and let $\hat{\tau} \in \hat{T}_\eta$ be such that $P(\hat{\tau}) = \tau$. We consider the set $S_N \setminus (\hat{\tau} + S_N)$. Since the codimension of Π^ζ is 1, we have that either $S_N \setminus (\hat{\tau} + S_N) \subset \{i \in \mathbb{Z}^d : \mathrm{dist}(i, \Pi^\zeta + N\zeta) < \eta\}$ or $S_N \setminus (\hat{\tau} + S_N) \subset \{i \in \mathbb{Z}^d : \mathrm{dist}(i, \Pi^\zeta) < \eta\}$. Hence, if $i, j \in S_N \setminus (\hat{\tau} + S_N)$, then $i - j \in \hat{T}_\eta$; by the definition of R_η, if $i \neq j$, then $\mathrm{dist}(i - j, 0) \geq R_\eta$, and we deduce that the distance between two points of $S_N \setminus (\hat{\tau} + S_N)$ is greater than R_η. If we set

$$B(\tau) = \{\alpha \in \mathbb{Z}^{d-1} : (P^{-1}(\alpha) \cap S_N) - \hat{\tau} \neq P^{-1}(\alpha - \tau) \cap S_N\}, \tag{3.27}$$

then $B(\tau) = P(S_N \setminus (\hat{\tau} + S_N))$. The distance between points in $B(\tau)$ is then greater than $2r_0 R_\eta$, where r_0 is a positive constant depending on ζ and such that $r_0 < \frac{1}{4}$. (Note that this argument is more delicate for higher-codimensional objects; see Remark 3.34.) Let Q_R denote the $d-1$-dimensional coordinate cube centered at 0 and with side length R. For any $\alpha \in \mathbb{Z}^{d-1}$ we consider the cube $\alpha + Q_{r_0 R_\eta}$. The family of such cubes for $\alpha \in B(\tau)$ is disjoint by the previous remark.

Now, in the set $B(\tau) \cap (\tau + Q^\nu_T)$ we consider the boundary set $J(\tau, T)$ defined by

$$J(\tau, T) = B(\tau) \cap \big(\tau + (Q^\nu_T \setminus Q^\nu_{T-2R_\eta})\big).$$

We note that

$$\#J(\tau, T) \leq \frac{c' T^{d-2}}{R_\eta^{d-2}}, \tag{3.28}$$

where c' is a positive constant independent of η, τ, and T. Moreover, we consider the set $I(\tau, T) \subset B(\tau) \cap (\tau + Q^\nu_{T-2R_\eta})$, where the energy of u^T is large; that is,

$$I(\tau, T) = \Big\{\alpha \in B(\tau) \cap (\tau + Q^\nu_{T-2R_\eta}) : E(u^T; \alpha - \tau + Q_{r_0 R_\eta}) > r_0 R_\eta\Big\}.$$

By the hypothesis on u^T we get that

$$\#I(\tau, T) \leq \frac{c'' T^{d-2}}{R_\eta}, \tag{3.29}$$

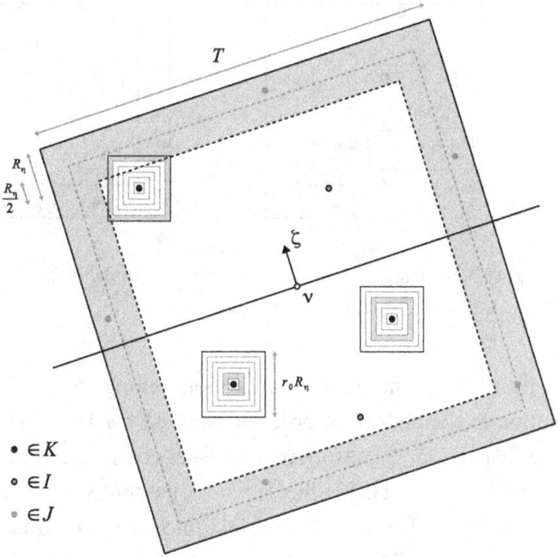

Figure 3.19 The sets $I(\tau,T)$, $J(\tau,T)$, and $K(\tau,T)$.

with c'' a positive constant independent of T, η, and τ. Let $K(\tau,T)$ be defined by

$$K(\tau,T) = \big(B(\tau) \cap (\tau + Q^{\nu}_{T-2R_\eta})\big) \setminus I(\tau,T).$$

For $\alpha \in K(\tau,T)$,

$$E(u^T; \alpha - \tau + Q_{r_0R_\eta}) \leq r_0R_\eta$$

and $\alpha + Q_{r_0R_\eta} \subset \tau + Q^{\nu}_{T-R_\eta}$ (see Fig. 3.19).

We set $C^0_{\alpha,\tau} = \alpha - \tau + Q_1$ and

$$C^k_{\alpha,\tau} = \alpha - \tau + (Q_{2k+1} \setminus Q_{2k-1}) \text{ for } 1 \leq k \leq \left\lfloor \frac{r_0R_\eta}{2} - 1 \right\rfloor.$$

The bound on $E(u^T; \alpha - \tau + Q_{r_0R_\eta})$ ensures that there exists $k(\alpha,\tau)$ such that $E(u^T; C^{k(\alpha,\tau)}_{\alpha,\tau}) = 0$, and since $C^{k(\alpha,\tau)}_{\alpha,\tau}$ is connected for $d > 2$ we deduce that u is constant on $S_N \cap (C^{k(\alpha,\tau)}_{\alpha,\tau} \times \mathbb{R})$ (see Fig. 3.20). This is the first argument where we use the assumption $d > 2$. The constant value is denoted by $u_{\alpha,\tau}$.

We define $u^{T,\tau}$ by setting

$$u^{T,\tau}_i = \begin{cases} 2\chi_{H^\nu}(P(i)) - 1 & \text{if } P(i) \in J(\tau,T), \\ 1 & \text{if } P(i) \in I(\tau,T), \\ u_{\alpha,\tau} & \text{if } P(i) \in \alpha + Q_{2k(\alpha,\tau)+1} \text{ for } \alpha \in K(\tau,T), \\ u^T_{i-\hat\tau} & \text{otherwise in } S_N. \end{cases} \tag{3.30}$$

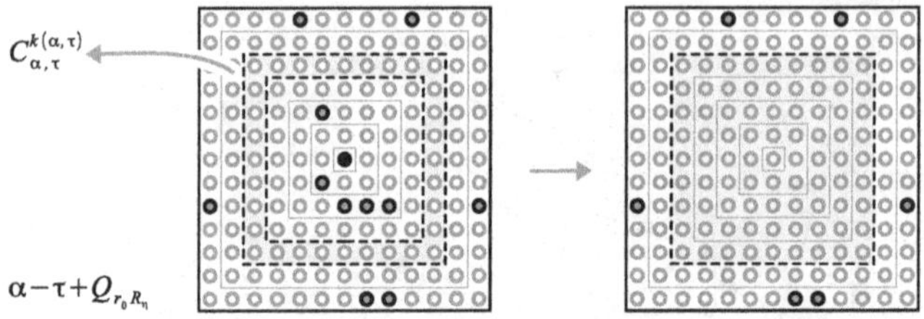

Figure 3.20 The function u close to an exceptional point.

With this definition, the points whose projection belongs to $I(\tau,T)$ or $J(\tau,T)$ give a contribution to the energy of order at most $\#I(\tau,T)$ or $\#J(\tau,T)$, respectively. Hence, (3.28) and (3.29) ensure that the total contribution of these points is of order at most $T^{d-2}R_\eta^{-1}$. This is the second argument where we use $d > 2$; more precisely, in the use of (3.28). The projection $\mathcal{P}(u^{T,\tau})$ is equal to $2\chi_{H^\nu}-1$ outside $\tau+Q^\nu_{T-2}$, and the estimate (3.26) is satisfied, concluding the proof. □

By using Proposition 3.30, we can prove the existence of the surface tension.

Proof of Proposition 3.29 We consider $S \gg T$ and the $d-1$-cube Q^ν_S. With fixed $\eta > 0$, let L_η be the inclusion length given by the relative density of T_η, and let $Q(k)$ with $k \in \{1,\dots,\lfloor\frac{S}{T+2L_\eta}\rfloor^{d-2}\}$ be disjoint cubes with a face orthogonal to ν and side length $T+2L_\eta$ included in $Q^\nu_{S-2}\cap\{|\langle x,\nu\rangle| < \frac{T}{2}+L_\eta\}$. The relative density ensures that for any k there exists a translation vector $\tau^k_\eta \in T_\eta$ such that $Q^\nu_T + \tau^k_\eta \subset Q(k)$.

Let $u^T : S_N \to \{-1,1\}$ be such that $\mathcal{P}(u) = 2\chi_{H^\nu} - 1$ on $\mathbb{Z}^{d-1} \setminus Q^\nu_{T-2}$ and $E(u^T; Q^\nu_T) = g_T(\nu)$. We can apply the result of Proposition 3.30 to u^T with each translation τ^k_η, obtaining a family of functions $u^k = u^{T,\tau^k_\eta} : S_N \to \{-1,1\}$ defined as in (3.30) satisfying the boundary condition $\mathcal{P}(u^k)(\alpha) = 2\chi_{H^\nu}(\alpha - \tau^k_\eta) - 1$ for any $\alpha \in \mathbb{Z}^{d-1} \setminus (\tau^k_\eta + Q^\nu_{T-2})$ and such that

$$E(u^k; \tau^k_\eta + Q^\nu_T) \le E(u^T; Q^\nu_T) + \frac{C}{R_\eta}T^{d-2}. \tag{3.31}$$

We define $u^S : S_N \to \{-1,1\}$ by setting

$$u^S_i = \begin{cases} u^k_i & \text{if } P(i) \in \tau^k_\eta + Q^\nu_T,\\ 2\chi_{H^\nu}(P(i)) - 1 & \text{otherwise in } S_N. \end{cases}$$

(See Fig. 3.21, where the projection of u^S on $\mathbb{R}^{d-1}$ is pictured.) With this definition, u^S is an admissible test function for $g_S(\nu)$. The contribution to the

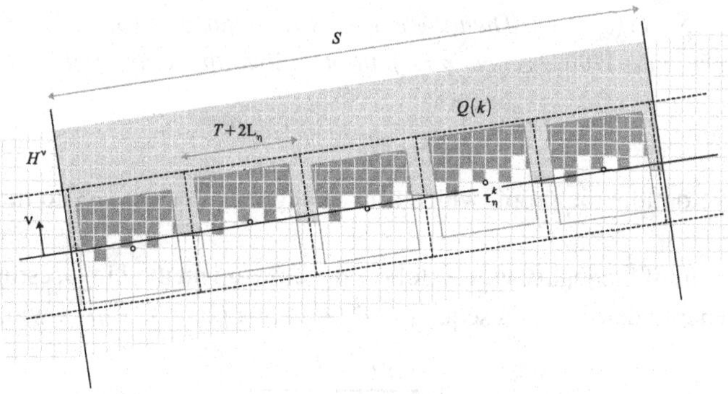

Figure 3.21 Projection of the construction of test sets.

energy $E(u^S;Q^\nu_S)$ outside the union of the cubes $\tau^k_\eta + Q^\nu_T$ is estimated by a constant times $(T+2L_\eta)^{d-2}S^{d-3}$, so that by (3.31) we get

$$E(u^S;Q^\nu_S) \le \left\lfloor \frac{S}{T+2L_\eta} \right\rfloor^{d-2} \left(E(u^T;Q^\nu_T) + c\frac{T^{d-2}}{R_\eta} + cL_\eta^{d-2} \right) + c(T+2L_\eta)^{d-2}S^{d-3},$$

where $c > 0$ is independent of S, T, η, and k. Thus, taking the upper limit as $S \to +\infty$, we have

$$\limsup_{S\to+\infty} \frac{1}{S^{d-2}} E(u^S;Q^\nu_S) \le \frac{1}{T^{d-2}} \left(E(u^T;Q^\nu_T) + c\frac{T^{d-2}}{R_\eta} + cL_\eta^{d-2} \right)$$

and, taking the lower limit as $T \to +\infty$, we obtain

$$\limsup_{S\to+\infty} \frac{1}{S^{d-2}} E(u^S;Q^\nu_S) \le \liminf_{T\to+\infty} \frac{1}{T^{d-2}} E(u^T;Q^\nu_T) + \frac{c}{R_\eta}.$$

Recalling that u^T is a minimizer of $g_T(\nu)$, and that u^S is an admissible test function for $g_S(\nu)$, we obtain

$$\limsup_{S\to+\infty} \frac{1}{S^{d-2}}\, g_S(\nu) \le \liminf_{T\to+\infty} \frac{1}{T^{d-2}}\, g_T(\nu) + \frac{c}{R_\eta}.$$

Letting $\eta \to 0$, since $R_\eta \to +\infty$ we obtain the existence of the limit. □

The definition of the energy density $\varphi_{\rm hom}$ allows us to prove a homogenization theorem for the ferromagnetic energy on quasicrystals.

Theorem 3.31 (Homogenization theorem for the ferromagnetic energy on quasicrystals) *Let ω be a bounded open subset of $\mathbb{R}^{d-1}$ with Lipschitz boundary,* *and* $\Omega_\varepsilon = \varepsilon S_N \cap (\omega \times \mathbb{R})$. *Then the energies* E_ε *defined in* (3.20) Γ*-converge with respect to the convergence in Definition* 3.26 *to the functional*

$$E(A) = \int_{\omega \cap \partial^* A} \varphi_{\hom}(\nu_A)\, d\mathcal{H}^{d-2}.$$

Proof We outline the proof, which uses an argument customary to thin-film theory.

For the liminf inequality we can follow the blow-up method in ω, estimating the limit energy density for a sequence $u^\varepsilon \to A$ at a point $x_0 \in \partial^* A$ with

$$\lim_{\varrho \to 0} \lim_{\varepsilon \to 0} \frac{E_\varepsilon(u^\varepsilon; Q^\nu_\varrho(x_\varepsilon))}{\varrho^{d-2}}$$

with $x_\varepsilon \to x_0$ and $x_\varepsilon \in \varepsilon T_\eta$ for some given η. After changing the boundary value close to the lateral boundary of $(Q^\nu_\varrho(x_\varepsilon) \times \mathbb{R}) \cap \varepsilon S_N$ we can then give a lower bound by using the formula defining $\varphi_{\hom}$, up to an error vanishing as $\eta \to 0$.

The upper bound can be directly constructed close to a hyperplane using the minima as in Proposition 3.30, and then we can proceed by density. □

Remark 3.32 (General energies) An analogous homogenization result holds for energies

$$E_\varepsilon(u) = \sum_{\varepsilon i, \varepsilon j \in \Omega_\varepsilon} \varepsilon^{d-2} a_{ij}(u_i - u_j)^2,$$

with a_{ij} periodic coefficients satisfying the hypotheses in Section 3.3. The formula defining $\varphi_{\hom}$ remains unchanged if we only have nearest-neighbor interactions, and must be modified as in (3.12) if the range of the interactions is R.

Remark 3.33 (Thin objects with rational subspaces) If the set in (3.19) is not the only point 0, then it contains a sublattice $\mathcal{L}'$ of $\mathbb{Z}^d$ of dimension $d' \le d-1$, along which directions the set S_N is periodic. In those directions we can use easier translation arguments. Note that in particular if $d' = d-1$, the space Π^ζ is a rational subspace and S_N consists in a number of copies of $\mathbb{Z}^{d-1}$.

Note that the statistical properties of quasicrystal as defined earlier are different from those of the *rational* thin films in that for quasicrystals the average number of points per given $d-1$ surface area on a volume of small dimension $N \in \mathbb{R}$ is proportional to N, while for rational thin objects it depends on $\lfloor N \rfloor$.

Remark 3.34 (Higher-codimensional quasicrystalline thin structures with polyhedral cross section) The arguments followed to prove the homogenization theorem for quasicrystalline thin films can be extended to more general discrete thin objects with polyhedral cross sections that can be modeled as follows.

Let n be such that $1 < n < d-1$, and let Π be a linear n-dimensional subspace of $\mathbb{R}^d$. We fix $K \in \mathbb{N}$, $N_1, \dots, N_K \geq 0$ and $\zeta_1, \dots, \zeta_K \in \Pi^\perp \cap S^{d-1}$ such that the polyhedral set

$$H = \{\zeta \in \Pi^\perp : \langle \zeta, \zeta_j \rangle \leq N_j \text{ for all } j = 1, \dots, K\}$$

is bounded. Then, we define the *discrete thin object*

$$S = \{x \in \mathbb{Z}^d : \langle x, \zeta_j \rangle \leq N_j \text{ for all } j = 1, \dots, K\}.$$

The hypotheses on S are the same as in the case $n = d-1$; that is,

(i) $\mathbb{Z}^d \cap \Pi^{\zeta_j} = \{0\}$, where $\Pi^{\zeta_j} = \{x \in \mathbb{R}^d : \langle x, \zeta_j \rangle = 0\}$ (irrationality);
(ii) S is connected in the sense of Definition 3.19.

Note that $\Pi \subset \Pi^{\zeta_j}$ for any j, hence $\mathbb{Z}^d \cap \Pi = \{0\}$.

In these hypotheses the homogenization theorem holds unchanged (with n in the place of $d-1$) for the energies

$$E_\varepsilon(u) = \sum_{\substack{\|i-j\|=1 \\ \varepsilon i, \varepsilon j \in \varepsilon S}} \varepsilon^{n-1}(u_i - u_j)^2$$

defined for $u : \varepsilon S \to \{-1, 1\}$.

The proof can be essentially followed as in the preceding case, except for some geometrical details in the construction of the almost translated functions $u^{T,\tau}$ in Proposition 3.30. Indeed, the points in $S \setminus (\hat{\tau} + S)$ are close to the boundary of S, which now is a connected set. Hence, even if their distance is larger than R_η their projections can be close (see Fig. 3.22). This implies that in $B(\tau)$ there may lie clusters of points. Since the size of these clusters can be

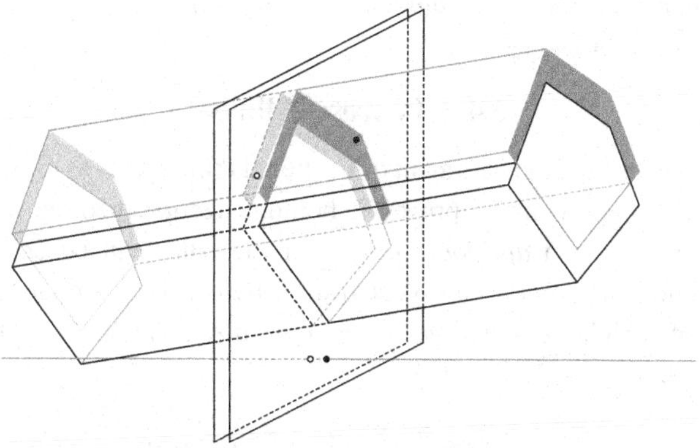

Figure 3.22 Exceptional points with neighboring projections.

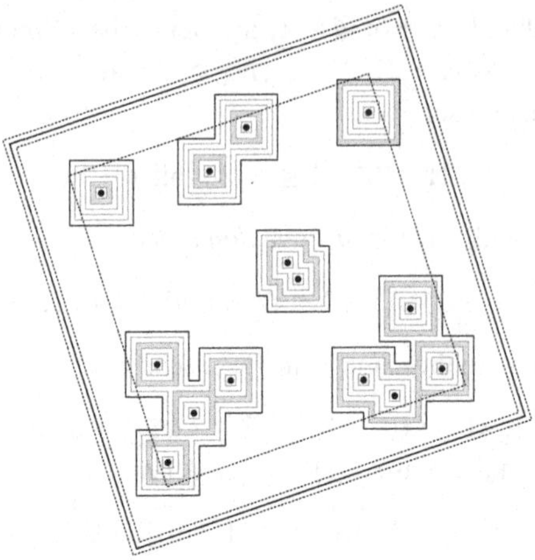

Figure 3.23 Clusters of exceptional points in higher codimension.

uniformly bounded, we may still define functions $u^{T,\tau}$ with the properties in Proposition 3.30 with a more complex procedure. The construction is pictured in Fig. 3.23.

The arguments of the proof of the homogenization of higher-codimensional quasicrystalline thin structures in Remark 3.34 are allowed by the fact that the cross section H of the thin object S is a polyhedral set.

Conjecture 3.35 (Homogenization with nonpolyhedral cross section) *We conjecture that the homogenization results for quasicrystalline thin structures hold for general cross sections and not only polyhedral.*

3.4.2 Penrose Tilings

The arguments used in the treatment of irrational thin objects as outlined in the previous section allow one to prove the homogenization theorem for systems of spin interactions on *aperiodic lattices* if the latter can be characterized as (a projection of) some thin object from a higher-dimensional space. The prototypes of such lattices are *Penrose lattices* in $\mathbb{R}^2$, which will be the object of this section.

Generation of Penrose Lattices by Projection

We define the *Penrose lattice* $\mathbb{P} \subset \mathbb{R}^2$ as follows. Let Π be the two-dimensional plane in $\mathbb{R}^5$ spanned by the vectors

$$v_1 = \sum_{k=1}^{5} \sin\left(\frac{2(k-1)\pi}{5}\right) e_k \quad \text{and} \quad v_2 = \sum_{k=1}^{5} \cos\left(\frac{2(k-1)\pi}{5}\right) e_k, \tag{3.32}$$

where e_k is the unit vector on the kth axis. We note that, considering the matrix M whose action is the permutation of all the coordinate axes in order, then Π is the plane of the vectors v such that the action of M on v is a rotation of $\frac{2}{5}\pi$. Then, we consider the set $\mathcal{I}$ of the points $z \in \mathbb{Z}^5$ such that $z + (0,1)^5 \cap \Pi \neq \emptyset$, and the function $\phi\colon \mathbb{Z}^5 \to \mathbb{R}^2$ defined as $\phi(z) = \sum_{k=1}^{5} z_k e^{\frac{ik\pi}{5}}$. We set $\phi(\mathcal{I}) = \mathbb{P}$.

This construction of $\mathcal{I}$ is very similar to that of a discrete thin object as in the previous section. The difference is that instead of taking points in $\mathbb{Z}^5$ whose distance vector from the plane Π lies in a given set, we take those in $\mathbb{Z}^5$ which are the vertex (with minimal coordinates) of a unit cube intersecting Π. Such a construction still gives a set satisfying analogous quasiperiodicity properties.

Remark 3.36 (Characterization of Penrose tilings) The tiling obtained by joining p and p' in $\mathbb{P}$ by an edge if and only if $\|p - p'\| = 1$ is a *Penrose tiling*. Hence, the preceding construction will be our definition of a *Penrose lattice*.

We let $\mathcal{T}$ denote the set of the *Penrose cells* of the tiling in $\mathbb{R}^2$; we get two possible shapes of rhombi for the cells $T \in \mathcal{T}$, each one with five possible orientations.

Discrete Perimeter on the Penrose Lattice

Given a discrete set $A \subset \mathbb{P} \subset \mathbb{R}^2$ and $\Omega \subset \mathbb{R}^2$ an open set, we define

$$\mathrm{Per}_{\mathbb{P}}(A;\Omega) = \#\left\{(i,j) : i \in A, j \notin A, \|i-j\| = 1, i,j \in \Omega\right\}, \tag{3.33}$$

and $\mathrm{Per}_{\mathbb{P}}(A) = \mathrm{Per}_{\mathbb{P}}(A;\mathbb{R}^2)$.

We identify a set $A \subset \mathbb{P}$ with a subset $i(A)$ of $\mathbb{R}^2$ in the following way. Given $p \in \mathbb{P}$, we consider the set $\{T^1,\ldots,T^{N(p)}\}$ of the Penrose cells with one vertex in p. We set

$$C(p) = p + \bigcup_{j=1}^{N(p)} \frac{T^j - p}{2}$$

so that we can define $i(A) = \bigcup_{p\in A} C(p)$ (see Fig. 3.24). Such a set will be called a *Penrose set*. Note that with this identification we have

$$\mathrm{Per}_{\mathbb{P}}(A) = \mathcal{H}^1(\partial(i(A))).$$

If $A \subset \varepsilon\mathbb{P}$, then we define the ε-Penrose set

$$i_\varepsilon(A) = \varepsilon\, i\Big(\frac{1}{\varepsilon}A\Big), \tag{3.34}$$

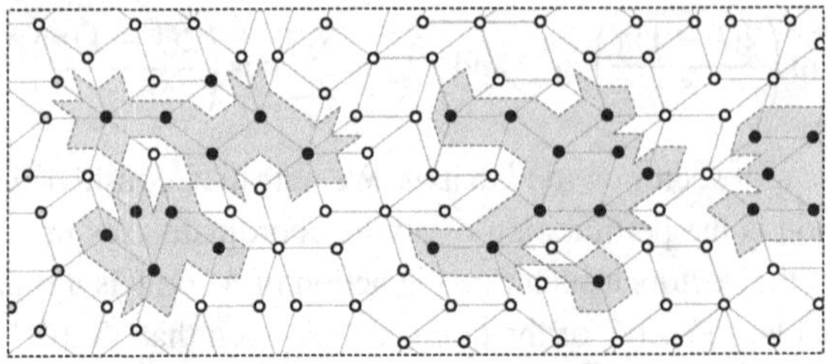

Figure 3.24 A discrete set A (black points) and the corresponding set $i(A)$ (shaded region).

and hence

$$\mathcal{H}^1\left(\partial(i_\varepsilon(A))\right) = \varepsilon\,\mathrm{Per}_{\mathbb{P}}\left(\frac{1}{\varepsilon}A\right). \tag{3.35}$$

Homogenization of Surface Energies on the Penrose Lattice

We will consider energies defined on functions $u \colon \Omega \cap \varepsilon\mathbb{P} \to \{-1, 1\}$ of the form

$$E_\varepsilon(u) = \frac{1}{8}\sum_{\langle i,j\rangle} \varepsilon(u_i - u_j)^2, \tag{3.36}$$

where $\langle i, j\rangle$ denotes the set of nearest neighbors in $\frac{1}{\varepsilon}\Omega \cap \mathbb{P}$; that is, such that $\|i - j\| = 1$. Note that the notion of nearest neighbors derives from that of nearest neighbors for the corresponding points in $\mathbb{Z}^5$, and not in the sense of the Voronoi cells in $\mathbb{R}^2$ (see Section 4.3). In the definition of the energy we have included the factor $\frac{1}{8}$ in order to directly interpret them as perimeters (see formula (3.37)).

Definition 3.37 (Convergence) A sequence $\{u^\varepsilon\}$ with $u^\varepsilon \colon \Omega \cap \varepsilon\mathbb{P} \to \{-1, 1\}$ *converges to* a set of finite perimeter A if the characteristic functions of the sets $i(A_\varepsilon)$, where $A_\varepsilon = \{\varepsilon i \in \Omega \cap \varepsilon\mathbb{P} \colon u_i^\varepsilon = 1\}$, converge to χ_A in $L^1_{\mathrm{loc}}(\Omega)$.

Remark 3.38 (Compactness) Let $\{u^\varepsilon\}$ be such that $\sup_\varepsilon E_\varepsilon(u_\varepsilon) < +\infty$; then, up to extraction of a subsequence, u^ε converge to some A in the sense of the preceding definition. Indeed, it is sufficient to remark that in this case we have

$$E_\varepsilon(u^\varepsilon) = \mathcal{H}^1\left(\partial(i_\varepsilon(A_\varepsilon))\right) \tag{3.37}$$

by (3.35), and use the compactness of families of sets of equibounded perimeter $i_\varepsilon(A_\varepsilon)$.

We then have the following homogenization theorem, whose proof can be obtained by following the almost-periodic arguments for the case of quasicrystals, up to minor modifications.

Theorem 3.39 (Homogenization of Penrose lattices) *Let E_ε be defined in* (3.36). *Then there exists a function $\varphi_{\rm hom}\colon S^1 \to \mathbb{R}$ such that E_ε Γ-converges as $\varepsilon \to 0$, with respect to the convergence in Definition* 3.37, *to the functional*

$$E(A) = \int_{\Omega\cap\partial^* A} \varphi_{\rm hom}(\nu)\, d\mathcal{H}^1. \tag{3.38}$$

The function $\varphi_{\rm hom}$ is characterized by the asymptotic formula

$$\varphi_{\rm hom}(\nu) = \lim_{T\to+\infty} \frac{1}{T} \min\Big\{\#\big\{(i,j)\colon i \in A\cap Q_T^\nu, j \notin A, \|i-j\| = 1\big\}\colon \\ A \subset \mathbb{P}, A = H^\nu \cap \mathbb{P} \text{ outside } Q_T^\nu\Big\}, \tag{3.39}$$

where $H^\nu = \{x\colon \langle x,\nu\rangle > 0\}$.

Conjecture 3.40 (Wulff shape of the Penrose perimeter) *We conjecture that the Wulff shape of E is a decagon. This is justified by the fact that $\mathbb{P}$ has a pentagonal symmetry*.

3.5 Thin Objects

As already observed in Remark 3.33 the study of thin objects is easier to formalize when the underlying lower-dimensional space is a sublattice of $\mathbb{Z}^d$. Following Remark 3.34, we can directly consider the case when this subspace has dimension $n \le d-1$. For the sake of simplicity we only consider the case when this subspace coincides with a coordinate n-dimensional space, and choose $S \subset \mathbb{Z}^d$ of the form

$$S = \mathbb{Z}^n \times H \ \text{ with } \ H \subset \mathbb{Z}^{d-n}.$$

We suppose that S is connected in the sense of Definition 3.19, fix periodic coefficients a_{ij} satisfying the conditions in the previous parts of this chapter, and, given a Lipschitz open set of $\mathbb{R}^n$, consider the functionals

$$E_\varepsilon(u) = \sum_{\varepsilon i,\varepsilon j\in\varepsilon S\cap(\omega\times\mathbb{R}^{d-n})} \varepsilon^{n-1} a_{ij}(u_i-u_j)^2$$

defined for $u\colon \varepsilon S\cap(\omega\times\mathbb{R}^{d-n}) \to \{-1,1\}$.

We can simplify the definition of dimensionally reduced convergence of functions as follows.

Definition 3.41 Let $u^\varepsilon : \varepsilon S \to \{-1,1\}$. The sequence $\{u^\varepsilon\}$ *converges to a set of finite perimeter* $A \subset \mathbb{R}^n$ if for all $h \in H$ the functions $u^\varepsilon_h : \varepsilon\mathbb{Z}^n \to \{-1,1\}$ defined by $u^\varepsilon_h(\varepsilon l) = u^\varepsilon(\varepsilon l, \varepsilon h)$ converge to A.

This definition is equivalent to Definition 3.26 in this simpler case. By Proposition 3.27 functionals E_ε are equicoercive for this convergence. The resulting homogenization theorem, obtained by rewriting the homogenization theorem for quasicrystals, reads as follows.

Theorem 3.42 *The* Γ*-limit of* E_ε *is of the form*

$$E(A) = \int_{\omega\cap\partial^* A} \varphi_{\rm hom}(\nu_A)\, d\mathcal{H}^{n-1},$$

and

$$\varphi_{\rm hom}(\nu) = \lim_{T\to+\infty} \frac{1}{T^{n-1}} \min\Big\{ \sum_{i,j\in\varepsilon(S\cap Q^{\hat\nu}_T)} a_{ij}(u_i - u_j)^2 :\ u\colon S\cap Q^{\hat\nu}_T \to \{-1,1\},$$
$$u = 2\chi_{H^{\hat\nu}} - 1 \text{ on } \mathbb{Z}^d \setminus Q^{\hat\nu}_{T-2}\Big\},$$

where $\hat\nu = (\nu_1, \ldots, \nu_n, 0, \ldots, 0)$.

Remark 3.43 (Dependence on thin-film thickness) We can apply the preceding result for thin films of thickness N; that is, when $n = d-1$ and $H = [0,N]$ with $N \geq 0$. Note that for $N = 0$ we simply have a bulk theory in $\mathbb{Z}^{d-1}$. Note, moreover, that even for homogeneous interactions the dependence of $\varphi_{\rm hom} = \varphi^N_{\rm hom}$ on N may be nontrivial. To see this, just take $d = 2$ and next-to-nearest-neighbor interactions with coefficients all equal to 1. Then $\varphi^N_{\rm hom}$ is a constant coefficient with value $6N+2$. Taking neighbors up to a distance R, we may obtain a more complex (nonlinear) behavior on N.

3.5.1 Chiral Thin Objects

The approach we outlined for thin films can be adapted to more general structures, involving periodicity conditions that can be described by some *chiral* parameter. We only include some simple examples of two-dimensional "nanotubes," with a one-dimensional limit, for which it is more convenient to use the notation of functions u rather than sets A.

Let $\mathcal{L}$ be a periodic lattice in $\mathbb{R}^2$, and suppose that $k_1, k_2 \in \mathbb{N}$ exist such that $k_2e_2 - k_1e_1 \in \mathcal{L}$, which in particular gives

$$\mathcal{L} - k_1e_1 = \mathcal{L} - k_2e_2. \tag{3.40}$$

The vector (k_1, k_2) will be called the *chirality* of the system.

We consider functions $u\colon \mathcal{L} \to \{-1, 1\}$ satisfying the periodicity assumption

$$u_{i+k_2e_2-k_1e_1} = u_i, \tag{3.41}$$

and coefficients a_{ij} periodic in the e_2 direction and satisfying the same assumption

$$a_{i+k_2e_2-k_1e_1\, j} = a_{ij}. \tag{3.42}$$

Theorem 3.44 *Let a_{ij} satisfy the preceding hypotheses. Let E_ε be defined by*

$$E_\varepsilon(u) = \sum_{i,j\in\mathcal{L}\cap([0,k_1)\times\mathbb{R})} a_{ij}(u_i - u_j)^2$$

on functions $u\colon \varepsilon\mathcal{L} \to \{-1, 1\}$ satisfying (3.41). Then E_ε Γ-converge to the functional

$$F(u) = K\#(S(u))$$

defined on piecewise-constant functions $u\colon \mathbb{R} \to \{-1, 1\}$, where

$$K = \lim_{T\to+\infty} \min\Big\{ \sum_{i,j\in\mathcal{L}\cap([0,k_1)\times\mathbb{R})} a_{ij}(u_i-u_j)^2 : u_i = -1 \text{ if } i_2 \leq -T,\ u_i = 1 \text{ if } i_2 \geq T\Big\}.$$

The theorem can be proved similarly as for the thin-film case.

Example 3.45 (Dependence on the chirality) We consider a nearest-neighbor interaction system with $a_{ij} = 1$. In this case the computation of K is reduced to a minimal-length problem with endpoints j and $j + k_2e_2 - k_1e_1$. Noting that for rational endpoints this is nothing but the problem in the definition of the homogenized energy density φ of the nearest-neighbor interactions on this lattice in $\mathbb{R}^2$, we deduce that

$$K = \varphi(k_2, k_1),$$

where φ is extended by 1-homogeneity. We compute K in terms of k_1 and k_2 and picture optimal transitions in some easy cases.

(i) (*square lattice*) In this case we have a multiplicity of minimal connections, and we trivially compute

$$K = k_1 + k_2$$

(see Fig. 3.25(b)).

(ii) (*triangular lattice*) In this case it is convenient to slightly modify the hypotheses of the theorem and suppose that $k_2 \in \sqrt{3}\mathbb{Z}$. The shape of the optimal transition depends on whether (k_2, k_1) lies in the angle between

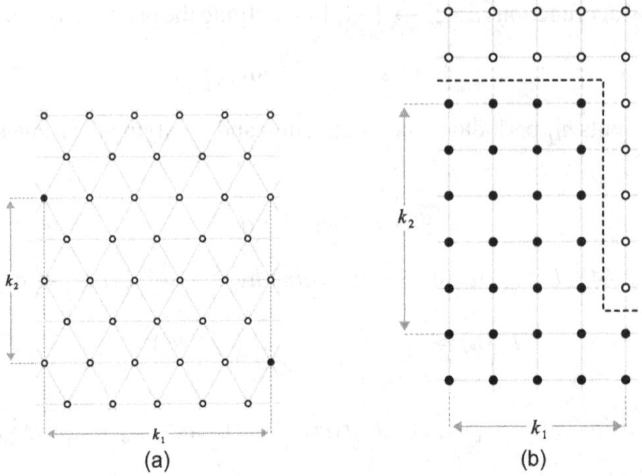

Figure 3.25 (a) identification of sites in a nanotube (the two black circles); (b) an optimal interface.

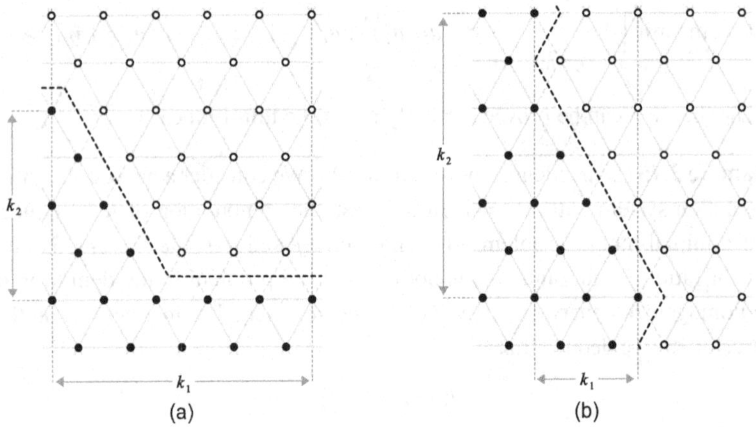

Figure 3.26 Optimal interfaces for different k_1 and k_2.

the vectors $(0,1)$ and $(\frac{\sqrt{3}}{2},\frac{1}{2})$ or between the vectors $(\frac{\sqrt{3}}{2},\frac{1}{2})$ and $(1,0)$ (parts (a) and (b) in Fig. 3.26, respectively). We then have

$$K = \begin{cases} 2k_1 + \frac{2}{\sqrt{3}}k_2 & \text{if } k_2 \leq \sqrt{3}k_1, \\ 4(1-\sqrt{3})k_1 + 4k_2 & \text{if } k_2 \geq \sqrt{3}k_1. \end{cases}$$

(iii) (*hexagonal lattice*) Also for the hexagonal lattice it is convenient to take $k_1 \in 2\mathbb{Z}$ and $k_2 \in \sqrt{3}\mathbb{Z}$, and the shape of the optimal transition depends

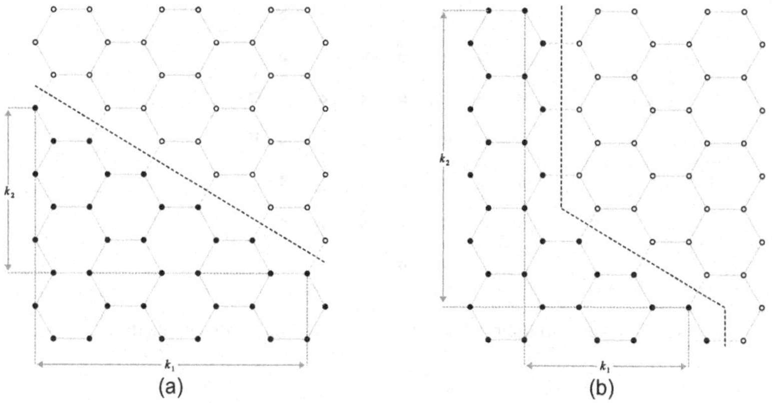

Figure 3.27 Optimal interfaces for different k_1 and k_2.

on whether (k_2, k_1) lies in the angle between the vectors $(0,1)$ and $(\frac{1}{2}, \frac{\sqrt{3}}{2})$ or between the vectors $(\frac{1}{2}, \frac{\sqrt{3}}{2})$ and $(1,0)$ (parts (a) and (b) in Fig. 3.27, respectively). We then have

$$K = \begin{cases} k_1 & \text{if } 2k_2 \leq \sqrt{3}k_1, \\ \frac{1}{\sqrt{3}}k_2 + \frac{1}{2}k_1 & \text{if } 2k_2 \geq \sqrt{3}k_1. \end{cases}$$

3.6 Optimal Bounds for Two-Phase Systems

The homogenization formula for periodic systems in Theorem 3.10 can be used to estimate the possible homogenized energy densities φ_{hom} under some *design constraint*. The simplest such constraint is to assume that the values of the connection strength are given, and that the percentage of connections taking a fixed value is also assigned, but the geometry of the connections is free (and periodic). This is a discrete analog of continuum G-closure problems, where often layered geometries are optimal in a given direction. A major difference in the discrete setting is that we can use layered geometries in more directions at the same time, as in Fig. 3.28, where dark and light gray segments represent different strengths of the interactions.

3.6.1 The General Case

In this section V denotes a given finite subset of $\mathbb{Z}^d$ containing the canonical orthonormal basis $\{e_1, \ldots, e_d\}$. For all $\xi \in V$, we fix α_ξ and β_ξ with $0 < \alpha_\xi <$

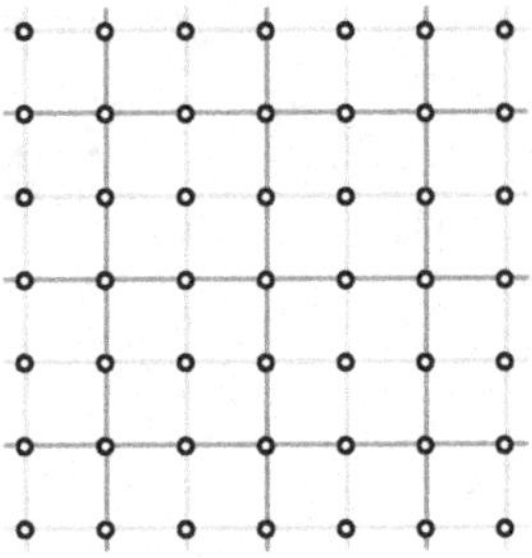

Figure 3.28 Layered interactions with period 2 in each coordinate direction.

β_ξ. We will consider systems of periodic coefficients a_{ij}, which are assumed to satisfy the *design constraint*

$$a_{i\,i+\xi} \in \{\alpha_\xi, \beta_\xi\}$$

for all $i \in \mathbb{Z}^d$ and $\xi \in V$. Denoted by K, the period of the system, we prescribe the *volume fraction* of β_ξ-bonds and the *total volume fraction of β-bonds*, defined as

$$\begin{aligned}\theta_\xi(\{a_{i\,i+\xi}\}) &= \frac{1}{K^d}\#\{i \in \mathbb{Z}^d : i \in \{1,\dots,K\}^d, a_{i\,i+\xi} = \beta_\xi\},\\ \theta(\{a_{i\,i+\xi}\}) &= \frac{1}{\#V}\sum_{\xi\in V}\theta_\xi(\{a_{i\,i+\xi}\}),\end{aligned} \tag{3.43}$$

respectively.

Definition 3.46 Let $\theta \in [0,1]$ and $\theta_\xi \in [0,1]$ be given for all $\xi \in V$. The set of *homogenized energy densities of mixtures of α_ξ and β_ξ bonds corresponding to V with volume fractions θ_ξ* (of β_ξ bonds) is defined as

$$\begin{aligned}H(\{\theta_\xi\}) = \big\{\varphi\colon \mathbb{R}^d \to [0,+\infty) : \text{there exist } \theta_\xi^k \to \theta_\xi \text{ and } \{a_{ij}^k\}\\ \text{with } \theta_\xi(\{a_{ij}^k\}) = \theta_\xi^k \text{ such that } \varphi_{\rm hom}^k \to \varphi\big\},\end{aligned} \tag{3.44}$$

where $\varphi_{\rm hom}^k$ is the homogenized energy density of the system $\{a_{ij}^k\}$ as in Definition 3.11. The set of *homogenized energy densities of mixtures of α and β bonds corresponding to V with volume fraction θ* (of β bonds) is defined as

$$\begin{aligned}H(\theta) = \big\{\varphi\colon \mathbb{R}^d \to [0,+\infty) : \text{there exist } \theta^k \to \theta \text{ and } \{a_{ij}^k\}\\ \text{with } \theta(\{a_{ij}^k\}) = \theta^k \text{ such that } \varphi_{\rm hom}^k \to \varphi\big\}.\end{aligned} \tag{3.45}$$

The following theorem completely characterizes the sets $H(\{\theta_\xi\})$ and $H(\theta)$.

Theorem 3.47 (Optimal bounds) *The elements of the set $H(\{\theta_\xi\})$ are all the even and convex positively homogeneous functions of degree one $\varphi\colon \mathbb{R}^d \to [0,+\infty)$ such that*

$$\sum_{\xi\in V} \alpha_\xi |\langle v,\xi\rangle| \le \varphi(v) \le \sum_{\xi\in V} (\theta_\xi\beta_\xi + (1-\theta_\xi)\alpha_\xi)|\langle v,\xi\rangle|, \tag{3.46}$$

and the elements of the set $H(\theta)$ are all φ such that (3.46) *holds for some $\theta_\xi \in [0,1]$ with $\frac{1}{\#V}\sum_{\xi\in V}\theta_\xi = \theta$.*

The proof of this theorem consists on one hand in proving that the bounds hold through a use of the homogenization formulas, and on the other hand to show their optimality by constructing suitable sequences $\{a^k_{ij}\}_k$. The latter construction involves a multiscale procedure using Γ-limit analyses on the continuum. We will give a short description of the arguments in the following section in a simplified two-dimensional setting.

3.6.2 Two-Dimensional Nearest-Neighbor Mixtures

In this section we give optimal bounds for the homogenization of two-dimensional periodic systems of the form

$$\sum_{\langle i,j\rangle} \varepsilon a_{ij}(u_i - u_j)^2,$$

where the bonds a_{ij} may take two positive values α and β with $0<\alpha<\beta$.

Such bounds are given in terms of the volume fraction θ of β-bonds as in (3.43), which now can be read as

$$\begin{aligned}
\theta_{\rm h}(\{a_{ij}\}) &= \frac{1}{K^2}\#\{i\in\mathbb{Z}^2 : i\in\{1,\dots,K\}^2, a_{i\,i+e_1}=\beta\},\\
\theta_{\rm v}(\{a_{ij}\}) &= \frac{1}{K^2}\#\{i\in\mathbb{Z}^2 : i\in\{1,\dots,K\}^2, a_{i\,i+e_2}=\beta\},\\
\theta(\{a_{ij}\}) &= \frac{1}{2}\Big(\theta_{\rm h}(\{a_{ij}\}) + \theta_{\rm v}(\{a_{ij}\})\Big).
\end{aligned} \tag{3.47}$$

To each such system $\{a_{ij}\}$ we associate the homogenized surface tension $\varphi = \varphi_{\rm hom}$ as in Definition 3.11. The claim of Theorem 3.47 in this case is that all possible such φ are the (positively homogeneous of degree one) convex functions such that

$$8\alpha(|\langle v,e_1\rangle| + |\langle v,e_2\rangle|) \le \varphi(v) \le 8(c_1|\langle v,e_1\rangle| + c_2|\langle v,e_2\rangle|), \tag{3.48}$$

where the coefficients c_1 and c_2 satisfy

$$c_1\le\beta,\quad c_2\le\beta,\qquad c_1+c_2 = 2(\theta\beta+(1-\theta)\alpha). \tag{3.49}$$

The discrete setting allows us to give a relatively easy description of the optimal bounds. On one hand, we use *bounds by projection*, where the homogenized surface tension is estimated from below by considering the minimal value of the coefficient on each section, and on the other hand *bounds by averaging*, where coefficients on a section are substituted with their average. The discrete setting allows us to construct (almost-)optimal periodic geometries, which optimize one type or the other of bounds in each direction.

We can picture the discrete bounds in terms of the Wulff shape of the related energies; that is, the solutions A_φ centered in 0 to the problem

$$\max\Big\{|A| : \int_{\partial A} \varphi(\nu)\, d\mathcal{H}^1(x) = 1\Big\}.$$

If $\varphi(\nu) = 8(c_1|\nu_1| + c_2|\nu_2|)$, then such a Wulff shape is the rectangle centered in 0 with one vertex in $(\frac{1}{64c_2}, \frac{1}{64c_1})$. A general φ satisfying (3.48) and (3.49) corresponds to a convex symmetric set contained in the square of side length $\frac{1}{32\alpha}$ (that is, the Wulff shape corresponding to $8\alpha(|\nu_1| + |\nu_2|)$) and containing one of such rectangles for c_1 and c_2 satisfying (3.49). The envelope of the vertices of such rectangles lies in the curve

$$\frac{1}{|x_1|} + \frac{1}{|x_2|} = 128(\theta\beta + (1-\theta)\alpha) \tag{3.50}$$

(see Fig. 3.29).

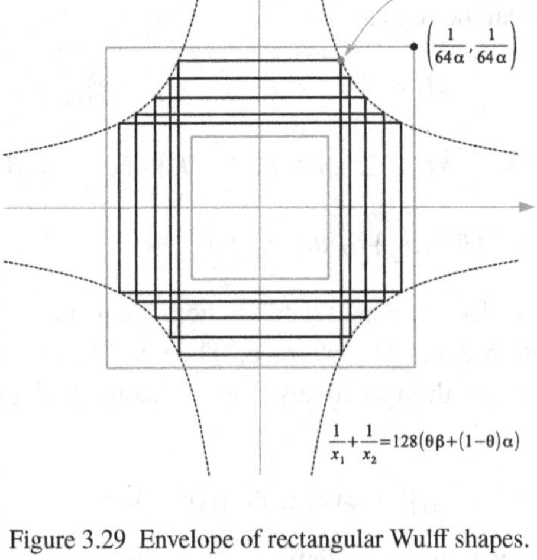

Figure 3.29 Envelope of rectangular Wulff shapes.

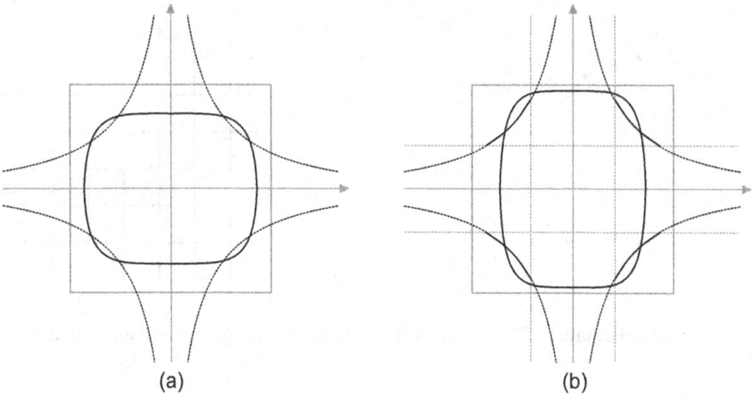

Figure 3.30 Possible Wulff shapes with $\theta \leq \frac{1}{2}$ (a) and $\theta \geq \frac{1}{2}$ (b).

In terms of such an envelope, the claim of Theorem 3.47 is that we can describe the possible Wulff shapes of φ as follows.

(i) If $\theta \leq \frac{1}{2}$, then it is any symmetric convex set contained in the square of side length $\frac{1}{32\alpha}$ and intersecting the four portions of the set of points satisfying (3.50) contained in that square (see Fig. 3.30(a)).
(ii) If $\theta \geq \frac{1}{2}$, then it is any symmetric convex set contained in the square of side length $\frac{1}{32\alpha}$ and intersecting the four portions of the set of points satisfying (3.50) with $|x_1| \geq \frac{1}{64\beta}$ and $|x_2| \geq \frac{1}{64\beta}$ contained in that square (see Fig. 3.30(b)). This second condition is automatically satisfied if $\theta \leq \frac{1}{2}$.

The proof in two dimensions is easier to visualize and formalize since boundaries of sets can be viewed as curves; in particular, in the discrete setting, as the union of line segments parallel to the axes. This fact allows one to characterize the homogenized energy density $\varphi_{\rm hom}$ defined in (3.12) by a *path-minimization formula*.

Noting that each pair of nearest neighbors $(i, j) \in \mathbb{Z}^2 \times \mathbb{Z}^2$ can be identified with the midpoint $\frac{i+j}{2}$, we consider the dual lattice of size $\frac{1}{\sqrt{2}}$ given by

$$\mathcal{Z} = \left\{\frac{i+j}{2} : i, j \in \mathbb{Z}^2 \text{ such that } \|i-j\| = 1\right\}. \tag{3.51}$$

We also identify $z \in \mathcal{Z}$ with the closed unit segment centered at z and orthogonal to $j - i$, denoted by $[i, j]^\perp$ (see Fig. 3.31). Now we give the definition of *path* in the scaled lattice $\varepsilon\mathcal{Z}$.

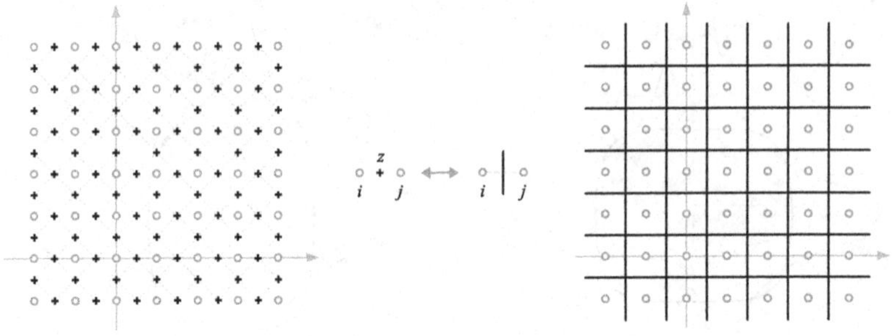

Figure 3.31 The lattice $\mathcal{Z}$ (cross nodes) and the identification of z and the segment $[i,j]^\perp$.

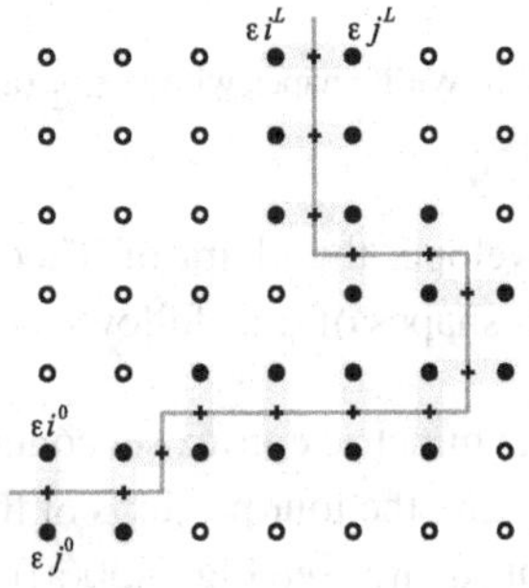

Figure 3.32 A path in $\varepsilon\mathcal{Z}$.

Definition 3.48 (Paths in $\varepsilon\mathcal{Z}$) A *path in* $\varepsilon\mathcal{Z}$ is a finite union

$$\sigma = \{\varepsilon z_l\} = \bigcup_{l=1}^{L} \varepsilon[i^l, j^l]^\perp$$

of segments with length ε parametrized by a sequence of points $z_l \in \mathcal{Z}$ (equivalently, by a sequence of pairs of nearest neighbors $(i^l, j^l) \in \mathbb{Z}^2 \times \mathbb{Z}^2$), such that the segment labeled by l has a common endpoint with the one labeled by $l+1$. The points εz_0 and εz_L are referred to as the *endpoints* of the path (see Fig. 3.32). If $\varepsilon = 1$, we say that σ is a *path in* $\mathcal{Z}$.

It is useful to define a projection from $\mathbb{R}^2$ to the lattice $\mathcal{Z}$, which is a Bravais lattice, generated by $\frac{e_1+e_2}{2}$ and $\frac{e_2-e_1}{2}$. In general, given a Bravais lattice $\mathcal{L} = \mathbb{Z}v_1 + \mathbb{Z}v_2$, we define the half-open cells corresponding to $\mathcal{L}$ by setting

$$C_k^{\mathcal{L}} = k + \left\{ tv_1 + sv_2 \colon t, s \in \left[-\frac{1}{2}, \frac{1}{2}\right) \right\}, \; k \in \mathcal{L},$$

as in (2.25).

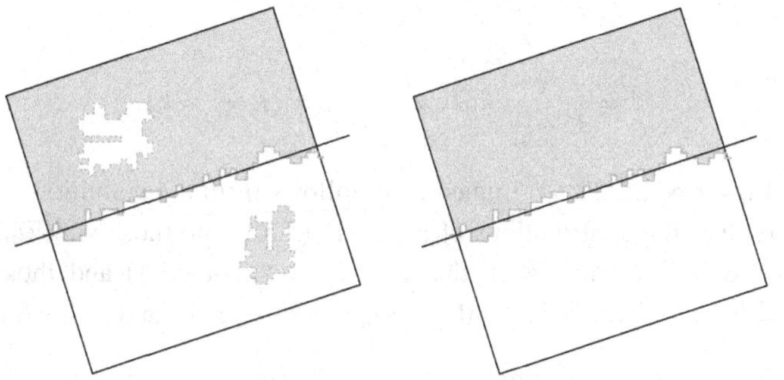

Figure 3.33 Reduction to minimization over paths.

Definition 3.49 (Projection on a Bravais lattice $\mathcal{L}$) Given $x \in \mathbb{R}^2$, $\pi_{\mathcal{L}}(x)$ denotes the (unique) $k \in \mathcal{L}$ such that $x \in C_k^{\mathcal{L}}$; we say that $\pi_{\mathcal{L}}(x)$ is the *projection of x on $\mathcal{L}$.*

Remark 3.50 (A path-minimization formula in dimension 2) In dimension 2, in the definition of φ_{hom} given in (3.12) we can consider only functions $v\colon \mathbb{Z}^2 \cap Q_T^\nu \to \{-1, 1\}$ such that $v = 2\chi_{H^\nu} - 1$ in $Q_T^\nu \setminus Q_{T-2R}^\nu$ and such that both the set

$$A = \bigcup_{v_i=1} \left(i + \left[-\frac{1}{2}, \frac{1}{2}\right]^2\right)$$

and its complement are connected. Hence, the boundary of A in the interior of the square Q_T^ν is (the restriction of) a path in $\mathcal{Z}$ (see Fig. 3.33). Since the boundary condition corresponds to the fact that the endpoints of the path are the projections of $-\frac{T}{2}\nu^\perp$ and $\frac{T}{2}\nu^\perp$ on $\mathcal{Z}$, we can minimize over such paths and get

$$\varphi_{\text{hom}}(\nu) = \lim_{T\to+\infty} \frac{8}{T} \min\Big\{ \sum_{l=1}^{L} a_{z_l} : \{z_l\} \text{ path in } \mathcal{Z} \text{ with endpoints, } \pi_{\mathcal{Z}}\big(-\tfrac{T}{2}\nu^\perp\big), \pi_{\mathcal{Z}}\big(\tfrac{T}{2}\nu^\perp\big); L \in \mathbb{N}\Big\}, \quad (3.52)$$

where $a_z = a_{ij}$ if $z \in \mathcal{Z}$ corresponds to the pair (i, j).

Proposition 3.51 (Lower bounds by projection) *Let φ_{hom} be the homogenized energy density of $\{a_{ij}\}$; then we have*

$$\varphi_{\text{hom}}(x) \geq 8c_1^p |\langle x, e_1\rangle| + 8c_2^p |\langle x, e_2\rangle|, \quad (3.53)$$

where

$$c_1^p = \frac{1}{K} \sum_{k=1}^{K} \min\{a_{ij} : \langle i, e_2\rangle = \langle j, e_2\rangle = k\} \quad (3.54)$$

and

$$c_2^p = \frac{1}{K}\sum_{k=1}^{K} \min\{a_{ij} \colon \langle i, e_1\rangle = \langle j, e_1\rangle = k\}. \tag{3.55}$$

Proof Lower bound (3.53) immediately follows from the definition of $\varphi_{\rm hom}$, by subdividing the contributions of $a_{i_{n-1}i_n}$ in (3.52) into those with $\langle i_n, e_2\rangle = \langle i_{n-1}, e_2\rangle$ (or equivalently such that $i_n - i_{n-1} \in \{-e_1, e_1\}$) and those with $\langle i_n, e_1\rangle = \langle i_{n-1}, e_1\rangle$ (or equivalently $i_n - i_{n-1} \in \{-e_2, e_2\}$), and estimating

$$a_{i_{n-1}i_n} \geq \min\{a_{ij} \colon \langle i, e_2\rangle = \langle j, e_2\rangle = \langle i_n, e_2\rangle\}$$

and

$$a_{i_{n-1}i_n} \geq \min\{a_{ij} \colon \langle i, e_1\rangle = \langle j, e_1\rangle = \langle i_n, e_1\rangle\},$$

respectively, in the two cases. □

Proposition 3.52 (Upper bounds by averaging) *Let $\varphi_{\rm hom}$ be the homogenized energy density of a_{ij}; then we have*

$$\varphi_{\rm hom}(x) \leq 8c_1^a|\langle x, e_1\rangle| + 8c_2^a|\langle x, e_2\rangle|, \tag{3.56}$$

where c_1^a is the average over horizontal bonds

$$c_1^a = \frac{1}{K^2}\sum_{k=1}^{K} \sum_{\substack{z\in[0,K)^2\cap\mathcal{Z} \\ \langle z, e_2\rangle = k}} a_z \tag{3.57}$$

and c_2^a is the average over vertical bonds

$$c_2^a = \frac{1}{K^2}\sum_{k=1}^{K} \sum_{\substack{z\in[0,K)^2\cap\mathcal{Z} \\ \langle z, e_1\rangle = k}} a_z. \tag{3.58}$$

Proof Let $n_1, n_2 \in \{1, \ldots, K\}$ be such that, setting

$$z_1^k = \Big(n_1 - \frac{1}{2}, k\Big), \quad z_2^k = \Big(k, n_2 - \frac{1}{2}\Big),$$

we have

$$\frac{1}{K}\sum_{k=1}^{K} a_{z_1^k} \leq \frac{1}{K^2}\sum_{k=1}^{K} \sum_{\substack{z\in[0,K)^2\cap\mathcal{Z} \\ \langle z, e_2\rangle = k}} a_z \quad \text{and} \quad \frac{1}{K}\sum_{k=1}^{K} a_{z_2^k} \leq \frac{1}{K^2}\sum_{k=1}^{K} \sum_{\substack{z\in[0,K)^2\cap\mathcal{Z} \\ \langle z, e_1\rangle = k}} a_z.$$

The upper bound is then obtained by considering only sets whose boundary lies in $(n_1 - \frac{1}{2}, n_2 - \frac{1}{2}) + K\mathbb{Z}^2$, and hence give minimal paths in $\mathcal{Z}$. □

Optimality of Bounds

The proof of the optimality of the bounds above is obtained by a multiscale construction involving the homogenization of continuum perimeter energies locally obtained as limits of discrete energies. We do not describe in detail this procedure, but give an idea of the "building blocks" of the discrete-to-continuum part of this process.

For simplicity we only treat the extreme case when $\theta = 1$, in which case we consider a family of periodic systems $\{a^k_{ij}\}_k$ with

$$\lim_{k\to+\infty} \theta_{\rm h}(\{a^k_{ij}\}) = 1 \qquad \text{and} \qquad \lim_{k\to+\infty} \theta_{\rm v}(\{a^k_{ij}\}) = 1,$$

and two paradigmatic geometries.

(i) (rectangular geometry) We construct systems $\{a^k_{ij}\}_k$ such that the φ obtained as limit of their homogenized energy densities is

$$\varphi(\nu) = 8\Big(\gamma_1|\langle \nu, e_1\rangle| + \gamma_2|\langle \nu, e_2\rangle|\Big),$$

whose Wulff shape is

$$W_\varphi = \Big[-\frac{1}{64\gamma_1}, \frac{1}{64\gamma_1}\Big] \times \Big[-\frac{1}{64\gamma_2}, \frac{1}{64\gamma_2}\Big],$$

with $\gamma_1, \gamma_2 \in [\alpha, \beta]$. We write $\gamma_i = t_i\alpha + (1 - t_i)\beta$.

Let $k \in \mathbb{N}$. The k-periodic coefficients a^k_z are given by

$$a^k_z = \begin{cases} \alpha & \text{if } \langle z, e_1\rangle = \frac{1}{2} \text{ and } 1 \le \langle z, e_2\rangle \le \lfloor kt_1\rfloor, \\ \alpha & \text{if } \langle z, e_2\rangle = \frac{1}{2} \text{ and } 1 \le \langle z, e_1\rangle \le \lfloor kt_2\rfloor, \\ \beta & \text{otherwise.} \end{cases}$$

Note that $\theta_{\rm h}(\{a^k_{ij}\})$ and $\theta_{\rm v}(\{a^k_{ij}\})$ tend to 1 as $k \to +\infty$. Moreover, since

$$\min\Big\{a^k_z : \langle z, e_1\rangle = k + \frac{1}{2}\Big\} = \begin{cases} \alpha & \text{if } 1 \le \langle z, e_2\rangle \le \lfloor kt_1\rfloor, \\ \beta & \text{otherwise} \end{cases}$$

and

$$\min\Big\{a^k_z : \langle z, e_2\rangle = k + \frac{1}{2}\Big\} = \begin{cases} \alpha & \text{if } 1 \le \langle z, e_1\rangle \le \lfloor kt_2\rfloor, \\ \beta & \text{otherwise,} \end{cases}$$

Proposition 3.51 gives the lower bound

$$\varphi^k_{\rm hom}(\nu) \ge \frac{8}{k}\Big((\lfloor kt_1\rfloor\alpha + (k - \lfloor kt_1\rfloor)\beta)|\langle \nu, e_1\rangle| + (\lfloor kt_2\rfloor\alpha + (k - \lfloor kt_2\rfloor)\beta)|\langle \nu, e_2\rangle|\Big).$$

The right-hand side gives also an upper bound, by considering only sets whose boundary lies in $(\frac{1}{2}, \frac{1}{2}) + k\mathbb{Z}^2$, as in the proof of Proposition 3.52. Finally, the claim is obtained by taking the limit as $k \to +\infty$.

(ii) (rhomboidal geometry) We construct systems $\{a^k_{ij}\}_k$ such that the φ obtained as limit of their homogenized energy densities is such that W_φ is the rhombus with vertices

$$\Big(\frac{1}{64\alpha},\frac{1}{64\alpha}\Big),\quad \Big(\frac{1}{64\beta},-\frac{1}{64\beta}\Big),\quad \Big(-\frac{1}{64\alpha},-\frac{1}{64\alpha}\Big),\quad \Big(-\frac{1}{64\beta},\frac{1}{64\beta}\Big).$$

With fixed $k \in \mathbb{N}$ we define k-periodic a^k_z as follows:

$$a^k_z = \begin{cases} \alpha & \text{if } z \text{ intersects the bisectrix line,} \\ \beta & \text{otherwise.} \end{cases}$$

Regrouping connections with $a^k_z = \alpha$ and $a^k_z = \beta$ in test paths for formula (3.52), we can estimate $\varphi^k_{\rm hom}(\nu)$ by the convex envelope of the function taking the value $\sqrt{2}\alpha$ in the direction $\nu_1 = \frac{e_1+e_2}{\sqrt{2}}$ and a value estimated by $\sqrt{2}\beta - o(1)$ as $k \to +\infty$ in the direction $\nu_2 = \frac{e_1-e_2}{\sqrt{2}}$. Since these two values are sharp in the directions ν_1 and ν_2, respectively, in the limit as $k \to +\infty$ we obtain the claim.

The proof of Theorem 3.47 in the two-dimensional case can be obtained by first approximating φ with a polyhedral Wulff shape combining the arguments described earlier. The proof in the general case follows by a similar argument, but is more technical since it does not use a minimal-path formula.

3.7 Random Systems

In this section we compute the Γ-limit of energies where we randomly choose coefficients. To this end we have to introduce some notions of percolation theory for what is called the *bond-percolation model*; that is, when the random choice is thought to be carried out on the connections. As a result we will consider coefficients $a_{ij} = a^\omega_{ij}$ on $\mathbb{Z}^d \times \mathbb{Z}^d$ that depend on the realization of a random variable. A different model, that can be treated similarly, is the *site-percolation model*. In our intuition it would correspond to choosing weak and strong *nodes* – and to define a weak connection as a connection between two nodes of which at least one is a weak node.

3.7.1 Random Mixtures

From now on we will restrict to the two-dimensional case $d = 2$. We recall definition (3.51) of the dual lattice of $\mathbb{Z}^2$

$$\mathcal{Z} = \Big\{\frac{i+j}{2} : i, j \in \mathbb{Z}^2 \text{ such that } \|i-j\| = 1\Big\}.$$

We identify each point $z = \frac{i+j}{2} \in \mathcal{Z}$ with the pair $(i,j) \in \mathbb{Z}^2 \times \mathbb{Z}^2$ or with the segment $[i,j]^\perp$ orthogonal to $[i,j]$ and with middle point z.

Let $0 < \alpha < \beta < +\infty$. We consider the simplest case of two kinds of nearest-neighbor connections with weights α and β. A choice of connections between nodes of $\mathbb{Z}^2$ is a function

$$\omega\colon \mathcal{Z} \to \{\alpha,\beta\}.$$

For any such ω we then define the coefficients

$$a_{ij}^\omega = \omega\Big(\frac{i+j}{2}\Big).$$

Let Ω be a bounded and Lipschitz open subset of $\mathbb{R}^2$, and consider for any $\varepsilon > 0$ the lattice $\mathcal{L}_\varepsilon = \varepsilon\mathbb{Z}^2 \cap \Omega$. For any $\omega \in \{\alpha,\beta\}^{\mathcal{Z}}$, we define the energies

$$E_\varepsilon^\omega(u) = \sum_{\langle i,j\rangle} \varepsilon a_{ij}^\omega (u_i - u_j)^2, \tag{3.59}$$

where as usual $\langle i,j\rangle$ indicates the set of pairs in $\mathbb{Z}^2$ such that $\varepsilon i, \varepsilon j \in \mathcal{L}_\varepsilon$ and $\|i-j\| = 1$.

We will consider ω with the property that

$$\omega\Big(\frac{i+j}{2}\Big) = a_{ij}^\omega = \begin{cases} \alpha & \text{with probability } 1-p, \\ \beta & \text{with probability } p, \end{cases} \tag{3.60}$$

with $p \in [0,1]$. This can be done rigorously by introducing some independent identically distributed random variables. For our presentation it suffices to describe the "almost-sure" properties of such ω with respect to a probability measure of the space $\{\alpha,\beta\}^{\mathcal{Z}}$.

Now, we want to prove some almost sure statements about the Γ-limit of the energies defined in (3.59).

Lower Bound by Blowup at $x_0 \in \partial^* A$

We prove the lower estimate by applying the blow-up method as in Section 3.2. Let $\{u^\varepsilon\}$ be such that $E_\varepsilon^\omega(u^\varepsilon)$ is equibounded and converges to a set A of finite perimeter, and such that the corresponding sequence of measures $\{\mu_\varepsilon^\omega\}$ weak* converge to a measure μ^ω. We restrict to points $x_0 \in \partial^* A$ suitable for the blowup of A and such that the measure-theoretical derivative of the limit measure μ^ω with respect to $\mathcal{H}^1 \llcorner \partial^* A$ exists, and follow the steps of the blow-up procedure in the case of periodic energies, obtaining

$$\lim_{\varepsilon\to 0} \frac{\mu_\varepsilon^\omega(Q_{\varrho_\varepsilon}^\nu(x_0))}{\varrho_\varepsilon} \geq \liminf_{\varepsilon\to 0} \frac{8}{\varrho_\varepsilon} \min\Big\{\sum_{l=1}^L \varepsilon a_{z_l}^\omega : L \in \mathbb{Z}, \{\varepsilon z_l\} \text{ path in } \varepsilon\mathcal{Z} \text{ joining}$$
$$\pi_{\varepsilon\mathcal{Z}}\Big(x_0 - \frac{\varrho_\varepsilon}{2}\nu^\perp\Big) \text{ and } \pi_{\varepsilon\mathcal{Z}}\Big(x_0 + \frac{\varrho_\varepsilon}{2}\nu^\perp\Big)\Big\},$$

where $\pi_{\varepsilon\mathcal{Z}}$ denotes the projection of $\mathbb{R}^2$ on $\varepsilon\mathcal{Z}$ as in Definition 3.49, and $a^\omega_z = a^\omega_{ij}$ if $z = \frac{i+j}{2} \in \mathcal{Z}$. We recall that points $z \in \mathcal{Z}$ are identified with segments in the definition of a path. We used the fact that in dimension 2 the lower bound can be described by optimizing over paths with endpoints close to $x_0 \pm \frac{\varrho_\varepsilon}{2}\nu^\perp$. Note that there exists a scale $\overline{\varrho}_\varepsilon \to 0$ such that in the blow-up procedure we can choose any infinitesimal sequence ϱ_ε satisfying $\varrho_\varepsilon \geq \overline{\varrho}_\varepsilon$.

By scaling ϱ_ε to $T_\varepsilon = \frac{\varrho_\varepsilon}{\varepsilon}$, we get

$$\lim_{\varepsilon\to 0} \frac{\mu_\varepsilon(Q^\nu_{\varrho_\varepsilon}(x_0))}{\varrho_\varepsilon} \geq \liminf_{\varepsilon\to 0} \frac{m^\omega(x_\varepsilon, x_\varepsilon + T_\varepsilon\nu^\perp)}{T_\varepsilon} \tag{3.61}$$

where $x_\varepsilon = \frac{x_0}{\varepsilon} - \frac{T_\varepsilon}{2}\nu^\perp = \frac{T_\varepsilon}{\varrho_\varepsilon} - \frac{T_\varepsilon}{2}\nu^\perp$ and

$$m^\omega(x,y) = 8\min\Big\{\sum_{l=1}^{L} a^\omega_{z_l} : \{z_l\} \text{ path in } \mathcal{Z} \text{ joining } \pi_{\mathcal{Z}}(x), \pi_{\mathcal{Z}}(y); L \in \mathbb{Z}\Big\}. \tag{3.62}$$

Until this point, we only used the blow-up method and the fact that we can see the minimum problem with Dirichlet boundary conditions as a minimum problem over paths in $\mathcal{Z}$ with fixed endpoints. In order to characterize almost surely the limit energy density we will use the following percolation result.

Theorem 3.53 (First-passage percolation) *Let $a^\omega_{ij} = \omega(\frac{i+j}{2})$ with ω, α, β, and p as in* (3.60). *Then the function m^ω defined in* (3.62) *satisfies the following properties:*

(i) *almost surely there exists the limit*

$$\lim_{T\to+\infty} \frac{m^\omega(0, T\nu^\perp)}{T} = \varphi^\omega(\nu);$$

(ii) *(translation invariance) almost surely there exists the limit*

$$\varphi^\omega(\nu) = \lim_{T\to+\infty} \frac{m^\omega(x_T, x_T + T\nu^\perp)}{T} \tag{3.63}$$

for any sequence $\{x_T\}$ such that $\|x_T\| \leq T^2$;

(iii) *(the limit is deterministic) there exists a function φ_p such that almost surely*

$$\varphi^\omega(\cdot) = \varphi_p(\cdot)$$

independently of ω.

Here and in the following, we only give a hint of the main arguments behind percolation results and to differences and correspondences with the periodic setting, referring to the literature for more details. As for the preceding result, we only mention that the existence of the limit comes from a *stationarity property* of the function m^ω seen as a subadditive "point process," for which a number of asymptotic results hold. Indeed, we have

$$m^\omega(x,y) \le m^\omega(x,z) + m^\omega(z,y)$$

for any $x, y, z \in \mathbb{R}^2$. Note that subadditivity is a key property also in the proof of the existence of the limit in the periodic homogenization case.

Note that translation invariance is different from the periodic case. Indeed, in that case it is possible to translate by a (multiple of the) period, and we can approximate any translation with an error of the order of the period, so that it is negligible when T diverges. Hence, the invariance holds without constraints on the growth of $\|x_T\|$. In the case of random coefficients, the same invariance property may not be true for some choice of x_T. A heuristic argument is that for any T there is a positive probability of having a large set where all connections are α-connections. Then, if we translate the minimum problem for $m^\omega(0, T\nu^\perp)$ to that set by a suitable x_T, we obtain $\alpha\|\nu\|_1$ as the minimum value. The same holds for β-connections, which contradicts translation invariance. The statement says that we may suppose almost surely that these sets are "far away" from 0.

Finally, the fact that the limit φ^ω is in fact deterministic depends on the *ergodicity property* of the point process m^ω.

Now we apply Theorem 3.53 to the sequence $\frac{m^\omega(x_\varepsilon, x_\varepsilon + T_\varepsilon\nu^\perp)}{T_\varepsilon}$ to obtain a lower bound. Note that in the blow-up procedure we can choose the infinitesimal sequence ϱ_ε large enough so as to have

$$\|x_\varepsilon\| = T_\varepsilon \frac{\|x_0\|}{\varrho_\varepsilon} \le T_\varepsilon^2.$$

Hence, by Theorem 3.53 we get that almost surely there exists the limit

$$\lim_{\varepsilon\to 0} \frac{m^\omega(x_\varepsilon, x_\varepsilon + T_\varepsilon\nu^\perp)}{T_\varepsilon} = \varphi_p(\nu).$$

By (3.61), we obtain the lower bound

$$\lim_{\varepsilon\to 0} \frac{\mu_\varepsilon^\omega(Q^\nu_{\varrho_\varepsilon}(x_0))}{\varrho_\varepsilon} \ge \varphi_p(\nu) \text{ for } \mathcal{H}^1\text{-almost all } x_0 \in \Omega \cap \partial^* A$$

for a set of ω with probability 1. Concluding the blow-up procedure by integrating, we deduce that for a set of realizations ω with probability 1 the following estimate holds:

$$\liminf_{\varepsilon\to 0} E_\varepsilon^\omega(u^\varepsilon) \ge \int_{\Omega\cap\partial^* A} \varphi_p(\nu)\, d\mathcal{H}^1.$$

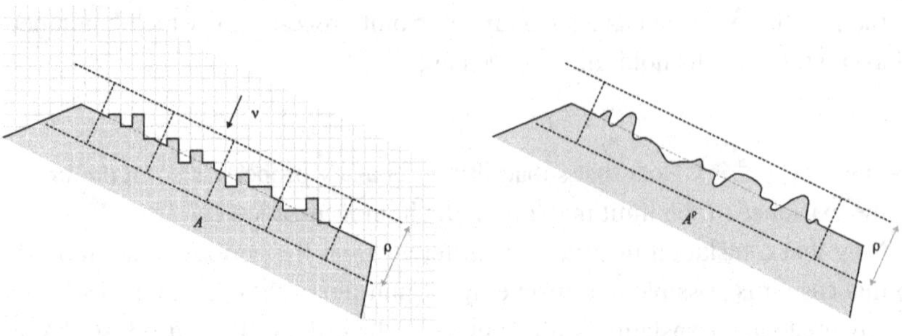

Figure 3.34 Construction of the recovery sequence.

Upper Bound

By a density argument, it is sufficient to construct a recovery sequence for a polyhedral set A.

We first assume that $\Omega \cap \partial A = L$, where L is a segment. Let ν be the normal to ∂A on L. Note that to apply Theorem 3.53(b) we have to take into account a boundedness requirement on the centers of the involved squares. Hence, differently from the periodic case, it is more convenient to subdivide the construction of the recovery sequence into two steps. First, we fix $\varrho > 0$ and consider a finite number of squares with side length ϱ covering the segment L except for a small neighborhood of the endpoints. We define a recovery sequence u_ε^ϱ by considering in each square $Q_\varrho^\nu(x_r^\varrho)$, $r \in \{1,\dots,R^\varrho\}$, (the scaling of) a solution of the minimum problem for $m^\omega(\frac{x_r^\varrho}{\varepsilon} - \frac{\varrho}{2\varepsilon}\nu^\perp, \frac{x_r^\varrho}{\varepsilon} + \frac{\varrho}{2\varepsilon}\nu^\perp)$. In the neighborhood of the endpoints we set u_ε^ϱ as the discretization of $2\chi_A - 1$. By construction it follows that

$$E_\varepsilon^\omega(u_\varepsilon^\varrho) = \varrho \sum_{r=1}^{R^\varrho} \frac{m^\omega(\frac{x_r^\varrho}{\varepsilon} - \frac{\varrho}{2\varepsilon}\nu^\perp, \frac{x_r^\varrho}{\varepsilon} + \frac{\varrho}{2\varepsilon}\nu^\perp)}{\frac{\varrho}{\varepsilon}} + c\varrho$$
$$= \mathcal{H}^1(L)\,\varphi_p(\nu) + c\varrho + o(1)_{\varepsilon\to 0}.$$

If A is an arbitrary polyhedral set, then we can repeat the construction for each of its sides (see Fig. 3.34), obtaining a family of spin functions again denoted by u_ε^ϱ. The characteristic functions of the union of the corresponding sets $A_\varepsilon(u_\varepsilon^\varrho)$ converge (up to subsequences) to some A^ϱ. For any fixed $\varrho > 0$ we get

$$\Gamma\text{-}\limsup_{\varepsilon\to 0} E_\varepsilon^\omega(A^\varrho) \le \int_{\Omega\cap\partial A} \varphi_p(\nu_A)\, d\mathcal{H}^1 + o(1)_{\varrho\to 0}.$$

This gives an upper bound for the upper Γ-limit on A^ϱ.

Now, we let $\varrho \to 0$. By semicontinuity, since $A^\varrho \to A$ as $\varrho \to 0$, we get the same upper bound on A, and then conclude by density for sets of finite perimeter.

This concludes the proof of the following Γ-convergence result.

Theorem 3.54 (Homogenization of random mixtures) *Let $a^\omega_{ij} = \omega(\frac{i+j}{2})$ with ω, α, β, and p as in (3.60). Then almost surely in ω there exists the limit*

$$\lim_{T\to+\infty} \frac{m^\omega(0, T\nu^\perp)}{T} = \varphi^\omega(\nu)$$

and

$$\Gamma\text{-}\lim_{\varepsilon\to 0} E^\omega_\varepsilon(A) = \int_{\Omega\cap\partial^* A} \varphi^\omega(\nu)\, d\mathcal{H}^1$$

on sets A of finite perimeter. Moreover, almost surely φ^ω depends only on p.

The formula defining φ^ω is usually called the *first-passage percolation formula* of the system.

Remark 3.55 We can extend the result to the case where the coefficients a^ω_{ij} are allowed to take values not only in $\{\alpha, \beta\}$, but in the interval $[\alpha, \beta]$. In this case, the notation becomes more complex since the result depends on the probability distribution and not only on the value p.

3.7.2 Extreme Cases

In order to treat the cases when $\alpha = 0$ or $\beta = +\infty$ we need some more refined geometric properties of path-connected sets of connections from percolation theory.

Given α and β such that $0 \le \alpha < \beta \le +\infty$, a realization $\omega \in \{\alpha, \beta\}^{\mathcal{Z}}$, and the corresponding system of coefficients a^ω_{ij}, we use the notation

$$\mathcal{Z}_\alpha = \bigcup_{a^\omega_{ij}=\alpha} [i,j]^\perp \quad \text{and} \quad \mathcal{Z}_\beta = \bigcup_{a^\omega_{ij}=\beta} [i,j]^\perp. \tag{3.64}$$

We again identify the segment $[i,j]^\perp$ with its center $z = \frac{i+j}{2} \in \mathcal{Z}$, so that $\mathcal{Z}_\alpha$ and $\mathcal{Z}_\beta$ can be viewed as subsets of $\mathcal{Z}$.

We give a result on the connected components of $\mathcal{Z}_\alpha$ and $\mathcal{Z}_\beta$ related to the cases $p < \frac{1}{2}$ and $p > \frac{1}{2}$, respectively. The homogenization results in the case $p = \frac{1}{2}$ can be obtained as a limit as $p \to \frac{1}{2}$ from Theorems 3.58 and 3.65, and will not be dealt with explicitly.

Theorem 3.56 (Bond percolation) *If $p < \frac{1}{2}$, then almost surely (in ω)*

(i) *there exists a unique infinite connected component of the set $\mathcal{Z}_\alpha$, which is denoted by W^ω (called the* weak cluster*);*

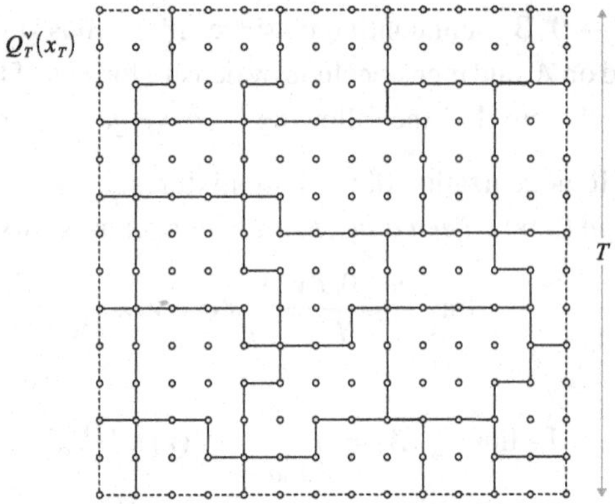

Figure 3.35 A grid of paths in the weak (or strong) cluster.

(ii) *there exist T_0 and η such that, for any $T > T_0$ and for any square $Q^\nu_T(x_T)$ such that $\|x_T\| \leq T^2$, there exist ηT disjoint paths in $Q^\nu_T(x_T) \cap W^\omega$ connecting pairs of opposite sides of $Q^\nu_T(x_T)$ (see Fig. 3.35).*

The corresponding statement holds for $p > \frac{1}{2}$ with β instead of α; the infinite connected component of the set $\mathcal{Z}_\beta$ is denoted by S^ω, and is called the strong cluster.

Theorem 3.56 ensures that, almost surely, if we consider a large enough square $Q^\nu_T(x_T)$ then we can think of the weak cluster (or the strong cluster if $p > \frac{1}{2}$) as a grid that we can use, in a sense, as a generalization of the grid given by the regular square lattice.

Remark 3.57 Note that a corresponding result holds if we take

$$\mathcal{Z}_\alpha = \bigcup_{a^\omega_{ij}=\alpha} [i,j] \quad \text{and} \quad Z_\beta = \bigcup_{a^\omega_{ij}=\beta} [i,j];$$

that is, if we consider connections between nodes in $\mathbb{Z}^2$.

Random Strong Inclusions

As in Section 3.7.1, we consider nearest-neighbor connections with weights α and β with independently assigned probability, but here we let the value β be "very large" and α normalized to 1; that is, we fix for $p \in [0,1]$,

$$a^\omega_{ij} = \omega\Big(\frac{i+j}{2}\Big) = \begin{cases} \alpha = 1 & \text{with probability } 1-p, \\ \beta = +\infty & \text{with probability } p \end{cases} \tag{3.65}$$

if $\|i-j\| = 1$ and 0 otherwise. We call this a *rigid system.*

We now consider the energies defined by

$$E_\varepsilon^\omega(u) = \sum_{\langle i,j\rangle} \varepsilon a_{ij}^\omega (u_i - u_j)^2, \tag{3.66}$$

with the coefficients a_{ij}^ω as in (3.65). We use the convention that $0 \cdot +\infty = 0$, so that, if $a_{ij}^\omega = \beta$, the choice $\beta = +\infty$ forces $u_i = u_j$ in order to have a bounded energy $E_\varepsilon^\omega(u)$.

Now we separately study the two cases $p > \frac{1}{2}$ and $p < \frac{1}{2}$, since we deal with different geometries.

Case $p > \frac{1}{2}$. In this case, we expect to have only isolated sets with finite energy, since we have an infinite connected component of bonds with value $+\infty$. By using the second statement in Theorem 3.56, we show that if these sets have equibounded perimeters, they necessarily converge to a set with perimeter equal to 0, so that the Γ-limit is finite (and it is equal to 0) only if $A = \emptyset$ or $A = \Omega$. It can be useful to show the convergence result in a more general case; that is, instead of assuming $\beta = +\infty$, which can be read as a constraint in the lattice, we introduce a dependence on ε allowing one to consider "very large" – finite or infinite – values of the strong coefficients. We assume, for i, j nearest neighbors in $\mathbb{Z}^2$,

$$a_{ij}^\omega = a_{ij}^{\varepsilon,\omega} = \begin{cases} \alpha = 1 & \text{with probability } 1 - p, \\ \beta = \beta_\varepsilon & \text{with probability } p, \end{cases} \tag{3.67}$$

with diverging β_ε, and show the following Γ-convergence result.

Theorem 3.58 (Γ-convergence with $p > \frac{1}{2}$) *Let E_ε^ω be defined as in (3.66) with the coefficients $a_{ij}^\omega = a_{ij}^{\varepsilon,\omega}$ given by (3.67), and let $\beta_\varepsilon \to +\infty$ as $\varepsilon \to 0$. Then, if $p > \frac{1}{2}$, we have that*

$$\Gamma\text{-}\lim_{\varepsilon\to 0} E_\varepsilon^\omega(A) = \begin{cases} 0 & \text{if } A = \emptyset \text{ or } A = \Omega, \\ +\infty & \text{otherwise.} \end{cases} \tag{3.68}$$

Proof It is sufficient to prove the lower estimate. To this end, we again apply the blow-up method. Let A be a set of finite perimeter such that $\mathcal{H}^1(\Omega \cap \partial^* A) \neq 0$. We choose a meaningful $x_0 \in \Omega \cap \partial^* A$ where we can blow up the set A, and consider the square $Q_{\varrho_\varepsilon}^\nu(x_0)$ and a path γ in $\varepsilon\mathcal{Z}$ with endpoints $\pi_{\varepsilon\mathcal{Z}}(x_0 - \frac{\varrho_\varepsilon}{2}\nu^\perp)$ and $\pi_{\varepsilon\mathcal{Z}}(x_0 + \frac{\varrho_\varepsilon}{2}\nu^\perp)$. Note that, if we define the sets $\mathcal{Z}_\gamma$ as in Remark 3.57, we can apply the second statement of Theorem 3.56 to $\mathcal{Z}_{\beta_\varepsilon}$ in the scaled squares $Q_{T_\varepsilon}^\nu(x_\varepsilon)$, where $T_\varepsilon = \frac{\varrho_\varepsilon}{\varepsilon}$ and $x_\varepsilon = \frac{x_0}{\varepsilon}$, since ϱ_ε can be chosen large enough with respect to ε. We then obtain that almost surely in ω, for T_ε large enough, there exist ηT_ε such disjoint paths in $\mathcal{Z}_{\beta_\varepsilon}$ inside $Q_{T_\varepsilon}^\nu(x_\varepsilon) \cap S^\omega$ connecting pairs of opposite sides of $Q^{\nu}_{T_\varepsilon}(x_\varepsilon)$, and hence the number of the segments in the union of these paths intersecting γ is at least $\eta\frac{\varrho_\varepsilon}{\varepsilon}$. For any such segment $[i,j]$, $a^{\varepsilon,\omega}_{ij} = \beta_\varepsilon$, and the blow-up method gives the estimate

$$\lim_{\varepsilon\to 0}\frac{\mu_\varepsilon(Q^{\nu}_{\varrho_\varepsilon}(x_0))}{\varrho_\varepsilon} \geq \liminf_{\varepsilon\to 0}\frac{m^{\omega}(x_\varepsilon, x_\varepsilon + T_\varepsilon\nu^{\perp})}{T_\varepsilon}$$

$$\geq \liminf_{\varepsilon\to 0}\frac{1}{T_\varepsilon}\eta\frac{\varrho_\varepsilon}{\varepsilon}\beta_\varepsilon = \lim_{\varepsilon\to 0}\eta\beta_\varepsilon = +\infty.$$

We can conclude that if $\mathcal{H}^1(\Omega\cap\partial^* A)\neq 0$, then the Γ-limit is $+\infty$. □

Case $p<\frac{1}{2}$. In this case, in order to apply the blow-up method to obtain a lower bound for the energies, we define a pseudometric by considering only paths which lie in the weak cluster. Recalling the definition of weak cluster W^ω in Theorem 3.56, which in this case is understood as a union of segments, we set

$$m^{\omega}_{\rm weak}(x,y) = 8\min\Big\{\sum_{l=1}^{M} a^{\omega}_{z_l}\colon \{z_l\}\text{ path in } \mathcal{Z}\text{ joining }\pi_w(x)\text{ and }\pi_w(y); M\in\mathbb{N}\Big\}$$

$$= 8\min\Big\{M\colon \{z_l\}\text{ path in } W^\omega\text{ joining }\pi_w(x)\text{ and }\pi_w(y)\Big\}, \qquad (3.69)$$

where $a^{\omega}_{z} = a^{\omega}_{ij}$ if $z = \frac{i+j}{2}$, and π_w is the projection on $\big((\frac{1}{2},\frac{1}{2})+\mathbb{Z}^2\big)\cap W^\omega$, with a choice given by the lexicographical order in case of nonuniqueness, and the convention that the minimum is 0 if $\pi_w(x)=\pi_w(y)$. Moreover, we let $L(\gamma)=M$ for a path $\gamma=\{z_l\}_{l=1}^{M}$ composed of M elements. The following result holds.

Theorem 3.59 (Existence of the asymptotic chemical distance) *There exists a function $\varphi=\varphi_p$ such that almost surely*

$$\varphi_p(\nu) = \lim_{T\to+\infty}\frac{m^{\omega}_{\rm weak}(x_T, x_T+T\nu^{\perp})}{T} \qquad (3.70)$$

if $\|x_T\|\leq T^2$. This function is called the asymptotic chemical distance.

Remark 3.60 (Metric interpretation of the chemical distance) The weak cluster can be seen as a connected submanifold of $\mathbb{R}^2$ equipped with the Euclidean distance. The scaled weak clusters then converge (e.g. in the sense of Gromov–Hausdorff convergence, or of the Γ-convergence of the related metrics) to the whole $\mathbb{R}^2$ equipped with the metric given by the asymptotic chemical distance; that is,

$$d_p(x,y) = \|x-y\|\,\varphi_p\Big(\frac{x-y}{\|x-y\|}\Big)$$

for $x\neq y$.

We start by fixing the coefficients a_{ij}^{ω} as in (3.65), that is, $\beta = +\infty$, and again apply the blow-up method to a sequence converging to a set A, to obtain a lower bound. Given a suitable $x_0 \in \partial^* A$, we consider the square $Q^{\nu}_{\varrho_\varepsilon}(x_0)$. Noting that the strong connections are never used in an optimal path, since $\beta = +\infty$, we have that $m^{\omega}(x,y) = m^{\omega}_{\rm weak}(x,y)$. Hence, we can repeat the blow-up procedure, obtaining

$$\liminf_{\varepsilon\to 0} \frac{\mu_\varepsilon(Q^{\nu}_{\varrho_\varepsilon}(x_0))}{\varrho_\varepsilon} \geq \liminf_{\varepsilon\to 0} \frac{m^{\omega}_{\rm weak}(x_\varepsilon, x_\varepsilon + T_\varepsilon \nu^\perp)}{T_\varepsilon}.$$

Then, by using Theorem 3.59, we obtain a lower bound for the Γ-limit. Since the upper estimate can be given exactly as in the proof of Theorem 3.54, we get the following Γ-convergence result.

Theorem 3.61 (Γ-convergence for rigid spins and $p < \frac{1}{2}$) *Let E^{ω}_{ε} be defined as in* (3.66) *with the coefficients a_{ij}^{ω} given by* (3.65). *Then, if $p < \frac{1}{2}$, almost surely we have that*

$$\Gamma\text{-}\lim_{\varepsilon\to 0} E^{\omega}_{\varepsilon}(A) = \int_{\Omega\cap\partial^* A} \varphi_p(\nu)\, d\mathcal{H}^1,$$

where φ_p is the asymptotic chemical distance defined in (3.70).

Now we suppose that $\beta = \beta_\varepsilon$ with $\beta_\varepsilon \to +\infty$; that is, we define the coefficients $a_{ij}^{\varepsilon,\omega}$ as in (3.67) and prove that the Γ-convergence result of Theorem 3.61 holds also in this case. Note that the upper estimate is the same as in the case $\beta = +\infty$; hence, it is sufficient to show the lower estimate.

If the strong connections have a finite value $\beta_\varepsilon < +\infty$, in the minimization of the energy it could be in principle more convenient to have a "short-cut" path containing some segment $[i,j]^\perp$ such that $a_{ij}^{\varepsilon,\omega} = \beta_\varepsilon$ instead of a longer path of weak connections (see Fig. 3.36). To show that this does not happen, we

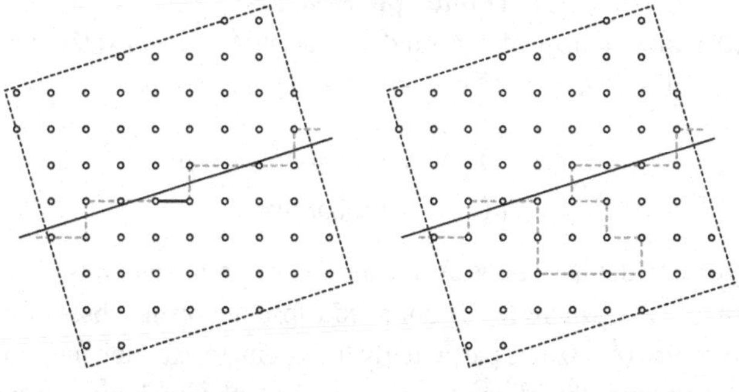

Figure 3.36 A "short-cut" path with a strong connection and a weak path.

have to use another percolation result, of which we only provide the statement without proof.

Lemma 3.62 (A lemma by H. Kesten) *For any $\delta > 0$ there exist $\eta > 0$ and T_0 such that for any $T > T_0$ and x_T satisfying $\|x_T\| \leq T^2$ the following property holds: if γ is a path between x_T and $x_T + T\nu^\perp$ with $8L(\gamma) < (\varphi_p(\nu) - \delta)T$, then there exist ηT strong connections in γ.*

We repeat the blow-up procedure at a suitable $x_0 \in \partial^* A$. We fix $\delta > 0$ and assume by contradiction that

$$\liminf_{\varepsilon\to 0} \frac{\mu_\varepsilon(Q^\nu_{\varrho_\varepsilon}(x_0))}{\varrho_\varepsilon} < \varphi_p(\nu) - \delta.$$

If we consider a sequence of paths γ_ε such that

$$\liminf_{\varepsilon\to 0} \frac{8L(\gamma_\varepsilon)}{\varrho_\varepsilon} \leq \liminf_{\varepsilon\to 0} \frac{\mu_\varepsilon(Q^\nu_{\varrho_\varepsilon}(x_0))}{\varrho_\varepsilon},$$

scaling as usual ϱ_ε to $T_\varepsilon = \frac{\varrho_\varepsilon}{\varepsilon}$, the result of the preceding lemma ensures that in each (scaled) path there are at least $\eta\frac{\varrho_\varepsilon}{\varepsilon}$ strong connections with coefficient β_ε. Then the energy of each path γ_ε is again greater than $\frac{\varepsilon}{\varrho_\varepsilon}\eta\frac{\varrho_\varepsilon}{\varepsilon}\beta_\varepsilon$, and we have

$$\liminf_{\varepsilon\to 0} \frac{\mu_\varepsilon(Q^\nu_{\varrho_\varepsilon}(x_0))}{\varrho_\varepsilon} \geq \liminf_{\varepsilon\to 0} \frac{8L(\gamma_\varepsilon)}{\varrho_\varepsilon} \geq \liminf_{\varepsilon\to 0} \eta\beta_\varepsilon.$$

Since $\beta_\varepsilon \to +\infty$, this gives a contradiction. Then

$$\liminf_{\varepsilon\to 0} \frac{\mu_\varepsilon(Q^\nu_{\varrho_\varepsilon}(x_0))}{\varrho_\varepsilon} \geq \varphi_p(\nu),$$

concluding the proof of the Γ-convergence result of Theorem 3.61 in the case $\beta_\varepsilon \to +\infty$.

Dilute Spin Systems

For a fixed realization ω of the random variable, we consider the nearest-neighbor energies defined in (3.66) with coefficients a^ω_{ij} as

$$a^\omega_{ij} = \begin{cases} 0 & \text{with probability } 1-p, \\ 1 & \text{with probability } p. \end{cases} \tag{3.71}$$

The energies are defined as usual on the set of spin functions $u\colon \mathcal{L}_\varepsilon(\Omega) = \varepsilon\mathbb{Z}^2 \cap \Omega \to \{-1, 1\}$, where Ω is a bounded Lipschitz open subset of $\mathbb{R}^2$.

With the choice $\alpha = 0$, it is particularly important to examine the geometry of the connections, since this allows to understand whether we have coerciveness or not. With respect to the rigid case, now it is necessary to look at the connections

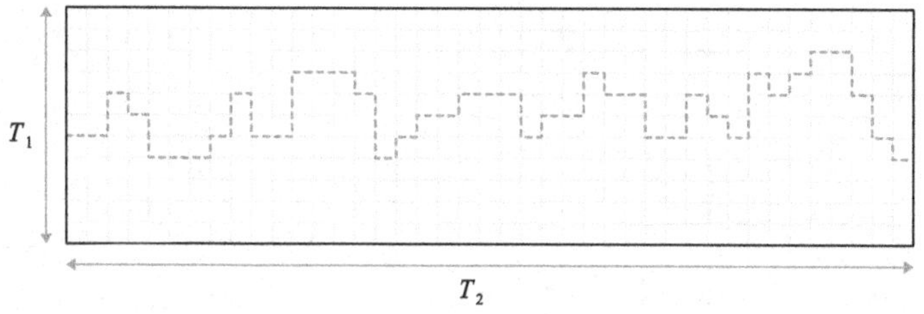

Figure 3.37 A path in the weak cluster in a rectangle.

in the dual lattice $\mathcal{Z}$ instead of the connections between the nodes of the lattice $\mathbb{Z}^2$, using the notation as in (3.64). In this context a subset of $\mathcal{Z}$ is connected if the union of the corresponding segments is connected. In this case the weak cluster W^ω given by Theorem 3.56 is a subset of $\mathcal{Z}_0$, and the strong cluster is a subset of $\mathcal{Z}_1$.

Now we separately consider the two cases $p < \frac{1}{2}$ and $p > \frac{1}{2}$.

Case $p < \frac{1}{2}$**.** We will use the geometric properties of the weak cluster stated in Theorem 3.56, and in particular a generalization given by the following remark.

Remark 3.63 From the second claim of Theorem 3.56, for any rectangle R with side lengths $T_1, T_2 > T_0$ and centered at x_R such that $\|x_R\| \leq (T_1 + T_2)^2$ there exist paths in the weak cluster contained in R joining each pair of opposite sides, as pictured in Fig. 3.37.

The preceding remark states that, for a fixed rectangle with controlled center and sides, we can always find a path with energy equal to 0; that is, there exists a spin function u such that the interface in the rectangle between the regions where $u = 1$ and $u = -1$ has energy 0. This allows one to prove the following proposition.

Proposition 3.64 (Recovery sequence for a square) *Almost surely in ω, for any square Q in $\mathbb{R}^2$ there exists a sequence $\{u^\varepsilon\}$ of spin functions such that $u^\varepsilon \to Q$ and $E^\omega_\varepsilon(u^\varepsilon) = 0$ for any ε small enough.*

Proof Let $Q = Q^\nu_\varrho(x_0)$. For any fixed $\delta \in (0, \frac{\varrho}{2})$ we consider the cube $Q_\delta = Q^\nu_{\varrho-2\delta}(x_0)$. The set $Q \setminus Q_\delta$ is then given by the union of four rectangles intersecting in the "corner squares," as pictured in Fig. 3.38. By scaling the square Q of a factor $\frac{1}{\varepsilon}$, for ε small enough we can apply the result of Remark 3.63 to each (scaled) boundary rectangle with side lengths $\frac{\varrho}{\varepsilon}$ and $\frac{\delta}{\varepsilon}$. Hence, in each rectangle R_k, $k = 1, \ldots, 4$, there exists a path γ_k in the weak cluster

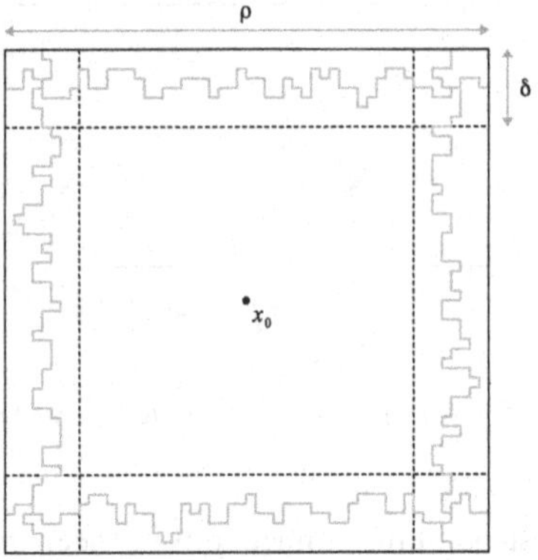

Figure 3.38 Construction of a recovery sequence in a square in the dilute case.

joining the sides with length $\frac{\delta}{\varepsilon}$. Now, we set A^δ_ε as the connected component of $\mathbb{R}^2 \setminus \bigcup_{k=1}^4 \gamma_k$ containing Q_δ, so that $Q_\delta \subset A^\delta_\varepsilon \subset Q$ and the boundary of A^δ_ε lies in the weak cluster. Setting

$$u_i^{\varepsilon,\delta} = \begin{cases} 1 & \text{if } \varepsilon i \in A^\delta_\varepsilon \cap \varepsilon\mathbb{Z}^2, \\ -1 & \text{otherwise,} \end{cases}$$

that is, $A_\varepsilon(u^{\varepsilon,\delta}) = A^\delta_\varepsilon$, we obtain by construction that $E^\omega_\varepsilon(u^{\varepsilon,\delta}) = 0$. Choosing a sequence δ_ε such that $\delta_\varepsilon \to 0$ and $\frac{\delta_\varepsilon}{\varepsilon} \to +\infty$ as $\varepsilon \to 0$, we define $u^\varepsilon = u^{\varepsilon,\delta_\varepsilon}$ and obtain that $u^\varepsilon \to Q$ and the energy of u^ε vanishes, concluding the proof. □

Since the energies are positive, Proposition 3.64 implies that the Γ-limit of E^ω_ε is 0 for any square, and hence for any finite union of squares. Then, by density, we have the following Γ-convergence result, which we only state for sets of finite perimeter.

Theorem 3.65 (Γ-convergence for dilute spin systems with $p < \frac{1}{2}$) *Let E^ω_ε be defined as in* (3.66) *with the coefficients a^ω_{ij} given by* (3.71). *Then, if $p < \frac{1}{2}$, almost surely we have that*

$$\Gamma\text{-}\lim_{\varepsilon\to 0} E^\omega_\varepsilon(A) = 0$$

for any A with finite perimeter.

Case $p > \frac{1}{2}$. We again use Theorem 3.56, in this case describing the geometry of the strong cluster.

Since $\alpha = 0$, we do not have coerciveness on the nearest-neighbor interactions. We already noticed that if $a_{ij} = 0$ for a pair (i, j) of nearest neighbors, we can recover coerciveness if we can find a path (with controlled length) joining i and j such that the bonds along this path have strictly positive coefficients. We can look at the problem of the asymptotic behavior of energies E_ε^ω by considering its deterministic counterpart, that is, the corresponding problem in the periodic framework (the perforated domains considered in Section 3.3.5).

We recall some definitions in the special case $d = 2$ (see Definition 3.19). Given a set $\mathcal{A} \subset \mathbb{Z}^2$, we say that it is *connected* if for $i, j \in \mathcal{A}$ there exists a path of nearest neighbors in $\mathcal{A}$ joining i and j; that is, there exists a subset $\{i_0, i_1, \dots, i_n\} \subset \mathcal{A}$ such that $i_0 = i$, $i_n = j$, and $\|i_k - i_{k-1}\| = 1$ for any $k \in \{1, \dots, n\}$. The *boundary of* $\mathcal{A}$ is defined as

$$\partial\mathcal{A} = \{i \in \mathcal{A} \colon \operatorname{dist}(i, \mathbb{Z}^2 \setminus \mathcal{A}) = 1\}.$$

Moreover, we define the *external boundary of* $\mathcal{A}$ as the boundary of its complement, that is,

$$\partial^{\mathrm{ext}}\mathcal{A} = \{i \in \mathbb{Z}^2 \setminus \mathcal{A} \colon \operatorname{dist}(i, \mathcal{A}) = 1\}.$$

In the following remark we revisit the compactness argument for perforated domains in $\mathbb{Z}^2$ in view of adapting it to the random case.

Remark 3.66 (Perforated domains in the periodic framework in $\mathbb{Z}^2$: compactness) Let the (symmetric) coefficients a_{ij} assume the values 0 and 1 and be periodic with a period K. We set

$$\mathcal{L} = \{i \in \mathbb{Z}^2 \colon j \in \mathbb{Z}^2 \text{ exists such that } \|i - j\| = 1 \text{ and } a_{ij} = 1\}$$

and assume that $\mathcal{L}$ is connected and infinite. Note that in dimension $d = 2$ this implies that each connected component of the complement is finite. We consider the energies given by

$$E_\varepsilon(u) = \sum_{\langle i,j \rangle} \varepsilon a_{ij}(u_i - u_j)^2.$$

Let $\{u^\varepsilon\}$ be such that $E_\varepsilon(u^\varepsilon)$ is equibounded. Let $\{C_\varepsilon^{k,+}\}_{k \in K_\varepsilon^+}$ and $\{C_\varepsilon^{k,-}\}_{k \in K_\varepsilon^-}$ be the sets of the connected components of the complement of $\varepsilon\mathcal{L}$ such that u^ε is identically equal to 1 or -1, respectively, in the external boundary of the component. Let $\{C_\varepsilon^{k,0}\}_{k \in K_\varepsilon^0}$ denote the sets of the remaining such connected components.

We modify u^ε by setting

$$\tilde{u}_i^\varepsilon = \begin{cases} u_i^\varepsilon & \text{if } \varepsilon i \in \varepsilon\mathcal{L}, \\ 1 & \text{if } \varepsilon i \in C_\varepsilon^{k,+} \text{ for } k \in K_\varepsilon^+, \\ -1 & \text{otherwise.} \end{cases}$$

With this definition, the energy does not increase. Now we show that the perimeters of the sets $A_\varepsilon(\tilde{u}^\varepsilon)$ are equibounded.

Indeed, it is sufficient to estimate $\#K^0_\varepsilon$. For each $C^{k,0}_\varepsilon$ there exist $\varepsilon i, \varepsilon j \in \partial^{\rm ext} C^{k,0}_\varepsilon$ such that $u^\varepsilon_i \neq u^\varepsilon_j$. Since $\mathcal{L}$ is connected, up to enlarging the period we can assume that each pair of points in $\mathcal{L} \cap [0,K]^2$ is connected by a path contained in $\mathcal{L} \cap [-K, 2K]^2$. Hence, for each $C^{k,0}_\varepsilon$ there exist i', j' nearest neighbors in the path connecting i and j with $a_{i'j'}(u^\varepsilon_i - u^\varepsilon_j)^2 = 4$, shared with an equibounded number of other such connected components. We then obtain $\#K^0_\varepsilon \leq CE_\varepsilon(u^\varepsilon)$. Since the additional contribution of each connected component $C^{k,0}_\varepsilon$ is at most $4\varepsilon K^2$, we get

$$\mathcal{H}^1(A_\varepsilon(\tilde{u}^\varepsilon)) \leq E_\varepsilon(u^\varepsilon) + C'K^2 E_\varepsilon(u^\varepsilon) \leq C''.$$

This implies, up to subsequences, the convergence of $\tilde{u}^\varepsilon$ to a set A with finite perimeter.

In the random case we can proceed similarly to the periodic case, but, since the connected components in the complement of the strong cluster may have arbitrarily large size, we have to provide a more refined estimate.

We introduce the set

$$\mathcal{L}^\omega = \{i \in \mathbb{Z}^2 \colon j \in \mathbb{Z}^2 \text{ exists such that } [i,j]^\perp \subset S^\omega\},$$

where S^ω is the strong cluster defined in Theorem 3.56, and consider the connected components of the complement of $\mathcal{L}^\omega$. Since $p > \frac{1}{2}$, we have that almost surely in ω each of these connected components is finite.

Now, let $\{u^\varepsilon\}$ be such that $E^\omega_\varepsilon(u^\varepsilon) \leq S < +\infty$ for all ε, where the energies are computed for nearest-neighbor pairs in the bounded Lipschitz open set Ω. As in the periodic framework, if we change the values of u^ε by setting $\tilde{u}^\varepsilon = 1$ in each connected component of the complement of $\varepsilon\mathcal{L}^\omega$ such that $u^\varepsilon = 1$ on its boundary, and -1 otherwise in the complement of $\varepsilon\mathcal{L}^\omega$, we do not change the values on $\varepsilon\mathcal{L}^\omega$ and the energy does not increase. Hence, we can assume that u^ε is constant on each such connected component.

If we consider a maximal connected component I^+ of the set $\{i \colon u^\varepsilon_i = 1\}$, the maximality of I^+ ensures that $u^\varepsilon = -1$ on the external boundary of εI^+. We show by contradiction that there exists $j \in \partial^{\rm ext} I^+$ such that $a^\omega_{ij} = 1$ for some $i \in I^+$, so that

$$\sum_{k \in I^+, l \in \partial^{\rm ext} I^+} a^\omega_{kl}(u^\varepsilon_k - u^\varepsilon_l)^2 \geq 4.$$

Indeed, assume that $a^\omega_{ij} = 0$ for all $j \in \partial^{\rm ext} I^+$ and $i \in I^+$. Let I^+_* be the subset of the external boundary of I^+ included in $\mathcal{L}^\omega$. If $I^+_* = \partial^{\rm ext} I^+$, this gives a contradiction with the hypothesis on u^ε since we have a connected component of the complement of $\varepsilon\mathcal{L}^\omega$ where the constant value on the external boundary is -1 and the internal value is 1. Otherwise, there exist i, j nearest neighbors in the same connected component of the complement of $\varepsilon\mathcal{L}^\omega$ with different values of u^ε, which again contradicts the hypothesis on u^ε. Let M^+_ε and M^-_ε denote the number of the maximal connected components where $u^\varepsilon_i = 1$ and $u^\varepsilon_i = -1$, respectively. Then, arguing in the same way for each maximal connected component where $u^\varepsilon_i = -1$, we obtain that

$$\#M^+_\varepsilon \leq \frac{S}{\varepsilon}, \quad \#M^-_\varepsilon \leq \frac{S}{\varepsilon}.$$

We now first prove that "small" connected components of $\{u^\varepsilon = 1\}$ and of $\{u^\varepsilon = -1\}$ have a small total measure, where we identify as usual a subset of $\varepsilon\mathbb{Z}^2$ with the union of the corresponding ε-squares. To that end we introduce $\delta > 0$ and note that each component of size (that is number of nodes) $O(\varepsilon^{-1+\delta})$ has measure $O(\varepsilon^{1+\delta})$, so that the total measure of such components is $O(\varepsilon^\delta)$.

As for the components with size much larger than $\varepsilon^{-1+\delta}$ and not greater than $\frac{1}{\varepsilon}$, we first note that the set of the boundary connections of such a component is of size much larger than $\varepsilon^{-\frac{1}{2}+\frac{\delta}{2}}$. A well-known fact from percolation theory is that almost surely each connected set of points $z \in \mathcal{Z}$ with $a^\omega_z = 0$ has size at most $O(|\log \varepsilon|)$. Hence, the set of the boundary connections of such a component has a number of connections z with $a^\omega_z = 1$ much larger than $e^{-\frac{1}{2}+\frac{\delta}{2}}|\log \varepsilon|^{-1}$. We then deduce that the energy contribution of such a component is much larger than

$$\varepsilon^{\frac{1}{2}+\frac{\delta}{2}}|\log \varepsilon|^{-1}.$$

Hence, the total number of the components with size much larger than $\varepsilon^{-1+\delta}$ is less than

$$S|\log \varepsilon|\varepsilon^{-\frac{1}{2}-\frac{\delta}{2}}.$$

Since we are considering components with size at most $\frac{1}{\varepsilon}$, the measure of each such component is at most ε, and the measure of their union is less than

$$S|\log \varepsilon|\varepsilon^{\frac{1}{2}-\frac{\delta}{2}},$$

so that it is negligible as $\varepsilon \to 0$.

Hence, as the L^1-convergence of the sequence $\{u^\varepsilon\}$ is concerned, we can assume that all connected components in $\{u^\varepsilon = 1\}$ and in $\{u^\varepsilon = -1\}$ have size at least $\frac{1}{\varepsilon}$. In order to estimate the perimeter of such components we will use the following percolation result, which we state without proof.

Lemma 3.67 (Percolation animal) *Let $p > \frac{1}{2}$. With fixed $M > 0$, almost surely there exist a deterministic positive constant κ and $\varepsilon_0 = \varepsilon_0(\omega) > 0$ such that for* *all connected sets contained in the square* $\left[-\frac{M}{\varepsilon}, \frac{M}{\varepsilon}\right]^2$ *and of size larger than* $\varepsilon^{-\frac{1}{2}}$ *with* $\varepsilon < \varepsilon_0$, *the proportion of strong links (such that* $a^\omega_{ij} = 1$*) in each such a set is at least* κ.

Lemma 3.67 ensures that for each $\Omega' \subset\subset \Omega$ we have

$$\mathcal{H}^1(\partial A_\varepsilon(u^\varepsilon) \cap \Omega') \leq \frac{1}{\kappa} E^\omega_\varepsilon(u^\varepsilon),$$

which gives the compactness of $\{u^\varepsilon\}$ (restricted to the strong cluster). We can state the result just proved as follows.

Proposition 3.68 (Compactness for dilute spins and $p > \frac{1}{2}$) *Let* $p > \frac{1}{2}$. *Let* Ω *be a bounded Lipschitz open subset of* $\mathbb{R}^2$. *Almost surely in* ω, *if* $\{u^\varepsilon\}$ *is such that* $E^\omega_\varepsilon(u^\varepsilon)$ *is equibounded, then there exists a set of finite perimeter* A *such that, up to subsequences,*

$$(u^\varepsilon - \chi_A)\chi_{Q_\varepsilon(\mathcal{L}^\omega)} \to 0$$

in $L^1_{\rm loc}(\Omega)$, *where* $Q_\varepsilon(\mathcal{L}^\omega) = \bigcup_{i \in \mathcal{L}^\omega} \varepsilon\big(i + \big[-\frac{1}{2}, \frac{1}{2}\big)^2\big)$.

Once the compactness is proved, the Γ-convergence result follows exactly as in the case $\alpha > 0$ for random mixtures in Theorem 3.54.

Theorem 3.69 (Γ-convergence for dilute spins and $p > \frac{1}{2}$) *Let* E^ω_ε *be defined as in* (3.66) *with the coefficients* a^ω_{ij} *given by* (3.71). *Then, if* $p > \frac{1}{2}$, *almost surely we have that*

$$\Gamma\text{-}\lim_{\varepsilon \to 0} E^\omega_\varepsilon(A) = \int_{\Omega \cap \partial^* A} \varphi_p(\nu)\, d\mathcal{H}^1,$$

where φ_p *is given by the first-passage percolation formula in Theorem 3.53.*

Bibliographical Notes to Chapter 3

Ferromagnetic energies, or Ising systems, are a classical subject of Statistical Mechanics, and a number of monographs are dedicated to them from that standpoint (e.g. Presutti, 2009). Their treatment using the variational approach presented in this chapter can be traced in the work of Caffarelli and de la Llave (2005) and in the first result using explicitly the terminology of Γ-convergence by Alicandro et al. (2006). Energies on vector spin functions, where the parameter takes its values in some unit sphere S^{m-1} for $m > 1$, are closer to bulk integral energies of the Ginzburg–Landau type and may produce lower-dimensional singularities such as vortices as in the case of the XY model studied by Alicandro et al. (2011b).

The superposition principle illustrated in the treatment of homogeneous energies follows earlier results on the approximation of free-discontinuity problems by Braides and Gelli (2002), and is also presented in the book by Braides and Solci (2021) in a two-dimensional setting. Zonoids are a subject of Convex Geometry, for which we refer to the book by Schneider (1988). Their connection with Ising systems is described by Braides and Chambolle (2023).

The blow-up method has been introduced by Fonseca and Müller (1992) for the study of lower-semicontinuous envelopes; its adaptation to the study of homogenization problems can be found in Braides et al. (2008). Periodic and almost-periodic homogenization results are proven in Braides and Piatnitski (2013). For the use of one-homogeneous extensions to derive a homogenization theorem for surface energies on the continuum we refer to Ambrosio and Braides (1990a) and the corresponding formula for convex homogenization by Braides and Chiadò Piat (1995). An analog in the case of discrete energies has been shown to give an alternate homogenization formula by Chambolle and Kreutz (2023), from which crystallinity follows. The existence of plane-like minimizers for periodic systems has been proved by Caffarelli and de la Llave (2005) (see also Caffarelli and de la Llave, 2001). The counterexample to existence of plane-like minimizers for an almost-periodic system can be found in Braides (2015). An analysis of geometric error terms in continuum approximations of discrete systems can be found in the paper by Rosakis (2014).

The crystallinity of Wulff shapes for finite-range ferromagnetic systems allows the study of variational evolutions for planar interfaces in terms of systems of ordinary differential equations as in the book by Braides and Solci (2021). For stability properties of crystalline Wulff shapes we refer to, for example, Figalli and Zhang (2022). Some interesting observations on the restrictions to evolution due to being a zonoid for Wulff shapes can be found in Elsey and Esedoglu (2018).

Discrete perforated domains have been studied by Braides et al. (2016a). Optimal bounds have been studied by Braides and Kreutz (2018a) in the spirit of the study of mixtures of conducting networks by Braides and Francfort (2004), for which, however, the optimality of the bounds has not been shown for all percentages. Quasicrystals have been studied by Braides et al. (2012), spin systems on Penrose lattices by Braides and Solci (2011). Discrete thin-object theories can be compared with continuum theories of thin objects starting from three-dimensional bodies through a dimension-reduction procedure. We refer to, for example, Le Dret and Raoult (1995), Braides et al. (2000), and Friesecke et al. (2006).

General references for almost-periodic functions are, for example, Besicovitch (1954) and Levitan and Zhikov (1982).

Variational percolation problems for spin systems have been considered by Braides and Piatnitski (2013) and Braides and Piatnitski (2012), following earlier results of the same authors for discrete free-discontinuity problems (Braides and Piatnitski, 2008). The case of rigid spin systems has been considered by Scilla (2013). The results presented use various first-passage percolation type formulas, such as in Theorem 3.53, which are often proved using subadditive point-process theorems (Akcoglu and Krengel, 1981). We refer to, for example, Boivin (1990) for first-passage percolation formulas for positive interactions, Garet and Marchand (2007) for the chemical distance, Wouts (2009), and Cerf and Théret (2011) for first-passage percolation formulas for dilute systems. Using Boivin's formula, we can treat the case where coefficients are distributed in an interval $[\alpha, \beta]$ and we can relax the hypothesis that a_{ij} be i.i.d. The results presented in this chapter are in dimension 2; in higher dimension, we have more percolation thresholds than just the value $\frac{1}{2}$. Moreover, in the dilute case one has to take into account the different connectedness properties of strong clusters at different values of the probability. For details we refer to the paper by Braides and Piatnitski (2012). Lemma 3.62 is included as an appendix in Braides and Piatnitski (2008) with a proof by H. Kesten. For general reference on Percolation, see Grimmett (1999) and Kesten (1982).

A metric interpretation of the convergence of clusters can be given following the idea of Gromov–Hausdorff convergence (Burago et al., 2022) or the convergence of metrics in the sense of Γ-convergence (Braides, 2002).

4

Compactness and Integral Representation

This chapter is devoted to the statement and proof of a compactness and integral-representation theorem for general interactions. We will leave the framework of the previous chapter, where energies are confined to spin functions and depend on pair interactions and the asymptotic behavior is described by perimeter energies characterized by homogenization formulas. The interactions will still be of the ferromagnetic type, in the sense that they favor uniform states, but we will not limit the number of such states. As a consequence, the continuum limit will be defined not on a single set but on a family of sets of finite perimeter. Moreover, in general we will not make any periodicity assumption, so that the limit energies may be inhomogeneous, even though some form of homogenization at a microscopic level does take place as we pass from discrete to continuum problems.

4.1 Some Examples of General Interactions

Since the generality of the compactness theorem that we are going to prove in the next sections will require a proper notation, in this section we consider some examples that explain the need of such a notation.

The first issue we face is how to include the possibility of obtaining nonhomogeneous limit energies. This requires us to consider ε-depending interaction energies, as shown by the following simple example.

Example 4.1 (Nonhomogeneous limit energies for ferromagnetic systems) Let $a\colon \mathbb{R} \to [0,+\infty)$ be a continuous function such that $a(x) \geq c > 0$ for any $x \in \mathbb{R}$. We fix $\Omega = \mathbb{R}$, and consider a discretization of the function a; that is, we consider interaction coefficients given by

$$a\Big(\frac{\varepsilon i + \varepsilon j}{2}\Big) = a^{\varepsilon}_{ij}.$$

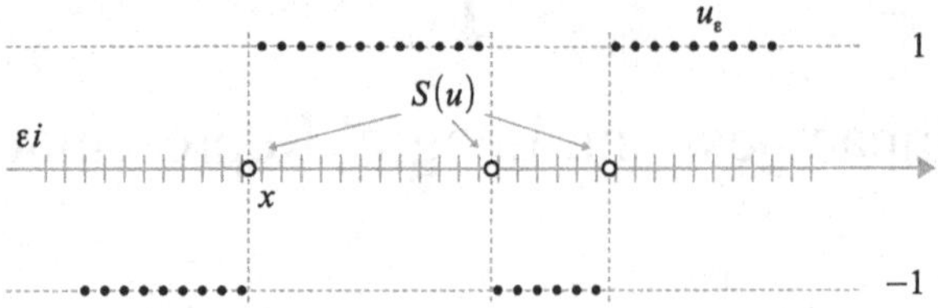

Figure 4.1 Discretization of piecewise constant u.

We define the nearest-neighbor energies

$$E_\varepsilon(u) = \sum_{|i-j|=1} a^\varepsilon_{ij}(u_i - u_j)^2.$$

The hypothesis $a(x) \geq c > 0$ gives compactness; hence, if $E_\varepsilon(u^\varepsilon)$ is equibounded, there exists $u\colon \mathbb{R} \to \{-1, 1\}$ such that $u^\varepsilon \to u$ in L^1, up to subsequences.

For any $x \in S(u)$, where there is a change of sign of the limit function, for ε small enough there is also a change of sign of u^ε at some $i = i_\varepsilon$ with $\varepsilon i_\varepsilon \to x$, which gives a contribution to the discrete energy of $8a^\varepsilon_{i\,i+1} = 8a(x) + o(1)_{\varepsilon\to 0}$ by the continuity of a, as pictured in Fig. 4.1. Hence, we have that

$$\Gamma\text{-}\lim_{\varepsilon\to 0} E_\varepsilon(u) = 8 \sum_{x\in S(u)} a(x),$$

which is a one-dimensional version of a nonhomogeneous perimeter functional.

This elementary case shows that even in a simple discretization argument we have to introduce coefficients depending on ε.

The general case of nonhomogeneous ferromagnetic energies can be then formalized as follows. We fix $\Omega \subset \mathbb{R}^d$ and look at energies given by

$$E_\varepsilon(u) = \sum_{\varepsilon i, \varepsilon j \in \mathcal{L}_\varepsilon} \varepsilon^{d-1} a^\varepsilon_{ij}(u_i - u_j)^2, \tag{4.1}$$

where $\mathcal{L}_\varepsilon = \varepsilon\mathbb{Z}^d \cap \Omega$ and $u\colon \mathcal{L}_\varepsilon \to \{-1, +1\}$. This will be a very particular case of the energies considered in the sequel, and at the same time will be a guideline for stating general conditions on the interactions that mirror and extend those in Chapter 3 for homogenous systems with coefficients a_{ij}, such that they Γ-converge (possibly up to subsequences) to an integral functional of the form

$$F(A) = \int_{\Omega\cap\partial^* A} \varphi(x, \nu)\, d\mathcal{H}^{d-1}$$

for some $\varphi\colon \Omega \times S^{d-1} \to [0, +\infty)$.

The following example shows that the analog of the conditions in Chapter 3 (in particular, the decay condition in Remark 3.1) must be suitably modified when considering ε-depending interactions.

Example 4.2 (A nonlocal Γ-limit) Let $d = 1$, $\Omega = \mathbb{R}$, and the family a^{ε}_{ij} be defined by

$$a^{\varepsilon}_{ij} = \begin{cases} 1 & \text{if } |i-j| = 1, \\ \varepsilon & \text{if } |i-j| = \lfloor \frac{1}{\varepsilon} \rfloor, \\ 0 & \text{otherwise.} \end{cases} \tag{4.2}$$

By separating the interactions between points at distance ε and points at distance of order 1, we write the energy as

$$E_{\varepsilon}(u) = \sum_{|i-j|=1} (u_i - u_j)^2 + 2\varepsilon \sum_{i\in\mathbb{Z}} \Big(u\Big(\varepsilon i + \varepsilon \Big\lfloor \frac{1}{\varepsilon} \Big\rfloor\Big) - u(\varepsilon i)\Big)^2.$$

Note that the second sum can be interpreted as the integral of the piecewise-constant interpolations of u, and it continuously converges to the (continuum) functional

$$2 \int_{\mathbb{R}} (u(x+1) - u(x))^2 \, dx.$$

Since Γ-convergence is stable under continuously converging perturbations, the Γ-limit is given by

$$F(u) = 8\#S(u) + 2 \int_{\mathbb{R}} (u(x+1) - u(x))^2 \, dx,$$

which is the sum of a perimeter functional and a nonlocal term. This example implies that the decay condition in Remark 3.1 is not sufficient to ensure that the limit is local, since it is satisfied by the coefficients a^{ε}_{ij} defined in (4.2). Indeed, for any $\varepsilon > 0$ and $i \in \mathbb{Z}$,

$$\sum_{j\in\mathbb{Z}} a^{\varepsilon}_{ij} |i-j| = 2 + 2\varepsilon \Big\lfloor \frac{1}{\varepsilon} \Big\rfloor \leq 4.$$

In this example the term $\varepsilon \lfloor \frac{1}{\varepsilon} \rfloor$ is always of order 1; more precisely, for any $i \in \mathbb{Z}$ and $R > 1$, we have

$$\sum_{\substack{j\in\mathbb{Z} \\ |i-j|\geq R}} a^{\varepsilon}_{ij} \| i-j \| \geq 1 - \varepsilon$$

for any ε such that $\lfloor \frac{1}{\varepsilon} \rfloor \geq R$. In order to obtain a local Γ-limit we will have to require conditions that forbid this type of behavior.

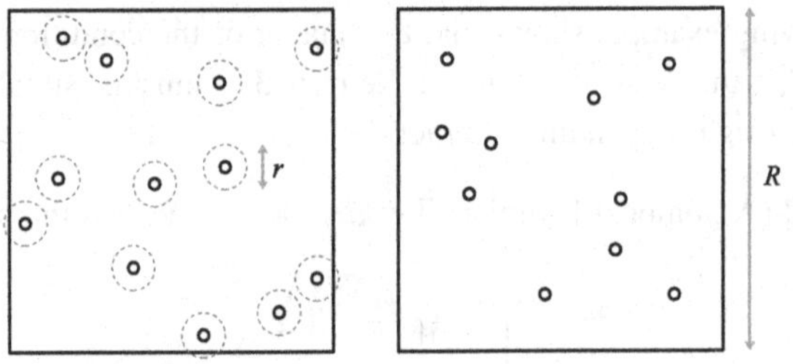

Figure 4.2 A disordered lattice.

A further issue is the possibility of substituting a Bravais lattice with a more disordered set. A minimal requirement on the set $\mathcal{L}$ in order to have a compactness and representation theorem is to be not too sparse and not too dense, while the precise geometric structure of $\mathcal{L}$ is not essential to prove those results.

Example 4.3 (Energies on disordered lattices) We consider a set $\mathcal{L} \subset \mathbb{R}^d$ such that:

(i) there exist $R > 0$ and $c > 0$ such that for any $x \in \mathbb{R}^d$,
$$c \leq \#(\mathcal{L} \cap Q_R(x)),$$
where $Q_R(x)$ is any cube centered in x with side length R;

(ii) there exists $r > 0$ such that $\|x - y\| \geq r$ for any $x, y \in \mathcal{L}$.

We refer to such a set as a disordered lattice (see Fig. 4.2).

Note that condition (ii) implies that in (i) we can assume $\#(\mathcal{L} \cap Q_R(x)) \leq \frac{1}{c}$. On $\mathcal{L}$ we may define ferromagnetic energies in such a way that functions with equibounded nearest-neighbor energies enjoy compactness properties. The only point to be made precise is the notion of nearest neighbors in this setting. To this end, for any $i \in \mathcal{L}$ we define the set

$$C_i = \{x \in \mathbb{R}^d : \|x - i\| \leq \|x - j\| \text{ for all } j \in \mathcal{L}\}.$$

These sets are called *Voronoi cells* of the lattice. Two points i and j in $\mathcal{L}$ are *nearest neighbors* if the corresponding Voronoi cells share a $d-1$-dimensional face; that is, $\mathcal{H}^{d-1}(\partial C_i \cap \partial C_j) > 0$ (see Fig. 4.3). We may then define

$$E_\varepsilon^{\mathcal{L}}(u) = \sum_{\langle i,j\rangle} \varepsilon^{d-1}(u_i - u_j)^2,$$

where as usual $u : \mathcal{L}_\varepsilon(\Omega) = \Omega \cap \varepsilon\mathcal{L} \to \{-1, 1\}$, $u_i = u(\varepsilon i)$, and $\langle i, j\rangle$ indicates the sum over nearest neighbors in $\mathcal{L}$. Properties (i) and (ii) presented earlier

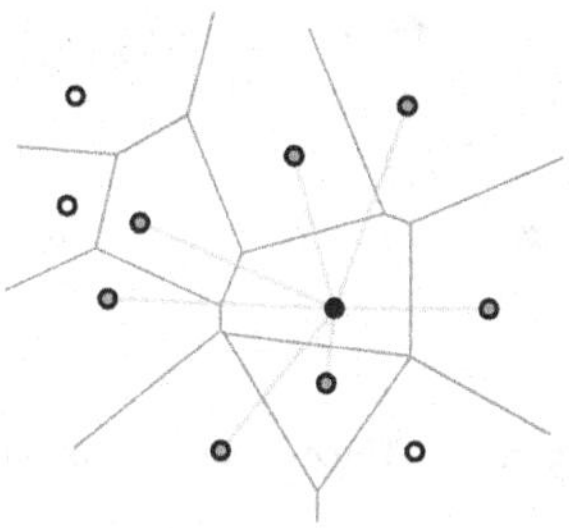

Figure 4.3 Nearest neighbors in a disordered lattice.

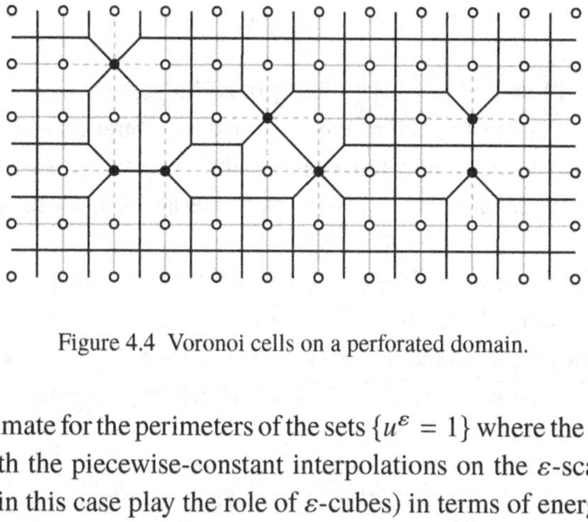

Figure 4.4 Voronoi cells on a perforated domain.

ensure an estimate for the perimeters of the sets $\{u^\varepsilon = 1\}$ where the functions are identified with the piecewise-constant interpolations on the ε-scaled Voronoi cells (which in this case play the role of ε-cubes) in terms of energies $E_\varepsilon^{\mathcal{L}}(u^\varepsilon)$.

Note in particular that perforated domains satisfy the condition required on $\mathcal{L}$. In Fig. 4.4 Voronoi cells on a perforated domain are pictured, where the black dots represent missing sites. This standpoint gives a different approach to compactness in that case.

We can enlarge the class of energies E_ε to functionals of the form

$$E_\varepsilon(u) = \sum_{i\in\mathcal{L}_\varepsilon(\Omega)} \varepsilon^{d-1}\phi_i^\varepsilon(\{u_j\}),$$

where ϕ_i^ε describes the way a point i interacts with all the other points in the lattice. In energies (4.1) as considered earlier we have

$$\phi_i^\varepsilon(\{u_j\}) = \sum_{j\in\mathcal{L}_\varepsilon(\Omega)} a_{ij}^\varepsilon(u_i - u_j)^2.$$

This approach allows one, for instance, to include the analysis of interactions of many-body type.

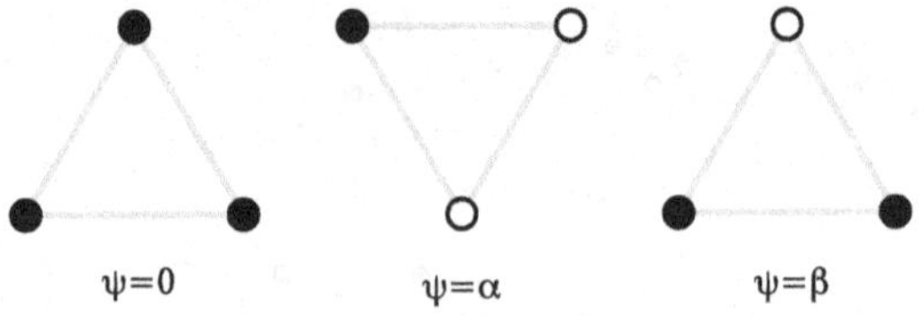

Figure 4.5 Three-point interactions.

Example 4.4 (Three-point interactions) We consider in $\mathbb{R}^2$ the triangular lattice $\mathbb{T}$, that is, the Bravais lattice given by

$$\mathbb{T} = \mathbb{Z}v_1 + \mathbb{Z}v_2,$$

where $v_1 = (1,0)$ and $v_2 = (\cos(\frac{\pi}{3}), \sin(\frac{\pi}{3}))$, and a generic triangular cell with vertices in $\mathbb{T}$ labeled as i, j, and k. We can consider interactions that are defined in terms of the values of spin functions on the three vertices of each triangle which cannot be reduced to sums of pair interactions, as in the following example.

Given a spin function $u\colon \mathbb{T} \to \{-1,1\}$ such that the values u_i, u_j, u_k on the vertices of the (considered) triangle are not all equal, we can assume that the two equal values are u_i and u_j and we have two possibilities, $i-j \in \{-e_1, e_1\}$ or $i-j \notin \{-e_1, e_1\}$. Now, we define a function ψ depending on the set $\{u_i, u_j, u_k\}$ by setting

$$\psi(\{u_i,u_j,u_k\}) = \begin{cases} 0 & \text{if } u_i = u_j = u_k, \\ \alpha & \text{if } u_i = u_j \neq u_k \text{ and } i-j \notin \{-e_1, e_1\}, \\ \beta & \text{if } u_i = u_j \neq u_k \text{ and } i-j \in \{-e_1, e_1\}, \end{cases}$$

where $0 < \alpha < \beta$ and we assume, as earlier, that the two equal values in the second and third cases are labeled as i and j (see Fig. 4.5). If we consider energies defined by functions ϕ_i^ε involving the interaction term ψ, we note that the configurations where the two coinciding values are on the horizontal side of the triangle are penalized, since $\beta > \alpha$, and this corresponds in the limit to a penalization of the horizontal interfaces.

It can be proven that the Γ-limit of these energies is given by

$$\int_{\Omega\cap\partial^* A} \varphi(\nu)\, d\mathcal{H}^1,$$

where the Wulff shape W_φ is a rhombus, as pictured in Fig. 4.6. Note that W_φ differs from the hexagonal Wulff shape corresponding to the nearest-neighbor interactions since in its boundary we cannot have horizontal edges.

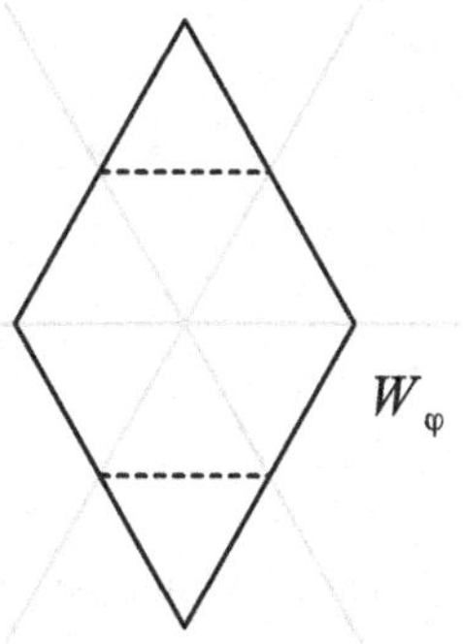

Figure 4.6 Wulff shape for three-point interactions.

A further issue is the extension to functions u taking values in a set of cardinality larger than two. In this case it may be convenient to embed this set in some $\mathbb{R}^k$.

Example 4.5 (Higher number of parameters and limits defined on partitions) We may consider energies on functions $u\colon \Omega \cap \varepsilon\mathbb{Z}^d \to Y$, where Y is a finite set (not necessarily embedded in a Euclidean space). In this case the limit of functions $\{u^\varepsilon\}$ must be understood as the limit of the sets (union of ε-cubes) where the piecewise-constant interpolations of $u^\varepsilon = x$ for each given $x \in Y$. If $\#Y > 2$ we expect the limit energy to be defined not on sets of finite perimeter but on *partitions into sets of finite perimeter* labeled by a subset of the elements of Y.

As a simple example we may consider $Y = \{1, 2, 3\} \subset \mathbb{R}$ and energies

$$E_\varepsilon(u) = \sum_{\langle i,j\rangle} \varepsilon^{d-1}\phi(u_i, u_j),$$

where $\phi(u, v) = 0$ if $u = v$ and $\phi(u, v) = 1$ if $u \neq v$. In this case, minimizers are the uniform states $u = x \in Y$, and the limit is defined on partitions (A_1, A_2, A_3). Note that ϕ is an analog of the ferromagnetic energy density, so that the Γ-limit can be computed likewise, and reads

$$F(A_1, A_2, A_3) = \sum_{k=1}^{3} \int_{\partial^* A_k} \|\nu\|_1 \, d\mathcal{H}^{d-1},$$

where ν denotes the normal to the corresponding $\partial^* A_k$.

Note that the same limit can be obtained by considering a set of vector parameters $Y = \{e_1, e_2, e_3\} \subset \mathbb{R}^3$ (and setting $A_k = A_{e_k}$), and the energies

$$E_\varepsilon(u) = \frac{1}{2} \sum_{\langle i,j\rangle} \varepsilon^{d-1} \|u_i - u_j\|^2,$$

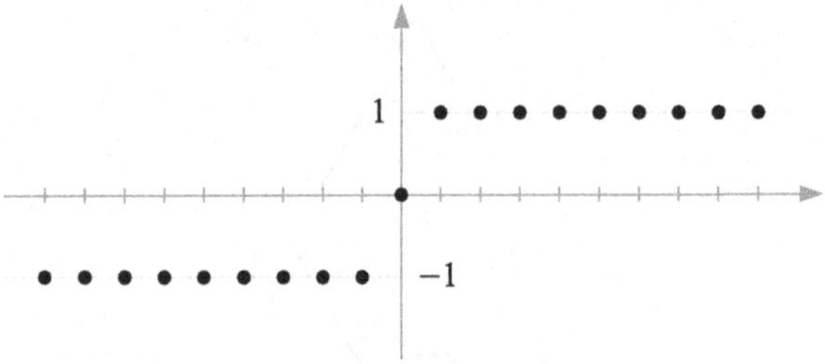

Figure 4.7 Wetting interface.

which are again a vector analog of the ferromagnetic energies and may arise in applications.

Example 4.6 (Interfacial energies with microstructure) Similarly to the example above, consider $Y = \{-1,0,1\} \subset \mathbb{R}$ and energies

$$E_\varepsilon(u) = \sum_{\langle i,j\rangle} \varepsilon^{d-1}(u_i - u_j)^2 .$$

Note that if we define $\phi(u,v) = (u-v)^2$, contrary to the preceding example, we have $\phi(0,1) \neq \phi(-1,1)$.

Again, the limits of E_ε are defined on partitions, but in this case the interfacial energy is not a sum of terms each depending only on one of the sets A_k. Indeed, note that in the one-dimensional case a recovery sequence for the triplet given by $A_{-1} = \emptyset$, $A_0 = (-\infty,0]$, $A_1 = (0,+\infty)$ is trivially $u^\varepsilon = \chi_{(0,+\infty)\cap\varepsilon\mathbb{Z}}$, for which the Γ-limit is 1. In contrast, for the triplet given by $A_{-1} = (-\infty,0]$, $A_0 = \emptyset$, $A_1 = (0,+\infty)$ a recovery sequence, pictured in Fig. 4.7, is given by $u^\varepsilon = \chi_{(0,+\infty)\cap\varepsilon\mathbb{Z}} - \chi_{(-\infty,0)\cap\varepsilon\mathbb{Z}}$ (taking the value $u_0^\varepsilon = 0$), for which the Γ-limit is 2. Note that, taking instead $u^\varepsilon = \chi_{(0,+\infty)\cap\varepsilon\mathbb{Z}} - \chi_{(-\infty,0]\cap\varepsilon\mathbb{Z}}$ (in this case $u_0^\varepsilon = -1$ and the value 0 is not taken), we get the value 4. The insertion of a third value between two given ones to lower the energy density is often referred to as *wetting*.

This example shows that in general the limit energy density may depend on the neighboring sets; that is, we may have

$$F(\{A_k\}) = \sum_{k\neq l} \int_{\partial^* A_k \cap \partial^* A_l} \varphi_{kl}(\nu)\, d\mathcal{H}^{d-1},$$

which shows the necessity of introducing more complex surface energies depending on partitions. Moreover, it also shows that in the determination of such φ_{kl} also values of u^ε different from k and l must be taken into account.

Example 4.7 (Limits parameterized by the ground states) We now give an example where the limit is parameterized on a subset of the set Y, considering a simple variation of the ferromagnetic energies presented earlier given by

$$E_\varepsilon(u) = \sum_{\langle i,j\rangle} \varepsilon^{d-1}(1 - u_i u_j)^2,$$

which are equivalent to the former when $u_i \in \{-1, 1\}$. When $Y = \{-1, 0, 1\}$ the two minimizers are the constants $u = 1$ and $u = -1$, so that the limit is again defined just on sets of finite perimeter as in the case $Y = \{-1, 1\}$, but in the computation of interfacial energies also the value 0 must be taken into account.

The use of vector parameters may be useful to treat some cases that have a finite number of (nonconstant) ground states, as in the following example.

Example 4.8 (Antiferromagnetic spin systems in $\mathbb{Z}^d$) For antiferromagnetic spin systems in $\mathbb{Z}^d$, we can rewrite the energies using a change of variables. We only consider the case $d = 1$ and $\Omega = \mathbb{R}$, where the nearest-neighbor antiferromagnetic energies can be written as

$$E_\varepsilon(u) = \sum_{i\in\mathbb{Z}} (u_i + u_{i+1})^2$$

defined on $u\colon \varepsilon\mathbb{Z} \to \{-1, 1\}$. The energy is only zero when $u_i = (-1)^i$ for all $i \in \mathbb{Z}$ or $u_i = -(-1)^i$ for all $i \in \mathbb{Z}$.

We may rewrite the energy as

$$E_\varepsilon(u) = \sum_{k\in\mathbb{Z}} \Big(\frac{1}{2}(u_{2k} + u_{2k+1})^2 + (u_{2k+1} + u_{2k+2})^2 + \frac{1}{2}(u_{2k+2} + u_{2k+3})^2\Big).$$

We now set $Y = \{-1, 1\}^2$ and define

$$\phi(v, w) = \frac{1}{2}(v_1 + v_2)^2 + (v_2 + w_1)^2 + \frac{1}{2}(w_1 + w_2)^2.$$

We have a one-to-one correspondence between functions $u\colon \varepsilon\mathbb{Z} \to \{-1, 1\}$ and $v\colon \varepsilon\mathbb{Z} \to X$, setting $v_k = (u_{2k}, u_{2k+1})$. Correspondingly, $E_\varepsilon(u)$ equals

$$\widetilde{E}_\varepsilon(v) = \sum_{k\in\mathbb{Z}} \phi(v_k, v_{k+1}), \tag{4.3}$$

which is an energy analog to the ones considered in Example 4.5. Since $\phi(v, w) = 0$ only when $v = w \in \{(-1, 1), (1, -1)\}$, the Γ-limit will be an energy parameterized on these two states.

Example 4.9 (Second-neighbor energies) We consider the second-neighbor energies

$$E_\varepsilon(u) = \sum_{i\in\mathbb{Z}} (u_{i+1} - u_{i-1})^2$$

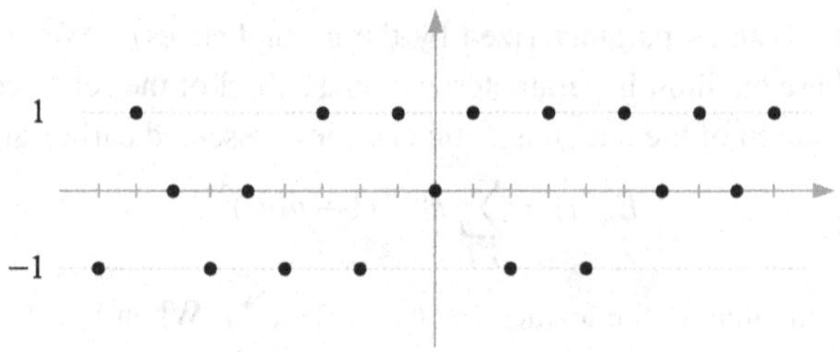

Figure 4.8 A "wild" ground state for ternary antiferromagnetic energies.

for $u\colon \varepsilon\mathbb{Z} \to \{-1,1\}$. In this case the sublattices of even and odd indices are decoupled, and we can separately consider the values of u_i to such lattices (see Example 6.7). Alternatively, we can introduce the variable $v\colon \varepsilon\mathbb{Z} \to Y$, where $Y = \{-1,1\}^2$ as in the previous example. In terms of this variable the equivalent energies are as in (4.3), where

$$\phi(v,w) = (v_1 - w_1)^2 + (v_2 - w_2)^2 = \begin{cases} 0 & \text{if } v = w, \\ 4 & \text{if } v \neq w \text{ but } v_1 = w_1 \text{ or } v_2 = w_2, \\ 8 & \text{if } v_1 \neq w_1 \text{ and } v_2 \neq w_2. \end{cases}$$

The limit energy is defined on functions $v\colon \mathbb{R} \to X$ and is of the form

$$\sum_{t\in S(v)} \phi(v^-(t), v^+(t)).$$

The last two examples show that introducing auxiliary (vector) parameters may be useful in treating problems with nonconstant ground states.

Finally, note that even simple energies may possess infinitely many ground states. For such energies a description of the Γ-limit in terms of interfaces is not possible.

Example 4.10 (Ternary antiferromagnetic energies) We consider the set of parameters $Y = \{-1,0,1\}$ and the nearest-neighbor energies

$$E_\varepsilon(u) = \sum_{i\in\mathbb{Z}} \phi(u_{i-1}, u_i),$$

defined for $u\colon \varepsilon\mathbb{Z} \to Y$, where

$$\phi(u,v) = \begin{cases} 0 & \text{if } u \neq v, \\ 1 & \text{if } u = v. \end{cases}$$

In this case ground states are all u with $u_{i-1} \neq u_i$, among which we have nonperiodic states and also u with arbitrary period. Note that all $u \in L^\infty(\mathbb{R})$

with $\|u\|_\infty \leq \frac{1}{2}$ can be obtained as weak limit of interpolations of ground states u^ε for E_ε, of which a pictorial representation is given in Fig. 4.8.

4.2 Partitions into Sets of Finite Perimeter

The generality of the approach in this chapter, which is necessary for future applications, will require the introduction of a more complex environment for the description of the Γ-limits of lattice energies. When dealing with functions taking values in a finite set of cardinality larger than two, it will be convenient to introduce functionals defined on partitions, which reduce to perimeter functionals in the case of partitions composed of two sets.

Definition 4.11 (Partition into sets of finite perimeter) Let Y be a finite set and Ω an open set. A *partition of* Ω *into sets of finite perimeter* (or *Caccioppoli partition*) indexed by Y is a collection of sets of finite perimeter $\{G^y\}_{y\in Y}$ with $G^y \subset \Omega$, $\bigcup_{y\in Y} G^y = \Omega$, and $|G^y \cap G^{y'}| = 0$ if $y \neq y'$.

Definition 4.12 (Convergence of partitions) We say that a sequence of partitions $\{\{G^y_n\}_{y\in Y}\}_n$ *converge to a partition* $\{G^y\}_{y\in Y}$ if for all $y \in Y$ the sequence of sets G^y_n converges to G^y; that is, $|G^y_n \triangle G^y| \to 0$. We say that the convergence is local if G^y_n converge to G^y locally in Ω.

Definition 4.13 (BV notation) Each partition of Ω into sets of finite perimeter is identified with a function $u\colon \Omega \to Y$ defined almost everywhere by setting $u(x) = y$ if $x \in G^y$. If Ω is bounded and Y lies in an Euclidean space, then such u is a function of bounded variation. With a little abuse of notation, the space of such partitions will be still denoted by $BV(\Omega; Y)$. We also keep the notation for the *jump set* of u

$$S(u) = \bigcup_{y\in Y} (\Omega \cap \partial^* G^y).$$

Moreover, we say that u_n *converge weakly** to u in $BV(\Omega; Y)$ if for each $y \in Y$ the partitions $\{G^y_n\}_{y\in Y}$ converge to $\{G^y\}_{y\in Y}$ and $\sup_n \mathcal{H}^{d-1}(\Omega \cap \partial^* G^y_n) < +\infty$. If u_n *converge weakly** to u in $BV(\Omega'; Y)$ for all $\Omega' \subset\subset \Omega$, then we say that u_n *converge weakly** to u in $BV_{\rm loc}(\Omega; Y)$. In this case u corresponds to the partition $\{G^y\}$. If Y lies in an Euclidean space, then $u_n \to u$ in $L^1_{\rm loc}(\Omega)$ and, if in addition Ω is bounded, then $u_n \to u$ in $L^1(\Omega)$.

The following result will allow us to recover integral functionals defined on partitions from their behavior as set functions.

Theorem 4.14 (Integral representation of energies on partitions) *Let Ω be an open set in $\mathbb{R}^d$, and Y a finite set. Let $\gamma_u(U)$ be defined for U open subset of Ω and $u \in BV(\Omega; Y)$ such that*

(i) *for all $u \in BV(\Omega; Y)$ the set function γ_u is the restriction to the family of open sets of a Borel measure;*
(ii) $\gamma_u(U)$ is local; *that is, for all open sets U we have $\gamma_u(U) = \gamma_v(U)$ whenever $u = v$ almost everywhere in U;*
(iii) *$u \mapsto \gamma_u(U)$ is L^1-lower semicontinuous for all U;*
(iv) *there exists c such that*

$$0 \leq \gamma_u(U) \leq c\mathcal{H}^{d-1}(S(u) \cap U).$$

Then there exists a Borel function $\varphi\colon \Omega \times Y \times Y \times S^{d-1} \to [0, +\infty)$ such that

$$\gamma_u(U) = \int_{U \cap S(u)} \varphi(x, u^+, u^-, \nu_u) d\mathcal{H}^{d-1}$$

and $\varphi(x, a, b, \nu)$ is given by

$$\begin{aligned}\varphi(x, a, b, \nu) &= \limsup_{\varrho \to 0} \frac{1}{\varrho^{d-1}} \inf_{\delta} \min\Big\{\gamma_w(Q^\nu_{\varrho+\delta}(x))\colon w \in BV(\Omega; Y) \\ &\qquad\qquad \text{such that } w = u^{\nu,x}_{a,b} \text{ on } \Omega \setminus Q^\nu_\varrho(x)\Big\} \\ &= \limsup_{\varrho \to 0} \frac{1}{\varrho^{d-1}} \sup_{\delta} \min\Big\{\gamma_w(Q^\nu_{\varrho}(x))\colon w \in BV(\Omega; Y) \\ &\qquad\qquad \text{such that } w = u^{\nu,x}_{a,b} \text{ on } \Omega \setminus Q^\nu_{\varrho-\delta}(x)\Big\},\end{aligned}$$

where for any a, $b \in Y$ we set

$$u^{\nu,x}_{a,b}(z) = \begin{cases} a & \text{if } \langle z - x, \nu\rangle > 0, \\ b & \text{if } \langle z - x, \nu\rangle \leq 0. \end{cases}$$

Remark 4.15 (BV-ellipticity) The lower semicontinuity of an integral functional on partitions entails that its integrand φ is *BV-elliptic*. If φ is homogeneous, that is, x-independent, then this condition can be interpreted that functions $u^{\nu,x}_{a,b}$ are minimizers with respect to compact perturbations; that is, for all $\varrho > 0$,

$$\begin{aligned}\varphi(a, b, \nu) = \frac{1}{\varrho^{d-1}} \min\Big\{\int_{S(u) \cap Q^\nu_\varrho(x)} \varphi(u^+, u^-, \nu_u) d\mathcal{H}^{d-1} : u \in BV(Q^\nu_\varrho(x); Y) \\ \text{such that } \{z\colon u(z) \neq u^{\nu,x}_{a,b}(z)\} \subset\subset Q^\nu_\varrho(x)\Big\}.\end{aligned}$$

This property implies that $\varphi(a, b, \cdot)$ is the restriction to S^{d-1} of a convex function positively homogeneous of degree one, and that $\varphi(\cdot, \cdot, \nu)$ is *subadditive*; that is,

$$\varphi(a, b, \nu) \leq \varphi(a, c, \nu) + \varphi(c, b, \nu).$$

As in the case of sets of finite perimeter it is very useful to prove a density result, which often allows one to reduce computations of Γ-limits to classes of smooth or polyhedral partitions.

Proposition 4.16 (Strong density of polyhedral partitions) *Let $u \in BV_{\rm loc}(\Omega; Y)$, and let $\Omega \subset \mathbb{R}^d$ be a Lipschitz set with $\partial\Omega$ compact. Then there is a sequence $\{u_j\} \subset BV(\Omega; Y)$ such that $S(u_j)$ is polyhedral (that is, it coincides with a finite union of simplexes up to a set of zero $\mathcal{H}^{d-1}$ measure), $u_j \to u$ in $L^1_{\rm loc}(\Omega; Y)$ and $Du_j \rightharpoonup Du$ as measures, and there are bijective maps $f_j \in C^1(\mathbb{R}^d; \mathbb{R}^d)$, with inverse also in C^1, which converge strongly in $W^{1,\infty}(\mathbb{R}^d; \mathbb{R}^d)$ to the identity map such that $|D(u \circ f_j) - Du_j|(\Omega) \to 0$.*

4.3 Admissible Lattices

We introduce the most general notion of lattice we will deal with and which includes in particular the case of periodic lattices. Such lattices have already been mentioned in Example 4.3 in Section 4.1, where also Voronoi cells are introduced.

Definition 4.17 (Admissible lattice) A countable set of points $\mathcal{L}$ in $\mathbb{R}^d$ is an *admissible lattice* if

(i) there exists a *maximal separation size* $R > 0$ such that $\inf_{x \in \mathbb{R}^d} \#\big(\mathcal{L} \cap B_R(x)\big) \geq 1$;

(ii) there exists a *minimal distance* $r > 0$ such that $\|i - j\| \geq r$ for all $i, j \in \mathcal{L}$ with $i \neq j$.

Roughly speaking, these assumptions rule out cluster points as well as arbitrarily big "holes" in the lattice.

We now define the set of nearest-neighboring points in $\mathcal{L}$. To this end, we associate to $\mathcal{L}$ a partition $\mathcal{P}(\mathcal{L})$ of $\mathbb{R}^d$ into sets, each of them containing one and only one point of $\mathcal{L}$. A natural choice for it is the *Voronoi tessellation* $\{C(i)\}_{i \in \mathcal{L}}$ associated to $\mathcal{L}$, whose *Voronoi cells* are defined as

$$C(i) = \{z \in \mathbb{R}^d \colon \ \|z - i\| \leq \|z - j\| \ \text{ for all } j \in \mathcal{L}\}. \tag{4.4}$$

By this definition, we have that the cells of a Voronoi tessellation are polyhedral sets.

We define the set of nearest-neighboring points in $\mathcal{L}$ as the sets of those points belonging to adjacent cells of $\mathcal{P}(\mathcal{L})$ according to the following definition. Note the different notation with respect to the cubic lattice due to the necessity of highlighting the dependence on the lattice.

Definition 4.18 (Nearest neighbors on admissible lattices)

(i) The set of *nearest-neighbor points* of a set $\mathcal{L}$ is defined by

$$\mathcal{NN}(\mathcal{L}) = \{(i,j) \in \mathcal{L} \times \mathcal{L} \colon \mathcal{H}^{d-1}(C(i) \cap C(j)) > 0\}.$$

(ii) Given $i \in \mathcal{L}$, the set of *nearest neighbors* of i is defined as

$$\mathcal{NN}(i) = \{j \in \mathcal{L} \colon (i,j) \in \mathcal{NN}(\mathcal{L})\}.$$

The following lemma describes the relevant properties of the Voronoi tessellation of an admissible lattice.

Lemma 4.19 *Let $\mathcal{L}$ be an admissible lattice with constants r, R as in Definition* 4.17. *Then there exist constants $M_1, M_2 > 0$ depending only on r, R such that, for all $i \in \mathcal{L}$,*

(i) $B_{\frac{r}{2}}(i) \subset C(i) \subset B_R(i)$;
(ii) $\#\{j \in \mathcal{L} \colon C(i) \cap C(j) \neq \emptyset\} \leq M_1$;
(iii) $\mathcal{H}^{d-1}(C(i) \cap C(j)) \leq M_2$ *for all* $j \in \mathcal{L} \setminus \{i\}$.

Proof (i) For $j \in \mathcal{L} \setminus \{i\}$ we have $\|i-j\| \geq r$, which implies $\|z-i\| \leq \|z-j\|$ for all $z \in B_{\frac{r}{2}}(i)$. By definition the first inclusion in (i) holds. Now suppose that there exists $z \in C(i)$ such that $\|z-i\| \geq R$. Since $\mathcal{L}$ is admissible, there exists $j \in \mathcal{L}$ such that $\|z-j\| < R$. It follows that $R \leq \|z-i\| \leq \|z-j\| < R$, leading to a contradiction.

(ii) Note that (i) implies that if $C(i) \cap C(j) \neq \emptyset$, then $\|i-j\| \leq 2R$. Using a covering argument, we deduce that it is enough to take $M_1 = \left(1 + \frac{4R}{r}\right)^d$.

(iii) By (i) the diameter of the set $C(i) \cap C(j)$ is bounded by $2R$ and the set is contained in a $(d-1)$-dimensional affine subspace so that we can take $M_2 = (2R)^{d-1}\omega_{d-1}$, where ω_{d-1} is the volume of the unit ball in $\mathbb{R}^{d-1}$. □

Remark 4.20 Note that the validity of claim (ii) ensures that the cardinality of the sets $\mathcal{NN}(i)$ of the nearest neighbors of points in $\mathcal{L}$ is equibounded.

4.4 Convergence of Discrete Functions

We now fix an admissible lattice $\mathcal{L}$ according to Definition 4.17, a finite set Y and an open set $\Omega \subseteq \mathbb{R}^d$. As in the previous chapters, for $\varepsilon > 0$ we set $\mathcal{L}_\varepsilon(\Omega) = \Omega \cap \varepsilon\mathcal{L}$ and for all i such that $\varepsilon i \in \mathcal{L}_\varepsilon(\Omega)$, we use the notation

$$\mathcal{L}^i_\varepsilon(\Omega) = \mathcal{L}_\varepsilon(\Omega) - \varepsilon i \tag{4.5}$$

for the corresponding translated lattice portion. For all such indices i, we consider functions

$$\phi^\varepsilon_i : \{z \colon \mathcal{L}^i_\varepsilon(\Omega) \to Y\} \to \mathbb{R}.$$

Given a function $u \colon \mathcal{L}_\varepsilon(\Omega) \to Y$, we recall the notation $u_i = u(\varepsilon i)$. For each such function u and for every i such that $\varepsilon i \in \mathcal{L}_\varepsilon(\Omega)$, we consider the function $\tau^i(u) \colon \mathcal{L}^i_\varepsilon(\Omega) \to Y$ defined by

$$\tau^i(u)(\varepsilon j) = \tau^i(u)_j = u_{i+j} = u(\varepsilon(i+j)). \tag{4.6}$$

In order not to overburden notation, we write $\{u_{i+j}\}_j$ in the place of $\tau^i(u)$. With this convention, we then introduce the family of functionals

$$E_\varepsilon(u) = E_\varepsilon(u;\Omega) = \sum_{\varepsilon i \in \mathcal{L}_\varepsilon(\Omega)} \varepsilon^{d-1} \phi^\varepsilon_i(\{u_{i+j}\}_j). \tag{4.7}$$

When we want to highlight the domain of ϕ^ε_i, we will also use the alternative notation

$$\phi^\varepsilon_i \left(\{u_{i+j}\}_{\varepsilon j \in \mathcal{L}^i_\varepsilon(\Omega)} \right)$$

in the place of $\phi^\varepsilon_i(\{u_{i+j}\}_j)$. We will study the Γ-convergence of E_ε with respect to the convergence defined as follows.

Definition 4.21 (Convergence on admissible lattices) Let $u^\varepsilon \colon \mathcal{L}_\varepsilon(\Omega) \to Y$ and $u \in BV(\Omega; Y)$. We say that u^ε *converge to* u in $BV_{\mathrm{loc}}(\Omega; Y)$ and we write $u^\varepsilon \to u$ if the piecewise-constant interpolations defined by $u^\varepsilon(x) = u^\varepsilon(\varepsilon i)$ for $x \in \varepsilon C(i)$ on each scaled Voronoi cell converge in the sense of Definition 4.13. If Y lies in a Euclidean space, this convergence implies the convergence of u^ε to u in $L^1_{\mathrm{loc}}(\Omega)$.

Our assumptions on the energy densities will enforce that sequences $\{u^\varepsilon\}$ with equibounded energy, that is, with $\sup_\varepsilon E_\varepsilon(u^\varepsilon;\Omega) < +\infty$, will be precompact with respect to the notion of convergence presented earlier. For later use we state the result in a localized form. To this end, for any open set U of Ω and $u \colon \mathcal{L}_\varepsilon(\Omega) \to Y$, we set

$$E_\varepsilon(u; U) = \sum_{\varepsilon i \in \mathcal{L}_\varepsilon(U)} \varepsilon^{d-1} \phi^\varepsilon_i(\{u_{i+j}\}_j). \tag{4.8}$$

We have the following compactness result, whose hypothesis is a more abstract and general version of the generalized coerciveness conditions introduced in Remark 3.3.

In this statement and in the following, we use the notation $Q_R(x)$ for the coordinate cube centered at x and with side length R, and we set $Q_R = Q_R(0)$ and $Q = Q_1$.

Theorem 4.22 (Compactness) *Suppose that there exist $M \in \mathbb{N}$, $C > 0$ such that, if $\varepsilon i \in \mathcal{L}_\varepsilon(\Omega)$ and u is not constant on $\varepsilon \mathcal{NN}(i) \cup \{\varepsilon i\}$, then there exists $j \in Q_M(i)$ with $\varepsilon j \in \mathcal{L}_\varepsilon(\Omega)$ such that*

$$\phi_j^\varepsilon(\{u_{j+k}\}_k) \geq C.$$

Let U be an open subset of Ω and let $\{u^\varepsilon\}$ be such that $\sup_\varepsilon E_\varepsilon(u^\varepsilon; U) < +\infty$. Then, up to subsequences, u^ε converge in $BV_{\rm loc}(U;Y)$ to some function in $BV(U;Y)$.

Proof Let $U' \subset\subset U$ so that $U' + Q_{\varepsilon M} \subset U$ for ε sufficiently small. By the hypothesis on ϕ_j^ε we have that, for all $v \in Y$,

$$\varepsilon^{d-1}\ \#\{i : \varepsilon i \in \mathcal{L}_\varepsilon(U'),\ u_i^\varepsilon = v \text{ and there exists } j \in \mathcal{NN}(i) \text{ such that } u_j^\varepsilon \neq v\}$$
$$\leq \frac{1}{C} \sum_{u_i^\varepsilon = v} \sum_{j \in Q_M(i) \cap \mathcal{L}} \varepsilon^{d-1} \phi_j^\varepsilon(\{u_{j+k}^\varepsilon\}_k) \leq C' \left(\frac{M}{r}\right)^d \frac{1}{C} E_\varepsilon(u^\varepsilon),$$

where we have taken into account that, by Definition 4.17 (ii), for any $j \in \mathcal{L}$,

$$\#\{i \in \mathcal{L} : j \in Q_M(i) \cap \mathcal{L}\} \leq C' \left(\frac{M}{r}\right)^d.$$

As a result, recalling that u^ε is identified with its piecewise-constant interpolation on the Voronoi cells of the lattice, by Lemma 4.19,

$$\mathcal{H}^{d-1}(U' \cap \partial\{x : u^\varepsilon(x) = v\}) \leq M_1 M_2\, C' \left(\frac{M}{r}\right)^d \frac{1}{C} E_\varepsilon(u^\varepsilon) < +\infty. \tag{4.9}$$

By the arbitrariness of $U' \subset\subset U$, estimate (4.9) implies that for all $v \in Y$ the family of sets $\{x \in U : u^\varepsilon(x) = v\}$ has locally equibounded perimeter in U, so that the corresponding characteristic functions are precompact in $BV_{\rm loc}(U)$ and, up to extraction of a subsequence, converge to some set of finite perimeter E_v. This implies that u^ε converge in $BV_{\rm loc}(U;Y)$ to $u = \sum_{v \in Y} v \chi_{E_v}$. The function u belongs to $BV(U;Y)$, since the estimate in (4.9) is independent of U'. □

We now note that the notion of convergence of a discrete system to a BV-function can be often given without resorting to interpolations.

Remark 4.23 (Convergence without lattice interpolation) Take a sequence of lattices $\mathcal{L}(\varepsilon)$ and functions $u^\varepsilon : \Omega \cap \mathcal{L}(\varepsilon) \to Y$. We say that such a sequence *converges to* $u \in BV(\Omega;Y)$ if there exist numbers $\delta_\varepsilon \to 0$ such that, having set

$$I_\varepsilon(y) = \{k \in \mathbb{Z}^d : u^\varepsilon(\zeta) = y \text{ for all } \zeta \in \mathcal{L}(\varepsilon) \cap \delta_\varepsilon(k + Q)\},$$

for all $y \in Y$ the sets

$$\bigcup_{k \in I_\varepsilon(y)} \delta_\varepsilon(k + Q)$$

converge in $L^1_{\rm loc}(\Omega)$ to $\{u = y\}$. Note that $\bigcup_{k\in\mathcal{I}_\varepsilon(y)} \delta_\varepsilon(k+Q)$ is the union of all cubes of the form $\delta_\varepsilon(k+Q)$ on which u^ε is identically y.

We can then derive a compactness criterion: suppose that δ_ε are such that

(i) if u is not constant on a cube Q^ε of side length δ_ε, then $E_\varepsilon(u;Q^\varepsilon)\geq C\delta_\varepsilon^{d-1}$, and
(ii) if u is not constant on the union of two cubes $Q^{1,\varepsilon}$ and $Q^{2,\varepsilon}$ of side length δ_ε intersecting on one side, then again $E_\varepsilon(u;Q^{1,\varepsilon}\cup Q^{2,\varepsilon})\geq C\delta_\varepsilon^{d-1}$;

then from each sequence u^ε such that $E_\varepsilon(u^\varepsilon;\Omega)\leq C<+\infty$ we can extract a subsequence converging to a function in $BV(\Omega;Y)$. Furthermore, if we also have that $E_\varepsilon(u;Q^\varepsilon)\geq C\delta_\varepsilon^{d-1}$, if u is not identically equal to $y\in Y_0\subset Y$ on a cube Q^ε of side length δ_ε, then $u\in BV(\Omega;Y_0)$.

4.5 The Localization Method

In order to prove compactness and integral-representation results for sequences of Γ-limits under general growth assumptions, it is often convenient to use a well-established method that can be summarized as follows.

(i) (*Localization of the energies*) We consider families $F_\varepsilon(u;U)$, where U is any open set of a fixed set Ω. We suppose that all energies are increasing functions of U. This is true for positive energy densities.
(ii) (*Extraction of a subsequence converging for a dense family of sets*) Apply the general compactness properties of Γ-convergence to extract a subsequence (labeled by ε_j) such that $F_{\varepsilon_j}(\cdot\,;U)$ Γ-converge to some $F_0(\cdot\,;U)$ for all U in a countable dense family $\mathcal{U}$. In other words, if we define

$$F'(u;U)=\Gamma\text{-}\liminf_{j\to+\infty} F_{\varepsilon_j}(u;U),\qquad F''(u;U)=\Gamma\text{-}\limsup_{j\to+\infty} F_{\varepsilon_j}(u;U),$$

then $F'(u;U)=F''(u;U)$ for $U\in\mathcal{U}$. We may take, for example, $\mathcal{U}$ equal to the class of all open sets that can be written as unions of d-dimensional rectangles in $\mathbb{R}^d$ with vertices having rational coordinates.
(iii) (*Existence of Γ-limits by inner regularity* – see Remark 4.26) Prove that for a dense family $\mathcal{R}$ of suitably regular open sets we have

$$F''(u;V)=\sup\{F''(u;U'):U'\subset\subset V\}. \tag{4.10}$$

From (4.10) it follows that

$$\begin{aligned}F''(u;V)&=\sup\{F_0(u;U'):U'\in\mathcal{U},\ U'\subset\subset V\}\\&=\sup\{F'(u;U'):\ U'\subset\subset V\}\leq F'(u;V)\end{aligned}$$

so that the Γ-limit $F_0(u; V)$ exists for all $V \in \mathcal{R}$. We will take as usual $\mathcal{R}$ equal to the family of open sets with Lipschitz boundary.

(iv) (*Representation*) Prove that the set function

$$\begin{aligned}\gamma_u(U) &= \sup\{F''(u; U') : U' \subset\subset U\} \\ &= \sup\{F_0(u; U') : U' \subset\subset U,\ U' \in \mathcal{U}\},\end{aligned} \tag{4.11}$$

which coincides with $F_0(u; U)$ if $U \in \mathcal{R}$, can be represented in an integral form.

The proof of (iv) will require the application of results which themselves assume that γ_u is a regular measure. To that end, a general criterion is the following.

Lemma 4.24 (De Giorgi–Letta measure criterion) *Let γ be an open-set function satisfying*

(i) *γ is an increasing set function;*
(ii) *γ is superadditive on disjoint sets;*
(iii) *γ is inner regular;*
(iv) *γ is subadditive.*

Then γ is the restriction to open sets of a Borel measure.

Note that, for $\gamma = \gamma_u$, properties (i) and (iii) of Lemma 4.24 follow immediately from the definition. The superadditivity (ii) of γ_u can be derived from the superadditivity of F_0; that is, that

$$F_0(u; U \cup V) \geq F_0(u; U) + F_0(u; V) \tag{4.12}$$

if U and V are disjoint open sets in $\mathcal{R}$. Indeed, it suffices to notice that the functions for a recovery sequence on $U \cup V$ can be used as test functions on U and V separately. From this property the superadditivity of γ_u follows from its definition.

In order to prove (iv) of Lemma 4.24 we need the following technical property, which will be proven in the following for discrete energies E_ε.

Definition 4.25 (Fundamental estimate) The energies F_ε satisfy the *fundamental estimate* if the following holds. Let U', U, and V be bounded open sets in Ω with $U' \subset\subset U$, $U' \in \mathcal{R}$, and let $\{u^{U,\varepsilon}\}$ and $\{u^{V,\varepsilon}\}$ be sequences such that

$$\sup_\varepsilon (F_\varepsilon(u^{U,\varepsilon}; U) + F_\varepsilon(u^{V,\varepsilon}; V \setminus \overline{U'})) \leq C < +\infty. \tag{4.13}$$

Then for all $\eta > 0$ there exists a sequence $\{v^\varepsilon\}$ such that

(i) $v^\varepsilon = u^{U,\varepsilon}$ on U',
(ii) $v^\varepsilon = u^{V,\varepsilon}$ on $V \setminus U$,
(iii) we have

$$F_\varepsilon(v^\varepsilon; U' \cup V) \leq F_\varepsilon(u^{U,\varepsilon}; U) + F_\varepsilon(u^{V,\varepsilon}; V \setminus \overline{U'}) \\ + C_{UV}\big(r(\varepsilon,\eta) + C(\eta)\|u^{U,\varepsilon} - u^{V,\varepsilon}\|_{L^1(U\setminus U')}\big), \quad (4.14)$$

where C_{UV} denotes a constant depending only on U and V, $C(\eta)$ a constant depending only on η, and $r(\varepsilon,\eta)$ denotes a function such that

$$\lim_{\eta\to 0}\limsup_{\varepsilon\to 0} r(\varepsilon,\eta) = 0.$$

Furthermore, if $u^{U,\varepsilon}$ and $u^{V,\varepsilon}$ both converge to u in $L^1(U \setminus \overline{U'})$, then also v^ε converge to u in $L^1(U \setminus \overline{U'})$.

Remark 4.26 Suppose that the fundamental estimate (Definition 4.25) holds and that

$$F''(u; U) \leq C\mathcal{H}^{d-1}(S(u) \cap U) \quad (4.15)$$

if $U \in \mathcal{R}$, where $\mathcal{R}$ is the family of sets with Lipschitz boundary. Then we can prove inner regularity (4.10) and that γ_u satisfies the locality and subadditivity properties.

As for the proof of (4.10), given $U'' \subset\subset U' \subset\subset U \subset\subset V$ with $U' \in \mathcal{R}$, choose $\{u^{U,\varepsilon_j}\} + j$ and $\{u^{V\setminus\overline{U''},\varepsilon_j}\}_j$ recovery sequences for the Γ-lim sup for u in U and $V \setminus \overline{U''}$, respectively, and apply the fundamental estimate in Definition 4.25 with $V \setminus \overline{U''}$ in place of V. Noting that $U' \cup (V \setminus \overline{U''}) = V$ and using the v^{ε_j} as in Definition 4.25 to test $F''(u; V)$, we get

$$F''(u; V) \leq F''(u; U) + F''(u; V \setminus \overline{U''})$$

by (4.14). By (4.15) the last term can be made arbitrarily small as U'' tends to V, upon requiring that $V \setminus \overline{U''} \in \mathcal{R}$, proving (4.10).

The *locality* of γ_u; that is, the property that $\gamma_u(V) = \gamma_{u'}(V)$ if $u = u'$ almost everywhere in V, can be obtained using the properties in Definition 4.25 as follows. We choose U, U', and V' Lipschitz sets with $U' \subset\subset U \subset\subset V' \subset\subset V$, and $u^{U,\varepsilon} \to u$, $u^{V,\varepsilon} \to u'$ such that

$$\limsup_{\varepsilon\to 0} F_\varepsilon(u^{U,\varepsilon}; U) = F''(u; U), \quad (4.16)$$

$$\limsup_{\varepsilon\to 0} F_\varepsilon(u^{V,\varepsilon}; V' \setminus \overline{U'}) = F''(u'; V' \setminus \overline{U'}). \quad (4.17)$$

Applying Definition 4.25 with the choice $V = V'$, observing that $v^\varepsilon \to u'$ and that the last two terms of estimate (4.14) are negligible, we obtain

$$F''(u'; V') \le F''(u; U) + C\mathcal{H}^{d-1}(S(u) \cap (V' \setminus \overline{U'})),$$

upon using (4.15). Letting U' invade V', we get

$$F''(u'; V') \le F''(u; V'),$$

and hence $\gamma_{u'}(V) \le \gamma_u(V)$. The locality of γ_u follows by exchanging the role of u and u'.

To prove that γ_u satisfies the subadditivity property (iv) in Lemma 4.24, we need to show that

$$\gamma_u(U \cup V) \le \gamma_u(U) + \gamma_u(V) \tag{4.18}$$

for any U and V open subsets of $\mathbb{R}^d$. Since $\mathcal{R}$ is a dense family, by (4.10) and the definition of γ_u it suffices to prove that, for all $U, V \in \mathcal{R}$,

$$F''(u; U \cup V) \le F''(u; U) + F''(u; V). \tag{4.19}$$

If we use Definition 4.25 with $\{u^{U,\varepsilon}\}$ and $\{u^{V,\varepsilon}\}$ recovery sequences for the Γ- lim sup for u in U and V, respectively, letting first ε and then η tend to 0 in (4.14), we infer that

$$F''(u; U' \cup V) \le F''(u; U) + F''(u; V). \tag{4.20}$$

Letting U' invade U, by (4.10) we get (4.19).

In the next section we state a compactness theorem, whose proof will be obtained by following the localization method. Summarizing, in order to make the procedure just described work, we will have to prove two items:

(i) the fundamental estimate (Definition 4.25);
(ii) the estimate from (4.15) for open sets with Lipschitz boundary.

4.6 The Compactness Theorem

In this section we state and prove a compactness and integral-representation result for Γ-limits of families of energies of the form (4.7) under very general assumptions on the interaction potentials

$$\phi_i^\varepsilon \colon \{z \colon \mathcal{L}_\varepsilon^i(\Omega) \to Y\} \to \mathbb{R}$$

in the light of the examples in Section 4.1. Namely, we assume that there exists $L > 0$ such that

$$0 \le \phi_i^\varepsilon(z) \le L \tag{4.21}$$

for all $\varepsilon > 0$, $\varepsilon i \in \mathcal{L}_\varepsilon(\Omega)$, and $z\colon \mathcal{L}^i_\varepsilon(\Omega) \to Y$, and that the following hypotheses are satisfied. For simplicity, we suppose that Ω is connected, upon considering each connected component in hypothesis (H1).

(H1) *(Existence of constant ground states)* there exists $Y_0 = \{v^1, \dots, v^K\} \subseteq Y$ such that for all $\varepsilon > 0$ sufficiently small we have $\phi^\varepsilon_i(\{u_{j+i}\}_j) = 0$ for all $\varepsilon i \in \mathcal{L}_\varepsilon(\Omega)$ if and only if $l \in \{1, \dots, K\}$ exists such that $u_j = v^l$ for all $\varepsilon j \in \mathcal{L}_\varepsilon(\Omega)$;

(H2) *(Coerciveness)* there exists $M \in \mathbb{N}$, $C > 0$ such that if u is not identically v^l on $\varepsilon NN(i) \cup \{\varepsilon i\}$ for some $l \in \{1, 2, \dots, K\}$, then there exists j in the cube $Q_M(i)$ with $\varepsilon j \in \mathcal{L}_\varepsilon(\Omega)$ such that

$$\phi^\varepsilon_j(\{u_{j+k}\}_k) \geq C;$$

(H3) *(Mild non-locality)* given $z, w\colon \mathcal{L}_\varepsilon(\Omega) \to Y$ and $m \in \mathbb{N}$ such that $z_j = w_j$ for all $j \in Q_m(i)$ with $\varepsilon j \in \mathcal{L}_\varepsilon(\Omega)$, then

$$|\phi^\varepsilon_i(\{w_{i+j}\}_j) - \phi^\varepsilon_i(\{z_{i+j}\}_j)| \leq c^\varepsilon_m \tag{4.22}$$

where the constants c^ε_m are such that

$$\limsup_{\varepsilon \to 0} \sum_{m=1}^{+\infty} c^\varepsilon_m m^{d-1} < +\infty \tag{4.23}$$

and that for all $\delta > 0$ there exists $R_\delta > 0$ such that

$$\limsup_{\varepsilon \to 0} \sum_{m > R_\delta} c^\varepsilon_m m^{d-1} < \delta. \tag{4.24}$$

We can take in particular $\Omega = \mathbb{R}^d$ and $\mathcal{L} = \mathbb{Z}^d$, in which case the hypotheses are formally simplified since every ϕ^ε_i is defined on the set $\{z\colon \varepsilon\mathbb{Z}^d \to Y\}$.

Theorem 4.27 (The Compactness Theorem) *Let Ω be an open subset of $\mathbb{R}^d$. For all $\varepsilon > 0$, let $\phi^\varepsilon_i\colon \{z\colon \mathcal{L}^i_\varepsilon(\Omega) \to Y\} \to [0, +\infty)$ satisfy* (4.21) *for all $\varepsilon i \in \mathcal{L}_\varepsilon(\Omega)$ and $\varepsilon > 0$. For all U open subsets of Ω and functions $u\colon \mathcal{L}_\varepsilon(\Omega) \to Y$, we define the energies*

$$E_\varepsilon(u; U) = \sum_{\varepsilon i \in \mathcal{L}_\varepsilon(U)} \varepsilon^{d-1} \phi^\varepsilon_i(\{u_{i+j}\}_j). \tag{4.25}$$

If (H1) *and* (H3) *are satisfied, then, upon extraction of a subsequence, there exists a bounded Borel function $\varphi\colon \Omega \times Y_0 \times Y_0 \times S^{d-1} \to [0, +\infty)$ such that for all U with Lipschitz boundary $E_\varepsilon(\cdot\,; U)$ Γ-converge with respect to the $BV_{\rm loc}(\Omega; Y)$ convergence on $BV(\Omega; Y_0)$ to the functional*

$$F_0(u; U) = \int_{U \cap S(u)} \varphi(x, u^+, u^-, \nu_u)\, d\mathcal{H}^{d-1}. \tag{4.26}$$

Furthermore, if (H2) *is also satisfied, then* $E_\varepsilon(\cdot\,;\Omega)$ Γ*-converge to* $F_0(\cdot\,;\Omega)$ *on the whole* $BV(\Omega;Y)$*, it is finite only for functions* $u \in BV(\Omega;Y_0)$ *with* $\mathcal{H}^{d-1}(S(u))$ *finite, and for almost every* $x \in \Omega$ *and* φ *satisfies*

$$\varphi(x,a,b,\nu) \geq c > 0 \tag{4.27}$$

for all $(a,b,\nu) \in Y_0 \times Y_0 \times S^{d-1}$ *with* $a \neq b$.

The proof of the theorem will be given at the end of the section, after some preparatory propositions.

Remark 4.28 (Simplified conditions for bounded domains) If Ω is a bounded set, then decay conditions (H3) can be improved to requiring that

$$\limsup_{\varepsilon\to 0} \sum_{m=1}^{+\infty} c_m^\varepsilon < +\infty \tag{4.28}$$

and that for all $\delta > 0$ there exists $R_\delta > 0$ such that

$$\limsup_{\varepsilon\to 0} \sum_{m>R_\delta} c_m^\varepsilon < \delta. \tag{4.29}$$

We first note that assumption (H3) allows the construction of the functions v^ε in the fundamental estimate. Such construction consists in defining v^ε separately on either side of an interface. While for pairwise interactions the argument relies in counting the interactions across the interface using a discrete coarea formula as in Remark 3.8, here we use (4.22) for an estimate of the energetic effect of the construction of v_ε in terms of the distance of points εi from the interface.

Proposition 4.29 *Under assumption* (H3)*, functionals* E_ε *satisfy the fundamental estimate as in Definition* 4.25.

Proof Following Definition 4.25, we fix U', U, and V bounded open sets in Ω with $U' \subset\subset U$, $U' \in \mathcal{R}$, and sequences $\{u^{U,\varepsilon}\}$ and $\{u^{V,\varepsilon}\}$ such that equiboundedness condition (4.13) be satisfied. Let $\eta > 0$ be fixed.

We first note that for ε small enough there exists

$$t_\varepsilon \in \Big(3\sqrt{d}\frac{\varepsilon}{\eta}, \mathrm{dist}(V\setminus U, U') - 3\sqrt{d}\frac{\varepsilon}{\eta}\Big)$$

such that, setting $A_\varepsilon = \{x \in \mathbb{R}^d\colon\ \mathrm{dist}(x,U') = t_\varepsilon\}$, the estimate

$$\|u^{U,\varepsilon} - u^{V,\varepsilon}\|_{L^1(A_\varepsilon + B_{3\sqrt{d}\frac{\varepsilon}{\eta}})} \leq \frac{\varepsilon}{\eta} C_{UV} \|u^{U,\varepsilon} - u^{V,\varepsilon}\|_{L^1(U\setminus U')} \tag{4.30}$$

holds, where B_r denotes the ball centered at 0 and with radius r and C_{UV} is a positive constant depending only on U and V. Indeed, setting $A(t) = \{x \in \mathbb{R}^d:$

$\operatorname{dist}(x,U')=t\}$, we may argue by contradiction and suppose that for all $N\in\mathbb{N}$, ε small enough, and t as given earlier we have

$$\|u^{U,\varepsilon}-u^{V,\varepsilon}\|_{L^1(A(t)+B_{3\sqrt{d}\frac{\varepsilon}{\eta}})}>\frac{\varepsilon}{\eta}N\|u^{U,\varepsilon}-u^{V,\varepsilon}\|_{L^1(U\setminus U')}.$$

Choosing a family of t with cardinality larger than $\operatorname{dist}(V\setminus U,U')\frac{\eta}{2\varepsilon}$ such that the sets $A(t)+B_{3\sqrt{d}\frac{\varepsilon}{\eta}}$ are pairwise disjoint, we get a contradiction.

Setting $U_\varepsilon=\{x\colon \operatorname{dist}(x,U')<t_\varepsilon\}$, we define

$$v^\varepsilon(x)=\begin{cases}u^{U,\varepsilon}(x) & \text{if } x\in U_\varepsilon,\\ u^{V,\varepsilon}(x) & \text{otherwise,}\end{cases}\tag{4.31}$$

and as usual we set $v_i^\varepsilon=v^\varepsilon(\varepsilon i)$. We have

$$E_\varepsilon(v^\varepsilon;V\cup U')=E_\varepsilon(v^\varepsilon;(V\cup U')\cap U_\varepsilon)+E_\varepsilon(v^\varepsilon;(V\cup U')\setminus U_\varepsilon).\tag{4.32}$$

Given i such that $\varepsilon i\in\mathcal{L}_\varepsilon(U_\varepsilon)$, set

$$m_\varepsilon(i)=\sup\{m\in\mathbb{N}\colon\ v^\varepsilon=u^{U,\varepsilon}\text{ on }\Omega\cap Q_{\varepsilon m}(\varepsilon i)\}.$$

By (H3) and the positivity of the energy, we have

$$\begin{aligned}E_\varepsilon(v^\varepsilon;(V\cup U')\cap U_\varepsilon)&\le E_\varepsilon(u^{U,\varepsilon};(V\cup U')\cap U)\\&\quad+\sum_{m=1}^{+\infty}\varepsilon^{d-1}c_m^\varepsilon\#\{i:\varepsilon i\in\mathcal{L}_\varepsilon(U_\varepsilon)\text{ and }m_\varepsilon(i)=m\}\\&\le E_\varepsilon(u^{U,\varepsilon};(V\cup U')\cap U)\\&\quad+C_{U'}\sum_{m>\frac{1}{\eta}}c_m^\varepsilon+\varepsilon^{d-1}\#I_\varepsilon^\eta\sum_{m\le\frac{1}{\eta}}c_m^\varepsilon+o(1)_{\varepsilon\to0},\end{aligned}\tag{4.33}$$

where $C_{U'}$ is a positive constant depending on the $d-1$-dimensional Hausdorff measure of the level sets of the signed distance from $\partial U'$ interior to U_ε, and

$$I_\varepsilon^\eta=\left\{i\colon\varepsilon i\in\mathcal{L}_\varepsilon(U_\varepsilon),\ m_\varepsilon(i)\le\frac{1}{\eta}\right\}.$$

If $m_\varepsilon(i)\le\frac{1}{\eta}$, then, by the definition of v^ε, $A_\varepsilon\cap Q_{\varepsilon+\frac{\varepsilon}{\eta}}(\varepsilon i)\neq\emptyset$, and by the definition of $m_\varepsilon(i)$ there exists $j(i)\in Q_{1+\frac{1}{\eta}}(i)$ such that $v_{j(i)}^\varepsilon\neq u_{j(i)}^{U,\varepsilon}$. Hence, $u_{j(i)}^{U,\varepsilon}\neq u_{j(i)}^{V,\varepsilon}$, and in particular we infer that

$$\int_{Q_{\frac{2\varepsilon}{\eta}}(\varepsilon i)}|u^{V,\varepsilon}-u^{U,\varepsilon}|\,dx\ge\int_{\varepsilon C(j(i))}|u^{V,\varepsilon}-u^{U,\varepsilon}|\,dx\ge C\varepsilon^d,$$

where $C(j(i))$ denotes the Voronoi cell of $j(i)$, as defined in (4.4). Hence, by (4.30),

$$\varepsilon^d \# I_\varepsilon^\eta \le C \sum_{i \in I_\varepsilon^\eta} \int_{Q_{\frac{2\varepsilon}{\eta}}(\varepsilon i)} |u^{V,\varepsilon} - u^{U,\varepsilon}|\, dx \le C \frac{1}{\eta^d} \int_{A_\varepsilon + B_{3\sqrt{d}\frac{\varepsilon}{\eta}}} |u^{V,\varepsilon} - u^{U,\varepsilon}|\, dx$$
$$\le \frac{\varepsilon}{\eta^{d+1}} C_{UV} \|u^{U,\varepsilon} - u^{V,\varepsilon}\|_{L^1(U\setminus U')}. \tag{4.34}$$

From (4.33), (4.34), and (4.23) we then get

$$E_\varepsilon(v^\varepsilon; (V \cup U') \cap U_\varepsilon) \le E_\varepsilon(u^{U,\varepsilon}; (V \cup U') \cap U) + C_{U'} \sum_{m > \frac{1}{\eta}} c_m^\varepsilon$$
$$+ \frac{1}{\eta^{d+1}} C_{UV} \|u^{U,\varepsilon} - u^{V,\varepsilon}\|_{L^1(U\setminus U')}$$
$$+ o(1)_{\varepsilon\to 0}. \tag{4.35}$$

With the same argument, we have

$$E_\varepsilon(v^\varepsilon; (V \cup U') \setminus U_\varepsilon) \le E_\varepsilon(u^{V,\varepsilon}; (V \cup U') \setminus \overline{U}') + C_{U'} \sum_{m > \frac{1}{\eta}} c_m^\varepsilon$$
$$+ \frac{1}{\eta^{d+1}} C_{UV} \|u^{U,\varepsilon} - u^{V,\varepsilon}\|_{L^1(U\setminus U')}$$
$$+ o(1)_{\varepsilon\to 0}, \tag{4.36}$$

where in this case $C_{U'}$ is a positive constant depending on the $d-1$-dimensional Hausdorff measure of the level sets of the signed distance from $\partial U'$ external to U_ε. We get (4.14) for E_ε by gathering inequalities (4.35), (4.36) together in (4.32) and taking into account (4.24) and (4.30). □

As remarked in Section 4.5, the second ingredient is an estimate from above, which is proved in the following proposition.

Proposition 4.30 *Let $U \subset \Omega$ be an open Lipschitz set. Under assumptions* (H1) *and* (H3) *there exists a constant $C > 0$ such that*

$$\Gamma\text{-}\limsup_{\varepsilon\to 0} E_\varepsilon(u; U) \le C \mathcal{H}^{d-1}(S(u) \cap U) \tag{4.37}$$

for all $u \in BV_{\rm loc}(\Omega; Y_0)$ with $\mathcal{H}^{d-1}(S(u) \cap U) < +\infty$.

Proof In the course of the proof we will use the property that

$$u \mapsto \Gamma\text{-}\limsup_{\varepsilon\to 0} E_\varepsilon(u; U)$$

is a lower-semicontinuous function with respect to the $BV_{\rm loc}(\Omega)$-convergence, so that we are allowed small variations of our target function.

By the density result in Proposition 4.16 it is enough to consider functions $u \in BV_{\rm loc}(\Omega; Y_0)$ with $S(u)$ a polyhedral set. Since the construction is local, we can deal with the case of u taking only two values, which we fix to be v^1 and v^2, both belonging to Y_0. We first consider the case when U is bounded.

We define v^ε on $\mathcal{L}_\varepsilon(\Omega)$ by setting $v^\varepsilon_i = u(\varepsilon i)$, and note that $v^\varepsilon \to u$. We set

$$\overline{m}_\varepsilon(i) = \sup\{m \in \mathbb{N}\colon\ v^\varepsilon \text{ is constant on } Q_{\varepsilon m}(\varepsilon i)\}, \tag{4.38}$$

and observe that, for any $m \in \mathbb{N}$,

$$\#\{i\colon\ \overline{m}_\varepsilon(i) = m\} \leq \frac{C}{\varepsilon^{d-1}}\mathcal{H}^{d-1}(S(u) \cap U),$$

where C is a positive constant independent of m. We estimate the energy of v^ε using (H3) as

$$\begin{aligned} E_\varepsilon(v^\varepsilon; U) &= \sum_{\varepsilon i \in U \cap \varepsilon\mathbb{Z}^d} \varepsilon^{d-1}\phi^\varepsilon_i(\{v^\varepsilon_{i+j}\}_j) \\ &\leq \sum_{\varepsilon i \in U \cap \varepsilon\mathbb{Z}^d} \varepsilon^{d-1} c^\varepsilon_{\overline{m}_\varepsilon(i)} \leq \sum_{m=1}^{+\infty} \varepsilon^{d-1}\#\{i : \overline{m}_\varepsilon(i) = m\} c^\varepsilon_m \\ &\leq C\Big(\mathcal{H}^{d-1}(S(u) \cap U) + o(1)_{\varepsilon\to 0}\Big) \sum_{m=1}^{+\infty} c^\varepsilon_m. \end{aligned}$$

If U is not bounded, the argument is more delicate; we only consider $U = \mathbb{R}^d$ not to overburden the notation. We decompose $S(u)$ as follows:

$$S(u) = \bigcup_{l \in \mathcal{I}} V_l, \tag{4.39}$$

where V_l is a closed $d-1$-dimensional polytope, $\mathcal{I}$ is a finite set of indices, and $V_l \cap V_{l'}$ is a $d-2$-dimensional polytope.

Let ν_l denote the exterior normal to $S(u)$ at V_l. Without loss of generality we may suppose that $S(u) \cap \varepsilon\mathbb{Z}^d = \emptyset$ and that $\langle \nu_l, e_k\rangle \neq 0$ for all $k \in \{1, 2, \ldots, d\}$ and $l \in \mathcal{I}$, so that for all $x \in \mathbb{R}^d$ there exists a unique $\overline{x} = \overline{x}_l \in \Pi^{\nu_l} = \{y\colon \langle y, \nu_l\rangle = 0\}$ such that

$$\|\overline{x} - x\|_\infty = \min\{\|y - x\|_\infty,\ y \in S(u)\}.$$

We define ξ_l as the unique vector such that $\langle \xi_l, e_k\rangle = \operatorname{sign}\langle \nu_l, e_k\rangle$ for all $k = 1, \ldots, d$. Then, $x - \overline{x}$ is parallel to ξ_l. We set $S_l = V_l + \xi_l\mathbb{R}$ and define the set of indices

$$\mathcal{J}_{l,\varepsilon} = \Big\{i\colon\ \varepsilon i \in S_l \cap \varepsilon\mathbb{Z}^d \text{ and } V_l \cap Q_{\varepsilon(\overline{m}_\varepsilon(i)+1)}(\varepsilon i) \neq \emptyset\Big\}. \tag{4.40}$$

We note that for any m,

$$\#\{i \in \mathcal{J}_{l,\varepsilon}\colon\ \overline{m}_\varepsilon(i) = m\} \leq \frac{C}{\varepsilon^{d-1}}(\mathcal{H}^{d-1}(V_l) + o(1)_{\varepsilon\to 0}), \tag{4.41}$$

where C is a (possibly different) positive constant independent of m.

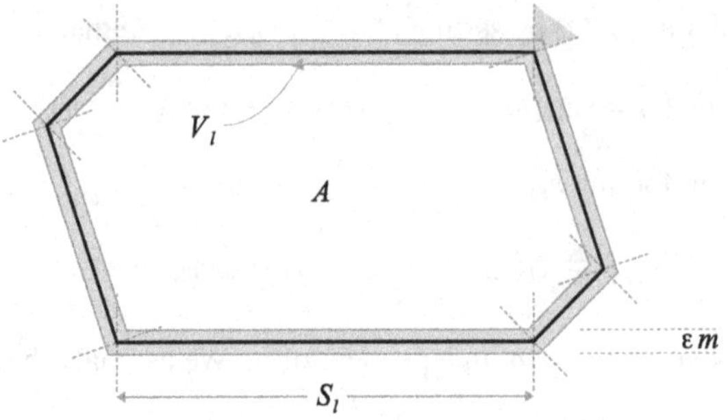

Figure 4.9 Different zones used in estimating the energy.

In this case the estimates will be separated into different zones: stripes perpendicular to each V_l, which we further subdivide into contributions close to the boundary (light gray zone in Fig. 4.9) or far from the boundary, and its complement that we further parameterize in terms of the $d-2$-dimensional polytopes of $S(u)$. In the two-dimensional picture in Fig. 4.9 this reduces to the contribution in each angle as the one highlighted in dark gray.

In this case, we estimate the energy of v^ε as

$$E_\varepsilon(v^\varepsilon) = \sum_{l\in\mathcal{I}}\sum_{i\in\mathcal{J}_{l,\varepsilon}} \varepsilon^{d-1}\phi_i^\varepsilon(\{v^\varepsilon_{i+j}\}_j) + \sum_{i\notin\bigcup_{l\in\mathcal{I}}\mathcal{J}_{l,\varepsilon}} \varepsilon^{d-1}\phi_i^\varepsilon(\{v^\varepsilon_{i+j}\}_j). \quad (4.42)$$

For fixed $l \in \mathcal{I}$ we have the same kind of estimate as earlier, that is,

$$\begin{aligned}\sum_{i\in\mathcal{J}_{l,\varepsilon}} \varepsilon^{d-1}\phi_i^\varepsilon(\{v^\varepsilon_{i+j}\}_j) &\le \sum_{i\in\mathcal{J}_{l,\varepsilon}} \varepsilon^{d-1} c^\varepsilon_{\overline{m}_\varepsilon(i)}\\ &\le \sum_{m=1}^{+\infty} \varepsilon^{d-1}\#\{i\in\mathcal{J}_{l,\varepsilon} : \overline{m}_\varepsilon(i)=m\}c^\varepsilon_m\\ &\le C(\mathcal{H}^{d-1}(V_l)+o(1)_{\varepsilon\to0})\sum_{m=1}^{+\infty} c^\varepsilon_m. \end{aligned}\quad (4.43)$$

We estimate the second term in (4.42) as

$$\begin{aligned}\sum_{i\notin\bigcup_{l\in\mathcal{I}}\mathcal{J}_{l,\varepsilon}} \varepsilon^{d-1}\phi_i^\varepsilon(\{v^\varepsilon_{i+j}\}_j) &\le \sum_{l\neq l'}\sum_{i\in C^n_{l,l',\varepsilon}} \varepsilon^{d-1}\phi_i^\varepsilon(\{v^\varepsilon_{i+j}\}_j)\\ &\quad+\sum_{l\neq l'}\sum_{i\in\mathbb{Z}^d\setminus C^n_{l,l',\varepsilon}} \varepsilon^{d-1}c^\varepsilon_{\overline{m}_\varepsilon(i)}, \end{aligned}\quad (4.44)$$

where

$$C^n_{l,l',\varepsilon} = \{i\in\mathbb{Z}^d : \operatorname{dist}_\infty(\varepsilon i, V_l\cap V_{l'}) \le \varepsilon n\}.$$

For the first sum we then use that $\phi_i^\varepsilon \leq L$ to get that

$$
\begin{aligned}
\sum_{l\neq l'}\sum_{i\in C^n_{l,l',\varepsilon}} \varepsilon^{d-1}\phi_i^\varepsilon(\{v^\varepsilon_{i+j}\}_j) &\leq \sum_{l\neq l'} CL\varepsilon^{d-1}\#C^n_{l,l',\varepsilon}\\
&\leq \sum_{l\neq l'} \varepsilon n^2 CL(\mathcal{H}^{d-2}(V_l\cap V_{l'})+o(1)_{\varepsilon n\to 0})\\
&\leq C\varepsilon n^2. \qquad (4.45)
\end{aligned}
$$

As for the second sum in (4.44), for all $i\in\mathbb{Z}^d$ we define

$$
R_\varepsilon(i)=\sup\big\{R\colon\ Q_{\varepsilon R}(\varepsilon i)\cap S(u)=\emptyset\big\}. \qquad (4.46)
$$

We note that, since $\bigcup_{l\in I}\mathcal{J}_{l,\varepsilon}=\varepsilon\mathbb{Z}^d\cap\bigcup_{l\in I}S_l$, for all $i\notin\bigcup_{l\in I}\mathcal{J}_{l,\varepsilon}$, we have that

$$
R_\varepsilon(i)=\mathrm{dist}_\infty(\varepsilon i,\partial V_l\cap\partial V_{l'})
$$

for some $l,l'\in I$, $l\neq l'$. We set

$$
I_{l,l',\varepsilon}=\Big\{i\notin\bigcup_{l\in I}\mathcal{J}_{l,\varepsilon}\colon\ R_\varepsilon(i)=\mathrm{dist}_\infty(\varepsilon i,V_l\cap V_{l'})\Big\}. \qquad (4.47)
$$

Note that for indices i in this set we have that $\lfloor R_\varepsilon(i)\rfloor\leq\overline{m}_\varepsilon(i)\leq\lfloor R_\varepsilon(i)\rfloor+1$ so that

$$
\begin{aligned}
\#\{i\in I_{l,l',\varepsilon}\colon\ \overline{m}_\varepsilon(i)=m\} &\leq C\frac{1}{\varepsilon^{d-1}}\mathcal{H}^{d-1}\{x\colon\ \mathrm{dist}(x,V_l\cap V_{l'})=\varepsilon(m+1)\}\\
&\leq \frac{1}{\varepsilon^{d-1}}C(\mathrm{diam}(V_l\cap V_{l'})+\varepsilon(m+1))^{d-1}\\
&\leq \frac{1}{\varepsilon^{d-1}}C(\mathrm{diam}(\{u=v^1\})+\varepsilon(m+1))^{d-1} \qquad (4.48)
\end{aligned}
$$

and

$$
\begin{aligned}
\sum_{i\in\mathbb{Z}^d\setminus C^n_{l,l',\varepsilon}} \varepsilon^{d-1}c_{\overline{m}_\varepsilon(i)} &\leq C\sum_{m\geq n}(\mathrm{diam}(\{u=v^1\})+\varepsilon(m+1))^{d-1}c^\varepsilon_m\\
&\leq C\Big(\frac{\mathrm{diam}\,\{u=v^1\}}{n}\Big)^{d-1}\sum_{m\geq n}m^{d-1}c^\varepsilon_m. \qquad (4.49)
\end{aligned}
$$

From these estimates we deduce that

$$
E_\varepsilon(v_\varepsilon)\leq C\sum_{l\in I}\mathcal{H}^{d-1}(V_l)+C\varepsilon n^2+C\sum_{m\geq n}m^{d-1}c^\varepsilon_m. \qquad (4.50)
$$

Taking the lim sup as $\varepsilon\to 0$, by (4.39) and the arbitrariness of n we obtain the desired estimate (4.37). □

We can now give the proof of the main compactness theorem.

Proof of Theorem 4.27 As a consequence of Proposition 4.29 and Proposition 4.30, under assumptions (H1) and (H3) the set function $\gamma_u(\cdot)$ defined by (4.11) for every $u \in BV(\Omega; Y_0)$ satisfies all the hypotheses of Theorem 4.14, from which (4.26) follows. The boundedness of φ is a consequence of Proposition 4.30. If (H2) is also satisfied, then, by Theorem 4.22, we have that if u^ε converge to u in $BV_{\rm loc}(\Omega; Y)$ and $E_\varepsilon(u^\varepsilon; \Omega)$ is uniformly bounded, then $u \in BV(\Omega; Y)$. By (H1) we obtain that in fact $u \in BV(\Omega; Y_0)$. Indeed, we only need to observe that, setting

$$I_\varepsilon = \{i : \varepsilon i \in \mathcal{L}_\varepsilon(\Omega); u_i^\varepsilon \notin Y_0\},$$

we have

$$\begin{aligned} |\{x \in \Omega' : u^\varepsilon(x) \notin Y_0\}| &\leq C\varepsilon^d \# I_\varepsilon \leq C\varepsilon^d \sum_{i \in I_\varepsilon} \sum_{j \in Q_M(i) \cap \mathcal{L}} \phi_j^\varepsilon(\{u_{j+k}^\varepsilon\}_k) \\ &\leq C\Big(\frac{M}{r}\Big)^d \varepsilon E_\varepsilon(u^\varepsilon) \leq \varepsilon C \end{aligned} \tag{4.51}$$

for all fixed $\Omega' \subset\subset \Omega$, where $r > 0$ is given by Definition 4.17. Letting ε tend to 0, we deduce that the limit u satisfies $u(x) \in Y_0$ almost everywhere by the arbitrariness of $\Omega' \subset\subset \Omega$. Finally, note that from (H2) we deduce that, for every $u \in BV(\Omega; Y_0)$,

$$E_\varepsilon(u; U) \geq C\mathcal{H}^{d-1}(S(u) \cap U);$$

hence, the same estimate holds also for $F_0(u; U)$. This allows us to deduce (4.27) as desired. □

4.7 Convergence of Minimum Problems

As a consequence of the Γ-convergence result in Theorem 4.27, in this section we derive the convergence of minimum problems involving the energies E_ε defined by (4.25) when boundary conditions or volume fractions of two minimal phases are prescribed.

4.7.1 Boundary-Value Problems

In order to treat minimum problems with Dirichlet-type boundary conditions for E_ε, it will be convenient to impose the boundary condition in a neighborhood of the boundary. To this end in what follows, for $U \subset \mathbb{R}^d$ a bounded Lipschitz set and $\delta > 0$ U_δ will denote the δ-interior of U defined by

$$U_\delta = \{x \in U : \operatorname{dist}(x, \mathbb{R}^d \setminus U) > \delta\}. \tag{4.52}$$

Proposition 4.31 (Convergence of boundary-value problems) *Let $U \subset \mathbb{R}^d$ be a bounded Lipschitz set and let $u_0 \in BV_{\rm loc}(\mathbb{R}^d; Y_0)$. Suppose that E_ε, defined by* (4.25), *satisfy* (H1)–(H3) *and that (upon extracting a subsequence) there exists*

$$F_0(u; U) = \Gamma\text{-}\lim_{\varepsilon\to 0} E_\varepsilon(u; U).$$

Let $u^{0,\varepsilon}: \Omega \cap \varepsilon\mathbb{Z}^d \to Y$ be such that $u^{0,\varepsilon} \to u^0$ and that

$$\inf_\delta \limsup_{\varepsilon\to 0} E_\varepsilon(u^{0,\varepsilon}; U \setminus \overline{U}_{2\delta}) = 0. \tag{4.53}$$

Then, setting

$$m_{\varepsilon,\delta} = \min\{E_\varepsilon(u; U) \colon \ u = u^{0,\varepsilon} \text{ on } \varepsilon\mathbb{Z}^d \setminus U_\delta\},$$
$$m_{0,\delta} = \min\{F_0(u; U) \colon \ u = u^0 \text{ on } \mathbb{R}^d \setminus U_\delta\},$$

we have

$$\inf_\delta m_{0,\delta} = \inf_\delta \liminf_{\varepsilon\to 0} m_{\varepsilon,\delta} = \inf_\delta \limsup_{\varepsilon\to 0} m_{\varepsilon,\delta}. \tag{4.54}$$

Proof Let $u^{\delta,\varepsilon}$ be a minimizer for $m_{\varepsilon,\delta}$. By Theorem 4.22 we have that (upon extracting a subsequence) $u^{\delta,\varepsilon} \to u^\delta$ with $u^\delta = u^0$ on $\mathbb{R}^d \setminus U_\delta$. Using the lower-semicontinuity inequality, we have

$$m_{0,\delta} \le F_0(u^\delta; \Omega) \le \liminf_{\varepsilon\to 0} E_\varepsilon(u^{\delta,\varepsilon}; \Omega) = \liminf_{\varepsilon\to 0} m_{\varepsilon,\delta}, \tag{4.55}$$

by which, taking the infimum over δ, one obtains the second inequality in (4.54).

Fix u a minimizer in $m_{0,\delta}$ and let u^ε be a recovery sequence for F_0 at u. By the properties in Definition 4.25 (which holds by Proposition 4.29) with $U' = U_{2\delta}$, $U = U_\delta$, $V = U$, $u^{U,\varepsilon} = u^\varepsilon$, and $u^{V,\varepsilon} = u^{0,\varepsilon}$, for fixed $\eta > 0$ there exists v^ε such that $v^\varepsilon = u^\varepsilon$ on $\varepsilon\mathbb{Z}^d \cap U_{2\delta}$, $v^\varepsilon = u^{0,\varepsilon}$ on $\varepsilon\mathbb{Z}^d \setminus U_\delta$, for $V' = V = U$,

$$\begin{aligned} E_\varepsilon(v^\varepsilon; U) &\le E_\varepsilon(u^\varepsilon; U) + E_\varepsilon(u^{0,\varepsilon}; U \setminus \overline{U}_{2\delta}) \\ &\quad + C(\eta)\|u^\varepsilon - u^{0,\varepsilon}\|_{L^1(U_\delta\setminus U_{2\delta})} + r(\varepsilon,\eta). \end{aligned} \tag{4.56}$$

As a consequence,

$$\begin{aligned} \limsup_{\varepsilon\to 0} E_\varepsilon(v^\varepsilon; U) &\le m_{0,\delta} + \limsup_{\varepsilon\to 0} E_\varepsilon(u^{0,\varepsilon}; U \setminus \overline{U}_{2\delta}) \\ &\quad + C(\eta)\limsup_{\varepsilon\to 0} \|u^\varepsilon - u^{0,\varepsilon}\|_{L^1(U_\delta\setminus U_{2\delta})} + o(1)_{\eta\to 0}. \end{aligned}$$

Noting that the applications

$$\delta \mapsto \limsup_{\varepsilon\to 0} \|u^\varepsilon - u^{0,\varepsilon}\|_{L^1(U_\delta\setminus U_{2\delta})}, \qquad \delta \mapsto \limsup_{\varepsilon\to 0} E_\varepsilon(u^{0,\varepsilon}; U \setminus \overline{U}_{2\delta})$$

are increasing, taking the infimum over δ and using (4.53), we finally obtain the first inequality in (4.54). □

4.7.2 Measure Constraints: the Wulff Problem

We now study the convergence of minimum problems for E_ε when the volume fraction of two minimal phases is prescribed.

We first consider the case of a bounded domain. Let v^l, $v^{l'} \in Y_0$, $p \in (0,1)$, $\delta > 0$ and consider the following minimum problems

$$m^\delta_{\varepsilon,p} = \min\Big\{E_\varepsilon(u;\Omega)\colon \big|\varepsilon^d\#\{i \in \Omega\cap\varepsilon\mathbb{Z}^d\colon\ u_i = v^l\} - p|\Omega|\big| \le \delta,$$
$$\big|\varepsilon^d\#\{i \in \Omega\cap\varepsilon\mathbb{Z}^d\colon\ u_i = v^{l'}\} - (1-p)|\Omega|\big| \le \delta\Big\}. \tag{4.57}$$

Proposition 4.32 *Let $\Omega \subset \mathbb{R}^d$ be a bounded Lipschitz set. Suppose that E_ε, defined by* (4.25), *satisfy* (H1)–(H3) *and that (upon extracting a subsequence) there exists*

$$F_0(u;\Omega) = \Gamma\text{-}\lim_{\varepsilon\to 0} E_\varepsilon(u;\Omega).$$

Then

$$\lim_{\delta\to 0}\liminf_{\varepsilon\to 0} m^\delta_{\varepsilon,p} = \lim_{\delta\to 0}\limsup_{\varepsilon\to 0} m^\delta_{\varepsilon,p} = m_p, \tag{4.58}$$

where $m_p = \min\{F_0(u;\Omega)\colon\ u \in BV(\Omega;\{v^l, v^{l'}\}),\ |\{u = v^l\}| = p|\Omega|\}$.

Proof Set

$$m^\delta_p = \min\Big\{F_0(u;\Omega)\colon\ u \in BV(\Omega;Y_0\}),\ \big||\{u = v^l\}| - p|\Omega|\big| \le \delta,$$
$$\big||\{u = v^{l'}\}| - (1-p)|\Omega|\big| \le \delta\Big\}.$$

The lower semicontinuity of F_0, (4.27), and Theorem 2.18 yield that

$$\liminf_{\delta\to 0} m^\delta_p = m_p. \tag{4.59}$$

Let $u^{\varepsilon,\delta}$ be a minimizer of $m^\delta_{\varepsilon,p}$. By Theorem 4.22, we have that (upon extracting a subsequence) $u^{\varepsilon,\delta} \to u \in BV(\Omega;Y_0)$, with $\big||\{u = v^l\}| - p|\Omega|\big| \le \delta$ and $\big||\{u = v^{l'}\}| - (1-p)|\Omega|\big| \le \delta$. Using the lower-semicontinuity inequality we have

$$m^\delta_p \le F_0(u;\Omega) \le \liminf_{\varepsilon\to 0} E_\varepsilon(u^{\varepsilon,\delta};\Omega) = \liminf_{\varepsilon\to 0} m^\delta_{\varepsilon,p},$$

which, by (4.59), implies that

$$m_p \le \liminf_{\delta\to 0}\liminf_{\varepsilon\to 0} m^\delta_{\varepsilon,p}.$$

Fix $u \in BV(\Omega; \{v^l, v^{l'}\})$ a minimizer of m_p and let $\{u^\varepsilon\}$ be a recovery sequence for F_0 at u. Note that, with fixed $\delta > 0$, u^ε is a test function for $m^\delta_{\varepsilon,p}$ for ε small enough. Hence,

$$\limsup_{\varepsilon\to 0} m^\delta_{\varepsilon,p} \leq \limsup_{\varepsilon\to 0} E_\varepsilon(u^\varepsilon; \Omega) = F_0(u; \Omega) = m_p.$$

The conclusion follows letting $\delta \to 0$. □

In the case of a homogeneous energy F_0, problems with a prescribed volume constraint in the whole space $\mathbb{R}^d$ carry all the necessary information to describe the energy density φ, as specified by the following remark.

Remark 4.33 (Wulff shape) Assume that F_0, defined in (4.26), is homogeneous; that is, that its surface density φ does not depend on x, and set $\overline{\varphi}(\nu) = \varphi(v^l, v^{l'}, \nu)$. Then the minimizers of

$$\min\{F_0(u; \mathbb{R}^d) \colon\ u \in BV_{\rm loc}(\mathbb{R}^d; \{v^l, v^{l'}\}),\ |\{u = v^l\}| = 1\}$$

are of the form $v^l \chi_U + v^{l'}(1 - \chi_U)$, where U agrees, up to translations and up to a set of measure zero, with $\gamma W_{\overline{\varphi}}$, where $W_{\overline{\varphi}}$ is the Wulff set

$$W_{\overline{\varphi}} = \{x \in \mathbb{R}^d : \langle x, \nu\rangle \leq \overline{\varphi}(\nu) \text{ for all } \nu \in S^{d-1}\}$$

and γ is the normalization constant such that $|\gamma W_{\overline{\varphi}}| = 1$.

4.8 Homogenization

We consider now energies E_ε of the form (4.25) in the particular case where, upon scaling, we can reduce to energy densities independent of ε and defined on the whole $\mathcal{L}$. For simplicity, we fix $\mathcal{L} = \mathbb{Z}^d$, so that

$$\mathcal{L}_\varepsilon(\Omega) = \Omega \cap \varepsilon\mathbb{Z}^d \text{ and } \mathcal{L}^i_\varepsilon(\Omega) = (\Omega \cap \varepsilon\mathbb{Z}^d) - \varepsilon i.$$

Since in our general framework energy densities ϕ^ε_i have domains depending on i, we have to take some care in order to state the required equivalence with energies independent of ε. To that end, we make the following assumptions.

(i) There exist constants t^ε_m such that

$$\limsup_{R\to+\infty} \limsup_{\varepsilon\to 0} \sum_{m\geq R} t^\varepsilon_m = 0 \tag{4.60}$$

and that for all $i \in \mathbb{Z}^d$, there exists a function $\phi_i \colon \{z \colon \mathbb{Z}^d \to Y\} \to [0, +\infty)$ such that

$$\left|\phi^\varepsilon_i(\{z_{i+j}\}_j) - \phi_i(\{\overline{z}_{i+\ell}\}_\ell)\right| \leq t^\varepsilon_{m_\varepsilon(i)} \tag{4.61}$$

for all functions $\overline{z}\colon \mathbb{Z}^d \to Y$ such that $\overline{z}_k = z_k = z(\varepsilon k)$ for all k such that $\varepsilon k \in \Omega \cap \varepsilon\mathbb{Z}^d$, where

$$m_\varepsilon(i) = \sup\{m \in \mathbb{N}\colon Q_{\varepsilon m}(\varepsilon i) \subset \Omega\}. \tag{4.62}$$

Note that in (4.61) we use a shortened notation, as in Section 4.4, in the place of $\phi_i^\varepsilon(\{z_{i+j}\}_{\varepsilon j \in \mathcal{L}_\varepsilon^i(\Omega)})$ and $\phi_i(\{\overline{z}_{i+\ell}\}_{\ell \in \mathbb{Z}^d})$, respectively.

(ii) If we set for any $\varepsilon > 0$ and $i \in \mathbb{Z}^d$,

$$\overline{\phi}_i^\varepsilon = \phi_i, \tag{4.63}$$

then $\overline{\phi}_i^\varepsilon$ satisfy the hypotheses of Theorem 4.27 with $\Omega = \mathbb{R}^d$.

Note that in the case $\Omega = \mathbb{R}^d$ these hypotheses amount to requiring that ϕ_i^ε can be written in terms of ε-independent energy densities as in (4.63). For $\Omega \neq \mathbb{R}^d$ and for long-range ferromagnetic interactions, for all i we simply have

$$\phi_i(\{z_\ell\}_{\ell\in\mathbb{Z}^d}) = \sum_{\ell\in\mathbb{Z}^d} a_{i\ell}(z_\ell - z_0)^2 \text{ and } \phi_i^\varepsilon(\{z_j\}_{\varepsilon j\in\mathcal{L}_\varepsilon^i(\Omega)}) = \sum_{\substack{j\in\mathbb{Z}^d\\ \varepsilon(i+j)\in\Omega}} a_{ij}(z_j - z_0)^2,$$

and (4.61) is the decay assumption on the system of the coefficients corresponding to (H3).

Proposition 4.34 (Homogenization formulas) *Suppose that ϕ_i^ε are as in Theorem 4.27. Suppose furthermore that ϕ_i satisfy* (4.63), (4.61) *and* (4.60), *and for all $\nu \in S^{d-1}$, $l, l' \in \{1, \dots, K\}$ and for all sequences $\{z_T\}_T \subset \mathbb{Z}^d$ with $\sup_T \frac{\|z_T\|}{T} < +\infty$, we have*

$$\inf_{\delta>0} \liminf_{T\to+\infty} \frac{1}{T^{d-1}} m_\delta^{l,l'}(T,\nu) = \inf_{\delta>0} \limsup_{T\to+\infty} \frac{1}{T^{d-1}} m_\delta^{l,l'}(T,\nu), \tag{4.64}$$

where

$$m_\delta^{l,l'}(T,\nu) = \inf\Big\{\Phi(u; Q_T^\nu(z_T)),\ u = u_{z_T}^{l,l',\nu} \text{ on } \mathbb{Z}^d \setminus Q_{(1-\delta)T}^\nu(z_T)\Big\}, \tag{4.65}$$

$$\Phi(u;U) = \sum_{i\in U\cap\mathbb{Z}^d} \phi_i(\{u_{i+j}\}_j), \tag{4.66}$$

and for $z \in \mathbb{R}^d$,

$$u_z^{l,l',\nu}(x) = \begin{cases} v^l & \text{if } \langle x - z, \nu\rangle \geq 0, \\ v^{l'} & \text{otherwise.} \end{cases} \tag{4.67}$$

Let $\varphi_{\rm hom}(v^l, v^{l'}, \nu)$ denote the common value in (4.64), *and suppose that it is independent of the sequence $\{z_T\}_T$. Then the function φ in Theorem 4.27 is* homogeneous; *that is, it is independent of x, we have that $\varphi(v^l, v^{l'}, \nu) =$*

$\varphi_{\hom}(v^l, v^{l'}, \nu)$, in particular φ does not depend on the subsequence, and the whole family E_ε Γ-converges.

Proof Let $\{\varepsilon_j\}$ be a sequence such that the claim of Theorem 4.27 holds, with limit energy F_0. By the derivation formulas in Theorem 4.14 we can take as the integrand of F_0 the function φ given by

$$\varphi(x_0, v^l, v^{l'}, \nu) = \lim_{\varrho\to 0}\inf_{\delta>0} \frac{1}{\varrho^{d-1}} \min\Big\{F_0(u; Q^\nu_\varrho(x_0)) : \\ u = u^{l,l',\nu}_{x_0} \text{ on } \mathbb{R}^d \setminus Q^\nu_{(1-\delta)\varrho}(x_0)\Big\}$$

for $x_0 \in \Omega$, $\nu \in S^{d-1}$ and $l, l' \in \{1, \dots, K\}$. We can suppose that in this formula the limit as $\varrho \to 0$ exists, up to passing to a subsequence.

For $i \in \mathbb{Z}^d$ we set $(u^{l,l',\nu}_z)_i = u^{l,l',\nu}_z(i)$. By the convergence of minima in Proposition 4.31 with $U = Q^\nu_\varrho(x_0)$, we then have

$$\varphi(x_0, v^l, v^{l'}, \nu) = \lim_{\varrho\to 0}\inf_{\delta>0}\liminf_{j\to+\infty} \frac{1}{\varrho^{d-1}} \inf\Big\{E_{\varepsilon_j}(u; Q^\nu_\varrho(x_j)) \colon u_i = (u^{l,l',\nu}_{x_j})_i \\ \text{if } \varepsilon_j i \in \varepsilon_j\mathbb{Z}^d \setminus Q^\nu_{(1-\delta)\varrho}(x_j)\Big\},$$

where $x_j \in \varepsilon_j\mathbb{Z}^d$ tends to x_0.

Let

$$\Phi_\varepsilon(u; U) = \varepsilon^{d-1} \sum_{i\in U\cap\varepsilon\mathbb{Z}^d} \phi_i(\{u_{i+j}\}_j).$$

If $\Omega = \mathbb{R}^d$, then $E_\varepsilon = \Phi_\varepsilon$. If Ω is not the whole $\mathbb{R}^d$, note that $m_\varepsilon(i) \geq R_j$ for any $\varepsilon_j i \in Q^\nu_\varrho(x_j) \cap \varepsilon_j\mathbb{Z}^d$, where $m_\varepsilon(i)$ is defined by (4.62) and R_j denotes the maximal integer such that $Q^\nu_{\varrho+\varepsilon R_j}(x_j) \subset \Omega$. Then by (4.61) we have

$$|E_{\varepsilon_j}(u; Q^\nu_\varrho(x_j)) - \Phi_{\varepsilon_j}(u; Q^\nu_\varrho(x_j))| \leq \varepsilon_j^{d-1} \sum_{m\geq R_j} t^{\varepsilon_j}_m \#\{i : m_{\varepsilon_j}(i) = m\} \\ \leq C_\Omega \sum_{m\geq R_j} t^{\varepsilon_j}_m$$

for some constant C_Ω. This last sum tends to 0 as $j \to +\infty$ by (4.60), since $R_j = O(\frac{1}{\varepsilon_j})$. In any case, we then have

$$\varphi(x_0, v^l, v^{l'}, \nu) = \lim_{\varrho\to 0}\inf_{\delta>0}\liminf_{j\to+\infty} \frac{1}{T_j^{d-1}} \inf\Big\{\Phi(u; Q^\nu_{T_j}(z_j)) \colon \\ u = u^{l,l',\nu}_{z_j} \text{ on } \mathbb{Z}^d \setminus Q^\nu_{(1-\delta)T_j}(z_j)\Big\},$$

where $T_j = \frac{\varrho}{\varepsilon_j}$ and $z_j = z_{T_j} = \frac{x_j}{\varepsilon}$. The claim is proved by (4.64). □

The previous proposition shows that the proof of a homogenization result can be decomposed into two steps:

(i) the translation invariance of the energy density φ;
(ii) the existence of the limits in homogenization formula (4.64) with $z_T = 0$.

Both properties are usually derived from some periodicity assumptions on the energy densities ϕ_k. Here we give a general criterion that may be verified separately under various hypotheses. Its proof is carried out by a usual subadditivity argument in homogenization theory, already introduced in the blow-up method of Section 3.3.1, for the use of which we have to estimate minimum problems on large cubes with minimum problems on smaller cubes, up to a negligible error. This can be done if a test function in a larger cube can be obtained by a patchwork procedure using minimal functions on translated smaller cubes, and we can estimate the energy of these translations with that of a single cube. The latter is precisely the assumption of the following proposition.

Proposition 4.35 (A homogenization criterion) *Suppose that the hypotheses of Theorem 4.27 are satisfied with $\Omega = \mathbb{R}^d$ and that ϕ_i^ε is independent of ε. Let $l, l' \in \{1, \dots, K\}$ and ν be fixed. For all $z \in \mathbb{R}^d$, $T > 0$, and $\delta > 0$, we set*

$$m(T,\delta,z) = \inf\Big\{\Phi(u; Q_T^\nu(z)) \colon\ u = u_z^{l,l',\nu} \text{ on } \mathbb{Z}^d \setminus Q^\nu_{(1-\delta)T}(z)\Big\}, \tag{4.68}$$

where Φ is defined in (4.66) and $u_z^{l,l',\nu}$ in (4.67).

Suppose that for all $\eta > 0$ there exist $L_\eta > 0$ and a subset $\mathcal{T}_\eta$ of $\mathbb{Z}^d$ such that

(i) $\mathcal{T}_\eta + [0, L_\eta]^d = \mathbb{R}^d$;
(ii) *for all $z \in \mathbb{Z}^d$ and $\tau \in \mathcal{T}_\eta$,*

$$m(T,\delta,z+\tau) \le m(T,\delta,z) + \omega(\eta)T^{d-1} + C(\eta,T), \tag{4.69}$$

where $\lim\limits_{\eta\to 0}\omega(\eta) = 0$ and $\lim\limits_{T\to+\infty}\frac{C(\eta,T)}{T^{d-1}} = 0$.

Then, equality (4.64) holds for all sequences $\{z_T\}_T \subset \mathbb{Z}^d$, with the common value of the limits independent of $\{z_T\}_T$.

Proof We use (4.69) to estimate $m(S,\delta,z_S)$ in terms of $m(T,\delta,z_T)$ for $S \gg T$. Up to a translation, we may suppose $z_S = 0$, and we then simply write $u^\nu = u^{0,\nu}$.

Let $\{\nu_1^\perp, \dots, \nu_{d-1}^\perp\}$ be an orthonormal basis of $\Pi^\nu = \{y \colon \langle y, \nu\rangle = 0\}$. For all indices $\hat{\imath} = (i_1, \dots, i_{d-1}) \in (T + L_\eta)\mathbb{Z}^{d-1}$, we set $z^{\hat{\imath}} = \sum_{k=1}^{n-1} i_k \nu_k^\perp$ and

$$Z_\delta(T,S) = \{\hat{\imath} \in (T+L_\eta)\mathbb{Z}^{d-1} \colon\ \lfloor z^{\hat{\imath}}\rfloor + Q_T^\nu \subseteq Q^\nu_{(1-\delta)S}\}.$$

For each $i \in (T + L_\eta)\mathbb{Z}^{d-1}$, let $\tau^{\hat{\imath}} \in \mathcal{T}_\eta \cap (\lfloor z^{\hat{\imath}}\rfloor - z_T + [0, L_\eta]^d)$, and a function $u^{\hat{\imath}}$ minimizing $m(T,\delta,z_T + \tau^{\hat{\imath}})$. Note that all cubes $Q_T^\nu(z_T + \tau^{\hat{\imath}})$ are disjoint.

We now define u^S as

$$u^S_j = \begin{cases} u^{\hat{i}}_j & \text{if } j \in Q^\nu_T(z_T + \tau^{\hat{i}}) \text{ and } \hat{i} \in Z_\delta(T,S), \\ u^\nu_i & \text{otherwise.} \end{cases}$$

In order to estimate $\Phi(u^S; Q^\nu_S)$ we fix $\beta \in \mathbb{N}$, for which we may suppose $\delta T \gg \beta$, and we rewrite it as the sum of three terms,

$$\Phi(u^S; Q^\nu_S) = \Sigma_1 + \Sigma_2 + \Sigma_3,$$

where

$$\Sigma_1 = \sum_{\hat{i} \in Z_\delta(T,S)} \Phi(u^S; Q^\nu_{T-\beta\sqrt{d}}(z_T + \tau^{\hat{i}})),$$
$$\Sigma_2 = \sum_{i \in Q^\nu_S \cap P_{\delta,\beta}(T,S)} \phi_i(\{u^S_{i+j}\}_j),$$
$$\Sigma_3 = \sum_{i \in Q^\nu_S \setminus P_{\delta,\beta}(T,S)} \phi_i(\{u^S_{i+j}\}_j),$$

and we have set

$$P_{\delta,\beta}(T,S) = \Big\{ i \in \mathbb{Z}^n \setminus \bigcup_{\hat{i} \in Z_\delta(T,S)} Q^\nu_{T-\beta\sqrt{d}}(z_T + \tau^{\hat{i}}) \colon \operatorname{dist}(i, \Pi^\nu) \le \beta \Big\}.$$

This subdivision is carried out in order to separate interactions in the interior of the cubes; that is, on points i such that the cube $Q_\beta(i)$ is still in the interior, in those close to the interface between v^l and $v^{l'}$ but outside those cubes, and finally on all other points i. For those we will take into account that in $Q_{\beta-L_\eta}(i)$ the function u^S is constant. The extra factor $\sqrt{d}$ just takes into account that the cubes Q^ν are not oriented in the coordinate directions.

We first estimate Σ_1. For all $\hat{i} \in Z_\delta(T,S)$ we have that

$$\begin{aligned} \Phi(u^S; Q^\nu_{T-\beta\sqrt{d}}(z_T + \tau^{\hat{i}})) &\le \Phi(u^{\hat{i}}; Q^\nu_{T-\beta\sqrt{d}}(z_T + \tau^{\hat{i}})) + CT^{d-1} \sum_{\alpha \ge \beta} c_\alpha \\ &\le \Phi(u^{\hat{i}}; Q^\nu_T(z_T + \tau^{\hat{i}})) + CT^{d-1} \sum_{\alpha \ge \beta} c_\alpha \\ &\le m(T,\delta,z) + \omega(\eta) T^{d-1} \\ &\quad + C(\eta,T) + T^{d-1} o(1)_{\beta \to +\infty}, \end{aligned}$$

where in the first inequality we have used the energy estimate in (H3), observing that for all $\hat{i} \in Z_\delta(T,S)$ the functions u^S and $u^{\hat{i}}$ agree on the cubes $Q^\nu_T(z_T + \tau^{\hat{i}})$; in the second inequality we have used the nonnegativity of ϕ_i; and in the third we have taken into account (4.69) and the summability property of c_α. Hence, noting that $\#Z_\delta(T,S) \le \frac{S^{d-1}}{T^{d-1}}$,

$$\Sigma_1 \le S^{d-1}\frac{m(T,\delta,z)}{T^{d-1}} + S^{d-1}\omega(\eta) + S^{d-1}\frac{C(\eta,T)}{T^{d-1}} + S^{d-1}o(1)_{\beta\to+\infty}. \tag{4.70}$$

Since $\phi_i \le L$, we estimate Σ_2 as

$$\Sigma_2 \le CL\,\beta\delta S^{d-1} + CL(\beta T^{d-2})\frac{S^{d-1}}{T^{d-1}}, \tag{4.71}$$

where we have separately taken into account the contribution in $Q^\nu_S \setminus Q^\nu_{(1-\delta)S}$ and on $Q^\nu_{(1-\delta)S}$. In order to estimate Σ_3 we observe that for all $i \in Q^\nu_S \setminus P_{\delta,\beta}(T,S)$ there exists $\alpha(i) \ge \beta - L_\eta$ such that the function u^S agrees with either v^l or $v^{l'}$ on all cubes $Q_{\alpha(i)}(i)$ and that $\#\{i \in Q^\nu_S \setminus P_{\delta,\beta}(T,S)\colon \alpha(i) = \alpha\} \le CS^{d-1}$. Therefore we have that

$$\Sigma_3 \le CS^{d-1}\sum_{\alpha\ge\beta-L_\eta} c_\alpha. \tag{4.72}$$

Summing up (4.70)–(4.72) we have

$$\frac{m(S,\delta,z_S)}{S^{d-1}} \le \frac{m(T,\delta,z)}{T^{d-1}} + \omega(\eta) + \frac{C(\eta,T)}{T^{d-1}} + C\beta\Big(\delta + \frac{1}{T}\Big) + r(\beta - L_\eta),$$

where $r(t)$ is infinitesimal as $t \to +\infty$. By taking first the limsup as $S \to +\infty$ and then the liminf as $T \to +\infty$, and using the hypothesis on $C(\eta,T)$, we then have

$$\limsup_{S\to+\infty}\frac{m(S,\delta,z_S)}{S^{d-1}} \le \liminf_{T\to+\infty}\frac{m(T,\delta,z)}{T^{d-1}} + \omega(\eta) + C\beta\,\delta + r(\beta - L_\eta).$$

We can take the limit as $\delta \to 0$, or equivalently the infimum, since the minimum β is increasing in δ, to deduce that

$$\lim_{\delta\to0}\limsup_{S\to+\infty}\frac{m(S,\delta,z_S)}{S^{d-1}} \le \lim_{\delta\to0}\liminf_{T\to+\infty}\frac{m(T,\delta,z)}{T^{d-1}} + \omega(\eta) + r(\beta - L_\eta).$$

We may finally let $\beta \to +\infty$ and subsequently $\eta \to 0$ to obtain the claim. □

Remark 4.36 (A more general statistical criterion) From the proof of the previous proposition, we note that the same argument can be followed word for word if for fixed $\eta > 0$ for all $S \gg T$, given z_T and z_S we can find a subset of indices $Z_\delta(T,S)$ and disjoint cubes $Q^\nu_T(w^{\hat{i}})$ with $\mathrm{dist}(w^{\hat{i}},\Pi^\nu)$ equibounded, $(Q^\nu_T(w^{\hat{i}}) \cap \Pi^\nu) \subset Q^\nu_{(1-\delta)S}(z^S)$, such that estimate (4.69) must hold for such cubes; that is,

$$m(T,\delta,z^{\hat{i}}) \le m(T,\delta,z_T) + \omega(\eta)T^{d-1} + C(\eta,T) \tag{4.73}$$

with the same assumptions on ω and $C(\eta, T)$, and

$$\limsup_{T\to+\infty}\limsup_{S\to+\infty}\frac{(1-\delta)^{d-1}S^{d-1}-T^{d-1}\#Z_\delta(T,S)}{S^{d-1}}=0;$$

that is, the percentage of $\Pi^\nu \cap Q^\nu_{(1-\delta)S}$ not covered by the union of those cubes is negligible on large cubes when T is also large. This property can be seen as a statistical assumption on the average distance between the cubes $Q^\nu_T(w^{\hat{i}})$. In the almost-periodic case this is satisfied since neighboring cubes are at an average distance at most $L_\eta\sqrt{d}$, so that

$$\#(Z_\delta(T,S)) \sim (1-\delta)^{d-1}\left(\frac{S}{T+L_\eta\sqrt{d}}\right)^{d-1}.$$

Finally, note that by Proposition 4.34 this assumption needs to be satisfied only for sequences $\{z_T\}_T$ with $\frac{\|z_T\|}{T}$ bounded.

4.9 The Pairwise Ferromagnetic Case

In this section we specialize the results of the previous part of this chapter to pairwise ferromagnetic interactions defined on a portion of the lattice $\mathcal{L}$ as

$$E_\varepsilon(u)=\sum_{\varepsilon i,\varepsilon j\in\mathcal{L}_\varepsilon(\Omega)}\varepsilon^{d-1}a^\varepsilon_{ij}(u_i-u_j)^2. \tag{4.74}$$

Comparing with Chapter 3, the interaction coefficients in (4.74) are allowed to depend on ε. Note, moreover, that the application of the Compactness Theorem also allows us to remove the finite-range hypothesis from the homogenization results in that chapter.

We consider the following assumptions, which will allow us to apply the Compactness Theorem 4.27.

(H$_f$1) (*Ferromagnetic energies*) we suppose that $a^\varepsilon_{ij}\geq 0$. Moreover, we assume that the nonrestrictive symmetry condition $a^\varepsilon_{ij}=a^\varepsilon_{ji}$ holds.

This hypothesis ensures that the constant states -1 and 1 are minima for the energy.

(H$_f$2) (*Uniform coerciveness*) there exists $M>0$, $c>0$ such that for all $\varepsilon i,\varepsilon j\in\mathcal{L}_\varepsilon(\Omega)$ with $j\in NN(i)$ there exist $j_0,\dots,j_N$ with $\varepsilon j_k\in\mathcal{L}_\varepsilon(\Omega)$ such that $j_0=i$, $j_N=j$ and $\max_{1\leq k\leq N}\|j_k-i\|\leq M$ such that

$$a^\varepsilon_{j_kj_{k-1}}\geq c>0.$$

This condition implies that the only two ground states are the constant states, and the domain of the Γ-limit is included in the family of sets of finite perimeter, in the sense of Chapter 3.

As in the homogeneous and periodic case, if we want the limit to be finite on sets of finite perimeter, we have to (locally) test the energies on half-spaces and require a bound, which in this case has to be uniform. We assume the following condition.

(H_f3) (*Finiteness condition*) If, for $\varepsilon > 0$ and $k \in \mathbb{N}$, we set

$$c^{\varepsilon,k} = \sup_{i\in\mathcal{L}} \sum_{\substack{\varepsilon j\in\mathcal{L}_\varepsilon(\Omega)\\ k\leq\|j-i\|_\infty<k+1}} a^\varepsilon_{ij}, \tag{4.75}$$

then we have

$$\limsup_{\varepsilon\to 0} \sum_{k=1}^{+\infty} k^d c^{\varepsilon,k} < +\infty.$$

Hypothesis (H_f3) is not sufficient to ensure that the limit be local, as shown in Example 4.2. To ensure the locality of the limit, we have to require that the tails of the sums given by the interactions are uniformly small. We assume the following condition.

(H_f4) (*Locality*) For any δ there exists $R_\delta > 0$ such that

$$\limsup_{\varepsilon\to 0} \sum_{k\geq R_\delta} k^d c^{\varepsilon,k} < \delta.$$

We then have the following compactness result, which we state for general Ω. Note that if Ω is bounded in (H_f3) and (H_f4), we can replace k with k^d.

Theorem 4.37 (Compactness and integral representation for ferromagnetic interactions) *Let Ω be a Lipschitz subset of $\mathbb{R}^d$. If (H_f1)–(H_f4) hold, then for any $\varepsilon_j \to 0$ there exists a subsequence $\{\varepsilon_{j_k}\}$ and a bounded Borel function $\varphi\colon \Omega\times S^{d-1} \to [0,+\infty)$ such that the functionals $E_{\varepsilon_{j_k}}$ defined in* (4.74) *Γ-converge to the functional F given by*

$$F(A) = \int_{\Omega\cap\partial^* A} \varphi(x,\nu)\,d\mathcal{H}^{d-1}$$

with respect to the convergence $u^\varepsilon \to A$ in Definition 2.22.

Proof Observe that E_ε is of the form (4.25) with $Y = Y_0 = \{-1, 1\}$ and $\phi^\varepsilon_i \geq 0$ defined by

$$\phi^\varepsilon_i(\{z_j\}_j) = \sum_{\varepsilon j\in\mathcal{L}^i_\varepsilon(\Omega)} a^\varepsilon_{i\,i+j}(z_j - z_0)^2. \tag{4.76}$$

Conditions (H_f1)–(H_f4) imply the assumptions of the Compactness Theorem 4.27, which then gives the claim. Indeed, condition (H_f3) gives the uniform boundedness of ϕ_i^ε, as remarked earlier, (H_f1) and (H_f2) ensure that the only ground states are the constant states. Moreover (H_f2) also yields that ϕ_i^ε satisfy the coerciveness assumption (H2). To check this, let $\varepsilon i, \varepsilon j \in \mathcal{L}_\varepsilon(\Omega)$ with $j \in NN(i)$ and $u_i \neq u_j$ and let $j_0, \dots, j_N$ as in (H_f2). Thus, necessarily $k \in \{1, \dots, N\}$ exists such that $u_{j_k} \neq u_{j_{k-1}}$. Hence,

$$\phi_{j_k}^\varepsilon(\{u_{j_k+j}\}_j) \geq 4a_{j_k j_{k-1}}^\varepsilon \geq 4c.$$

Finally, if w and z are such that $z_j = w_j$ for all $j \in Q_m(i) \cap \mathcal{L}$, we have

$$\left|\phi_i^\varepsilon(\{w_{i+j}\}_j) - \phi_i^\varepsilon(\{z_{i+j}\}_j)\right| \leq \left| \sum_{\varepsilon j \in \mathcal{L}_\varepsilon(\Omega)} a_{ij}^\varepsilon \Big((w_j - w_i)^2 - (z_j - z_i)^2\Big)\right| \leq c_m^\varepsilon,$$

where

$$c_m^\varepsilon = 4 \sup_{i \in \mathcal{L}} \sum_{\substack{\varepsilon j \in \mathcal{L}_\varepsilon(\Omega) \\ \|j-i\|_\infty \geq m}} a_{ij}^\varepsilon, \tag{4.77}$$

proving (4.22) with bounds satisfying (4.23) and (4.24) by (H_f3) and (H_f4). Indeed observe that

$$c_m^\varepsilon \leq 4 \sum_{k \geq m} c^{\varepsilon,k},$$

where $c^{\varepsilon,k}$ is defined in (4.75). Hence, for any $\alpha > 0$ we have

$$\sum_{m=1}^{+\infty} m^\alpha c_m^\varepsilon \leq 4 \sum_{m=1}^{+\infty} m^\alpha \sum_{k \geq m} c^{\varepsilon,k} = 4 \sum_{k=1}^{+\infty} c^{\varepsilon,k} \sum_{1 \leq m \leq k} m^\alpha \leq C \sum_{k=1}^{+\infty} k^{\alpha+1} c^{\varepsilon,k}.$$

We may show that

$$\sum_{m > M_\delta} m^\alpha c_m^\varepsilon \leq C \sum_{k > M_\delta} k^{\alpha+1} c^{\varepsilon,k}$$

in an analogous way. □

As an application of Theorem 4.37 we can improve the homogenization Theorem 3.10 by removing the finite-range assumption.

Theorem 4.38 (Homogenization theorem) *Let $\mathcal{L} = \mathbb{Z}^d$ and let $a_{ij}^\varepsilon = a_{ij}$ satisfy hypotheses (H_f1)–(H_f4), with a_{ij} periodic; that is, there exists $K \in \mathbb{N}$ such that*

$$a_{i+Ke_l\, j+Ke_l} = a_{ij} \textit{ for all } i, j \in \mathbb{Z}^d \textit{ and } l \in \{1, \dots, d\}. \tag{4.78}$$

Then the limit F is independent of the subsequence and is of the form

$$F(A) = \int_{\Omega\cap\partial^* A} \varphi_{\text{hom}}(\nu)\, d\mathcal{H}^{d-1},$$

with φ_{hom} *satisfying*

$$\varphi_{\text{hom}}(\nu) = \lim_{\delta\to 0}\liminf_{T\to+\infty} \frac{1}{T^{d-1}} \min\Big\{ \sum_{i,j\in Q_T^\nu} a_{ij}(v_i - v_j)^2 \colon v\colon \mathbb{Z}^d \cap Q_T^\nu \to \{-1,1\},$$
$$v = 2\chi_{H^\nu} - 1 \text{ in } Q_T^\nu \setminus Q_{(1-2\delta)T}^\nu \Big\}.$$

Proof We can apply Proposition 4.34 with ϕ_i defined by

$$\phi_i(\{z_n\}_{n\in\mathbb{Z}^d}) = \sum_{n\in\mathbb{Z}^d} a_{in}(z_n - z_0)^2. \tag{4.79}$$

Indeed arguing as in the proof of Theorem 4.37 we can show that (4.61) is satisfied with $t_m^\varepsilon = c_m^\varepsilon$, where the constants c_m^ε are defined in (4.77) and satisfy (4.24). Hence, in particular (4.60) is satisfied. In order to apply Proposition 4.34, we have to prove that (4.64) holds for $\Phi(u;U)$ defined in (4.66) with ϕ_i given by (4.79). To this end, it suffices to apply Proposition 4.35. Note that, by (4.78), the minimum problem defined in (4.68) satisfies

$$m(T,\delta,z+\tau) = m(T,\delta,z)$$

for any $z \in \mathbb{Z}^d$ and $\tau \in (K\mathbb{Z})^d$. Hence, (4.69) holds with $\mathcal{T}_\eta = (K\mathbb{Z})^d$ and $L_\eta = K$. □

4.10 An Application: Design of Networks

The Compactness Theorem allows one to formalize problems of network design as the characterization of energy densities satisfying some constraints. In this section we briefly describe one of such problems.

Similarly to Section 3.6, we consider a given finite subset V of $\mathbb{Z}^d$ containing $e_1,\dots,e_d$. For all $\xi \in V$ we fix α_ξ and β_ξ with $0 < \alpha_\xi < \beta_\xi$. We will consider systems of coefficients a_{ij}^ε, which are assumed to satisfy the *design constraint*

$$a_{i\,i+\xi}^\varepsilon \in \{\alpha_\xi, \beta_\xi\}$$

for all $i \in \mathbb{Z}^d$ and $\xi \in V$. Contrary to Section 3.6.1, the systems are not assumed to be periodic, but the distribution of α_ξ and β_ξ-connections is supposed to be arbitrary. Note that, taking $V = \{e_1,\dots,e_d\}$ and $\alpha_\xi = \alpha$ and $\beta_\xi = \beta$, the preceding assumption corresponds to having arbitrary mixtures of two types of nearest-neighbor connections.

Up to subsequences, we can suppose that for every $\xi \in V$ the *percentage of β_ξ-connections* has a weak limit $\theta_\xi \in L^1(\Omega)$ that we call the *limit percentage of β_ξ-connections*; that is, the measures

$$\mu_\varepsilon^\xi = \sum_{i \in I_\varepsilon^\xi} \varepsilon^d \delta_{\varepsilon i}, \text{ where } I_\varepsilon^\xi = \left\{ i \in \mathbb{Z}^d \cap \frac{1}{\varepsilon}\Omega : a_{i\,i+\xi}^\varepsilon = \beta_\xi \right\} \tag{4.80}$$

converge weakly* to a limit measure μ^ξ, which is absolutely continuous with respect to the Lebesgue measure, with a density that is denoted by θ_ξ. Constraints on the percentage of β_ξ-connections (and hence of α_ξ-connections) can be translated into constraints on θ_ξ. Using the Compactness Theorem 4.37 we can state the design problem as follows.

Design Problem We now examine the problem of determining all energy densities $\varphi = \varphi(x, \nu)$ that can be obtained as limits of energies (4.74) with given limit percentages θ_ξ of β_ξ-connections for all $\xi \in V$.

We call this a "design problem" in that it requires us to design optimal geometries that generate a given φ, which we think itself is optimal for some problem that has to be solved under the preceding constraint.

The solution of this problem requires some technical results on the limit of perimeter energies on the continuum that are not central to the scope of this book. We are going to state the result without entering in the proofs. The key argument is that the analysis can be reduced locally to periodic systems, which are described by the bounds in Section 3.6.1. We recall that, given $\theta_\xi \in [0, 1]$ for all $\xi \in V$, the set $H(\{\theta_\xi\})$ is the closure of the limits of periodic systems with percentages of β_ξ-connections tending to θ_ξ (see Definition 3.46). The first result is a localization theorem that characterizes the bounds of the energy densities in the design problem.

Theorem 4.39 (A "Dal Maso–Kohn localization principle") *Let a_{ij}^ε be a system of coefficients with limit percentages of β_ξ-connections given by $\theta_\xi \in L^1(\Omega)$. Then we have $\varphi(x, \cdot) \in H(\{\theta_\xi(x)\})$ for almost every $x \in \Omega$ and for all $\xi \in V$.*

The proof of this result can be obtained by noting that the minimum problems in the blow-up formula giving $\varphi(x_0, \nu)$ can be interpreted as those in a homogenization formula related to a periodic discrete lattice with percentage of β_ξ-connections close to $\theta_\xi(x_0)$ for almost every x_0. The following result shows that Theorem 4.39 is sharp.

Theorem 4.40 (Design theorem) *For all $\xi \in V$ let $\theta_\xi : \Omega \to [0, 1]$ be measurable and let $\varphi : \Omega \times \mathbb{R}^d \to [0, +\infty)$ be positively 1-homogeneous and even in* *the second variable such that the trivial bounds* (3.6) *are satisfied for all* $x \in \Omega$ *and* $\varphi(x,\cdot) \in H(\{\theta_\xi(x)\})$ *for almost every* $x \in \Omega$. *Then there exists a system of coefficients* a^ε_{ij} *such that* E_ε *in* (4.74) Γ*-converge to* F, *given by*

$$F(A) = \int_{\Omega\cap\partial^* A} \varphi(x,\nu(x))d\mathcal{H}^{d-1}$$

and $\theta_\xi(\{a^\varepsilon_{ij}\})$ *converge to* θ_ξ *as* $\varepsilon \to 0$ *for all* $\xi \in V$.

The reason why we have to specify the requirement that the trivial bounds (3.6) be satisfied pointwise by φ is that the constraint $\varphi(x,\cdot) \in H(\{\theta_\xi(x)\})$ holds only almost everywhere in Ω, so that the value of φ may not be constrained on a set of positive $\mathcal{H}^{d-1}$-measure, which is meaningful for the energy F. Note that this result is the same as the optimality Theorem 3.47 if the function $x \mapsto \varphi(x,\nu)$ and the limit percentages θ_ξ are constant.

Bibliographical Notes to Chapter 4

Functionals on partitions of sets of finite perimeter have been studied by Ambrosio and Braides (1990a,b) as part of the program for the study of free-discontinuity problems proposed by De Giorgi (2006). The density of polyhedral partitions has been proved in Braides et al. (2017).

The localization method and the subsequent development of integral-representation techniques has been devised by De Giorgi (1975) (see also the selected papers of De Giorgi, 2006) in order to provide compactness theorems independent of regularity hypotheses or restrictive structure assumptions on the functionals. The measure criterion has been proved by De Giorgi and Letta (1977). The localization method is the core of the monograph of Dal Maso (1993). An account can also be found in the books by Braides and Defranceschi (1998) and Braides (2002), and in the handbook by Braides (2006). Its application to a discrete-to-continuum analysis was first carried out by Alicandro and Cicalese (2004) for limits defined on Sobolev functions. For spin systems with pairwise ferromagnetic interactions, an analogous result has been proved by Alicandro and Gelli (2016). The version included here, in particular the formulation of the hypotheses for integrands depending on all sites, is original and generalizes a homogenization result by Braides and Cicalese (2017) (see also Braides and Kreutz, 2018b; Bach et al., 2020).

Since spaces of functions of bounded variation $BV(\Omega, Y_n)$ related to partitions parameterized on a set $Y_n \subset \mathbb{R}^m$ with $n \in \mathbb{N}$ become dense in the space of functions of bounded variation if we let Y_n invade $\mathbb{R}^m$, by diagonal arguments we can obtain functionals on BV as limits of discrete systems. We do not pursue this issue since rigorous statements would involve distracting technicalities. We only note that in the vector case $m > 1$ the situation can be even more complex as in the work by Cicalese et al. (2022), involving functionals on Cartesian currents (Giaquinta et al., 1998).

The homogenization approach to optimal design of networks has been analyzed by Braides and Kreutz (2018a). The Dal Maso–Kohn localization principle refers to an unpublished note; a version of this result is proved by Raitums (2001). An account of the homogenization method in Optimal Design for elliptic equations is given by Tartar (2000) (see also the book by Allaire, 2002).

5
Random Lattices

In this chapter we abandon the setting of interactions between nodes of a fixed deterministic lattice and consider random environments. Contrary to the percolation problems in Section 3.7, where the randomness governed the behavior of the connections in a fixed lattice, in this chapter we focus on discrete sets with a random distribution of sites $\mathcal{L}^\omega$, which will depend on the realization ω of a suitable random variable. For these sets geometric and metric properties of connections play a crucial role. We will only deal with pairwise ferromagnetic spin energies and treat two types of random assumptions, linking them to the approaches described in the previous chapters. If the random lattices are admissible, we can draw an analogy with the homogenization of random mixtures in Section 3.7.1, while nonadmissible lattices are reminiscent of dilute spin systems in the supercritical case, as in Section 3.7.2.

First, in Section 5.1 we follow an approach based on the compactness theorem in Chapter 4. To that end, we consider stochastic lattices $\mathcal{L}^\omega$ that are almost surely admissible lattices in the sense of Definition 4.17, and interaction coefficients depending on the distance of the corresponding sites. By scaling $\mathcal{L}^\omega$ and the corresponding energies we can directly conclude that a Γ-limit exists at fixed ω by the results in Chapter 4. If suitable stationarity and ergodicity properties hold, then we can almost surely deduce that almost surely the energy density of the limit is traslation invariant and independent of ω. To that end, we define an energy density through a stochastic asymptotic homogenization formula, whose existence as a limit and almost-sure independence from ω are obtained using ergodic theorems for discrete subadditive stochastic processes. Those stochastic results allow an almost-sure version of the subadditive argument used for proving homogenization formulas for periodic media.

In general, the assumption that $\mathcal{L}^\omega$ be almost surely admissible may be a modeling restriction. A paradigmatic model case in which this hypothesis is not satisfied is that of Poisson random sets, which, loosely speaking, are random point sets $\mathcal{L}^\omega$ in $\mathbb{R}^d$ such that we expect that a measurable set A in $\mathbb{R}^d$ contain a number of points of $\mathcal{L}^\omega$ proportional to its measure, with a Poisson distribution. Note that this characterization is invariant by translations and rotations, so that the limit energy, if it exists, is a multiple of the Euclidean perimeter. In Section 5.2 we relax the hypothesis of being admissible lattices. The weaker conditions that we require restrict the domain of applicability of the results mainly to dimension two. We study the model case of Poisson random sets in the plane. To prove homogenization, we follow the blow-up procedure in Section 3.2 and reformulate the asymptotic properties of the ferromagnetic energies in terms of the asymptotic behavior of minimal paths in the corresponding Delaunay triangulation. This asymptotic behavior can then be related to Bernoulli Percolation results such as those used in Section 3.7 through a comparison lemma between large sets V of Voronoi cells and unions of squares intersection V, for which compactness arguments similar to those for admissible lattices can be used.

5.1 Stochastic Lattices

By using the compactness theorems of Chapter 4, in this section we extend the results of Section 3.7 to the case in which the spin variables are defined on a random point set obtained as the outcome of a random variable fulfilling suitable assumptions.

In what follows we assume that $(O, \mathcal{F}, \mathbb{P})$ is a probability space where $\mathcal{F}$ is a complete σ-algebra. We first introduce the notion of *stochastic lattice.*

Definition 5.1 (Admissible stochastic lattice) A random variable $\mathcal{L}\colon O \to (\mathbb{R}^d)^{\mathbb{N}}$ is called a *stochastic lattice*. The stochastic lattice $\mathcal{L}$ is said to be *admissible* if $\mathcal{L}(\omega)$ is an admissible lattice according to Definition 4.17 with the maximal separation size R and the minimal distance r independent of ω $\mathbb{P}$-almost surely.

In the rest of the section, R and r will denote the maximal separation size and the minimal distance of the lattice, respectively.

For fixed $\varepsilon > 0$, $\omega \in O$ and U open subset of Ω, we set $\mathcal{L}_\varepsilon^\omega(U) = \varepsilon\mathcal{L}(\omega) \cap U$ and consider energies defined on functions $u\colon \mathcal{L}_\varepsilon^\omega(\Omega) \to \{-1, 1\}$ of the form

$$E_\varepsilon^\omega(u; U) = \sum_{\varepsilon i, \varepsilon j \in \mathcal{L}_\varepsilon^\omega(U)} \varepsilon^{d-1} a_{ij}^\omega (u_j - u_i)^2. \tag{5.1}$$

Nearest-neighbor and long-range interactions are separated in the definition of the coefficients a_{ij}^ω, which have the form

$$a^{\omega}_{ij} = \begin{cases} a^{\text{nn}}(j-i) & \text{if } (i,j) \in \mathcal{NN}(\omega), \\ a^{\text{lr}}(j-i) & \text{otherwise,} \end{cases} \tag{5.2}$$

where the set of nearest neighbors $\mathcal{NN}(\omega)$ is as in Definition 4.18 with $\mathcal{L}(\omega)$ in the place of $\mathcal{L}$, and the functions $a^{\text{nn}}, a^{\text{lr}}$ defined on $\mathbb{R}^d$ satisfy the following assumptions.

($\text{H}_{fs}1$) (*ferromagnetic energies*) The functions a^{nn} and a^{lr} are nonnegative.

($\text{H}_{fs}2$) (*coerciveness*) There exists $C > 0$ such that

$$\frac{1}{C} \leq a^{\text{nn}}(x) \leq C \tag{5.3}$$

for all $x \in \mathbb{R}^d$.

($\text{H}_{fs}3$) (*finiteness condition and locality*) There exists a nonincreasing function $J_{\text{lr}} \colon [0, +\infty) \to [0, +\infty)$ with

$$\int_{\mathbb{R}^d} J_{\text{lr}}(\|x\|)\|x\|^d \, dx = J < +\infty \tag{5.4}$$

such that

$$a^{\text{lr}}(x) \leq J_{\text{lr}}(\|x\|) \tag{5.5}$$

for all $x \in \mathbb{R}^d$.

Note that for any $\omega \in O$ the functional E^{ω}_{ε} is of the form (4.74) with $a^{\varepsilon}_{ij} = a^{\omega}_{ij}$. Condition ($\text{H}_f2$) is satisfied by assumption (5.3). Moreover, by (5.5) and since J is decreasing, the constants $c^{\varepsilon,k}$ defined in (4.75) satisfy $c^{\varepsilon,k} \leq CJ_{\text{lr}}(k)k^{d-1}$ for all $k \in \mathbb{N}$ and $\varepsilon > 0$. Hence (5.4) implies that (H_f3) and (H_f4) are satisfied.

Remark 5.2 (Coerciveness) Given $\omega \in O$ such that $\mathcal{L}(\omega)$ is admissible, we may apply Theorem 4.37 and obtain that for every $\varepsilon_n \to 0$ there exist a subsequence (not relabeled) and a function $\varphi^{\omega} \colon \Omega \times S^{d-1} \to [0, \infty)$ such that for all $u \in BV(\Omega, \{-1, 1\})$ and U regular open subsets of Ω, we have

$$\Gamma\text{-}\lim_{n\to+\infty} E^{\omega}_{\varepsilon_n}(u; U) = \int_{S(u)\cap U} \varphi^{\omega}(x, \nu_u) \, d\mathcal{H}^{d-1}. \tag{5.6}$$

Note that equivalently we can consider energies as defined on sets of finite perimeter, setting $u = 2\chi_A - 1$, as in Chapter 3, so that (5.6) may be rewritten as

$$\Gamma\text{-}\lim_{n\to+\infty} E^{\omega}_{\varepsilon_n}(A, U) = \int_{U\cap\partial^* A} \varphi^{\omega}(x, \nu) \, d\mathcal{H}^{d-1}.$$

We now describe how the preceding Γ-limits depend on the randomness of the problem. Note that if $\mathcal{L}(\omega) = \mathbb{Z}^d$, the functionals E^ω_ε satisfy the hypotheses of Theorem 3.1, taking into account the possibility of extensions to systems with infinite range of interactions satisfying (3.5), and the homogenization result stated there holds. In particular, φ^ω does not depend on x. We will see that an analogous result holds in our stochastic setting if the periodicity of the deterministic lattice is replaced with the stationarity of our stochastic lattice, according to the definition introduced in the following discussion. Moreover, the limit energy turns out to be deterministic under some ergodicity properties of the stochastic lattice.

We may conveniently rewrite energies E^ω_ε by introducing an auxiliary deterministic square lattice, which will allow us to provide estimates on the discrete energies that are uniform with respect to the stochastic variable. Let r be the minimal distance of the lattice as in Definition 4.17. If we set $r' = \frac{r}{\sqrt{d}}$, for all $\alpha \in r'\mathbb{Z}^d$ it holds $\#(\mathcal{L}(\omega) \cap (\alpha + [0, r')^d)) \leq 1$. Setting

$$\mathcal{Z}_{r'}(\omega) = \{\alpha \in r'\mathbb{Z}^d \colon \ \#(\mathcal{L}(\omega) \cap (\alpha + [0, r')^d)) = 1\},$$

for $\alpha \in \mathcal{Z}_{r'}(\omega)$ let i_α denote the only point in the intersection $\mathcal{L}(\omega) \cap (\alpha + [0, r')^d)$, and, for $\xi \in r'\mathbb{Z}^d$, $U \subset \mathbb{R}^d$, and $\varepsilon > 0$, define

$$R^\xi_\varepsilon(U) = \{\alpha \colon \ \alpha, \alpha + \xi \in \mathcal{Z}_{r'}(\omega); \ \varepsilon i_\alpha, \varepsilon i_{\alpha+\xi} \in U\}. \tag{5.7}$$

We can now rewrite the energy as

$$E^\omega_\varepsilon(u; U) = \sum_{\xi \in r'\mathbb{Z}^d} \sum_{\alpha \in R^\xi_\varepsilon(U)} \varepsilon^{d-1} a^\omega_{i_\alpha i_{\alpha+\xi}} (u_{i_{\alpha+\xi}} - u_{i_\alpha})^2 \tag{5.8}$$

and use this characterization to obtain a uniform upper bound for $E^\omega_\varepsilon(u; U)$ in terms of $\mathcal{H}^{d-1}(S(u) \cap U)$.

In the following lemma, for the sake of notational simplicity we use a slight abuse of notation with respect to that in (4.52).

Lemma 5.3 *Let U be a convex subset of Ω and let*

$$U_\varepsilon = \{x \in U \colon \ \mathrm{dist}\,(x, \partial U) > 2R\varepsilon\}.$$

Then there exists a positive constant $C = C(r, R)$ such that, for any $u \colon \mathcal{L}^\omega_\varepsilon(\Omega) \to \{-1, 1\}$ and $\xi \in r'\mathbb{Z}^d$, we have

$$\sum_{\alpha \in R^\xi_\varepsilon(U_\varepsilon)} \varepsilon^{d-1} a^\omega_{i_\alpha i_{\alpha+\xi}} (u_{i_{\alpha+\xi}} - u_{i_\alpha})^2 \leq C\mathcal{H}^{d-1}(S(u) \cap U) J_{\mathrm{lr}}(\|\hat{\xi}\|)\|\xi\|, \tag{5.9}$$

where $\hat{\xi} \in \xi + [-r', r']^d$ is such that $\|\hat{\xi}\| = \mathrm{dist}\,([0, r')^d, [0, r')^d + \xi)$. In *particular,*

$$E_\varepsilon^\omega(u;U_\varepsilon) \le C\mathcal{H}^{d-1}(S(u)\cap U) \sum_{\xi\in r'\mathbb{Z}^d} J_{\rm lr}(\|\hat{\xi}\|)\|\xi\|. \tag{5.10}$$

Proof By ($H_{fs}2$) and ($H_{fs}3$), we get that

$$\sum_{\alpha\in R_\varepsilon^\xi(U_\varepsilon)} \varepsilon^{d-1} a^\omega_{i_\alpha i_{\alpha+\xi}} (u_{i_{\alpha+\xi}} - u_{i_\alpha})^2 \le CJ_{\rm lr}(\|\hat{\xi}\|) \sum_{\alpha\in R_\varepsilon^\xi(U_\varepsilon)} \varepsilon^{d-1}(u_{i_{\alpha+\xi}} - u_{i_\alpha})^2. \tag{5.11}$$

Now, if $u_{i_\alpha} \neq u_{i_{\alpha+\xi}}$, then there exists a pair $(i,j)\in \mathcal{NN}(\omega)$ such that $i,j\in [\alpha,\alpha+\xi]+B_{2R}$ and $u_i \neq u_j$. Moreover, setting

$$I_\varepsilon^\xi(i,j) = \{\alpha\in R_\varepsilon^\xi(U_\varepsilon)\colon\ i,j\in[\alpha,\alpha+\xi]+B_{2R}\},$$

by Lemma 4.19 we infer that $\#I_\varepsilon^\xi(i,j)\le C(R,r)\|\xi\|$. Then we get

$$\begin{aligned}
&\sum_{\alpha\in R_\varepsilon^\xi(U_\varepsilon)} \varepsilon^{d-1} a^\omega_{i_\alpha i_{\alpha+\xi}} (u_{i_{\alpha+\xi}} - u_{i_\alpha})^2 \\
&\quad \le CJ_{\rm lr}(\|\hat{\xi}\|) \sum_{\substack{\varepsilon i,\varepsilon j\in U\\ (i,j)\in\mathcal{NN}(\omega)}} \varepsilon^{d-1}\#I_\varepsilon^\xi(i,j)(u_j-u_i)^2 \\
&\quad \le C\mathcal{H}^{d-1}(S(u)\cap U)\, J_{\rm lr}(\|\hat{\xi}\|)\|\xi\|,
\end{aligned}$$

by (5.11) and again by Lemma 4.19, and the claim. □

5.1.1 Stationary and Ergodic Stochastic Lattices

We now introduce the relevant notions in the stochastic framework that allow one to prove a homogenization result.

Definition 5.4 (Additive group action) A family $\{\tau_z\}_{z\in\mathbb{Z}^d}, \tau_z\colon \mathcal{O}\to\mathcal{O}$ is an *additive group action* on $\mathcal{O}$ if $\tau_{z_1+z_2} = \tau_{z_2}\circ\tau_{z_1}$ for all $z_1,z_2\in\mathbb{Z}^d$.

Such an additive group action is called *measure preserving* if $\mathbb{P}(\tau_z B) = \mathbb{P}(B)$ for all $B\in\mathcal{F}$, $z\in\mathbb{Z}^d$. Moreover, a measure-preserving additive group action is called *ergodic* if $\mathbb{P}(B)\in\{0,1\}$ for all $B\in\mathcal{F}$ such that $\tau_z(B)=B$ for all $z\in\mathbb{Z}^d$.

Definition 5.5 (Stationary and ergodic stochastic lattices) A stochastic lattice $\mathcal{L}\colon\mathcal{O}\to(\mathbb{R}^d)^{\mathbb{N}}$ is said to be *stationary* if there exists a measure-preserving group action $\{\tau_z\}_{z\in\mathbb{Z}^d}$ on $\mathcal{O}$ such that $\mathcal{L}(\tau_z\omega) = \mathcal{L}(\omega)+z$ for $\mathbb{P}$-almost every $\omega\in\mathcal{O}$. Moreover, $\mathcal{L}$ is said to be *ergodic* if $\{\tau_z\}_{z\in\mathbb{Z}^d}$ is ergodic.

We now introduce a notation that is customary in the treatment of stochastic processes. Since we will use this notion to characterize surface integrands, we confine our framework to $\mathbb{R}^{d-1}$. Given $a,b\in\mathbb{R}^{d-1}$, $[a,b)$ denotes the corresponding rectangle in $\mathbb{R}^{d-1}$; that is,

$$[a,b) = \{x \in \mathbb{R}^{d-1}: \; a_n \leq x_n < b_n \text{ for all } n = 1,\ldots,d-1\}.$$

The family of such rectangles is denoted by

$$\mathcal{I} = \{[a,b): \; a,b \in \mathbb{Z}^{d-1}, a \neq b\}.$$

Now, we can introduce the notion of discrete subadditive stochastic process.

Definition 5.6 (Discrete subadditive stochastic process) A function $\mu\colon \mathcal{I} \to L^1(O)$ is a *discrete subadditive stochastic process* if the following properties hold $\mathbb{P}$-almost surely.

(i) For every $I \in \mathcal{I}$ and for every finite partition $\{I_k\}_{k\in K} \subset \mathcal{I}$ of I, we have

$$\mu(I,\omega) \leq \sum_{k\in K} \mu(I_k,\omega).$$

(ii) We have

$$\inf\left\{\frac{1}{|I|}\int_O \mu(I,w)\,\mathrm{d}\mathbb{P}(\omega): \; I \in \mathcal{I}\right\} > -\infty.$$

Our analysis relies on the following ergodic theorem, which we state without proof.

Theorem 5.7 (Akcoglu–Krengel) *Let $\mu\colon \mathcal{I} \to L^1(O)$ be a discrete subadditive stochastic process. If μ is stationary with respect to a measure-preserving group action $\{\tau_z\}_{z\in\mathbb{Z}^{d-1}}$; that is, for all $I \in \mathcal{I}$ and for all $z \in \mathbb{Z}^{d-1}$ we have $\mu(I+z,\omega) = \mu(I,\tau_z\omega)$ almost surely, then there exists $\Phi\colon O \to \mathbb{R}$ such that, for $\mathbb{P}$-almost every ω,*

$$\lim_{k\to+\infty} \frac{\mu([-k,k)^{d-1},\omega)}{(2k)^{d-1}} = \Phi(\omega).$$

Furthermore, if $\{\tau_z\}_{z\in\mathbb{Z}^{d-1}}$ is ergodic, then Φ is almost surely constant.

5.1.2 Stochastic Homogenization

If $\mathcal{L}$ is a stationary admissible stochastic lattice according to the preceding definition, the following stochastic homogenization result holds.

Theorem 5.8 (Stochastic homogenization) *Let $\mathcal{L}$ be a stationary admissible stochastic lattice and let a_{ij}^{ω} satisfy (5.2) with a^{nn} and a^{lr} satisfying ($\mathrm{H}_{fs}1$), ($\mathrm{H}_{fs}2$), and ($\mathrm{H}_{fs}3$). Then, for $\mathbb{P}$-almost every ω and for all $\nu \in S^{d-1}$, we have*

$$\inf_{\delta>0}\liminf_{T\to+\infty} \frac{m_\delta(\omega)(T,\nu)}{T^{d-1}} = \inf_{\delta>0}\limsup_{T\to+\infty} \frac{m_\delta(\omega)(T,\nu)}{T^{d-1}} \in \mathbb{R}, \tag{5.12}$$

where

$$m_\delta(\omega)(T,\nu) = \inf\Big\{ \sum_{i,j\in\mathcal{L}(\omega)\cap Q_T^\nu} a_{ij}^\omega (u_j - u_i)^2 : \ u = u^\nu \ on\ \mathcal{L}(\omega)\setminus Q^\nu_{(1-\delta)T}\Big\}, \tag{5.13}$$

and

$$u^\nu(x) = \begin{cases} 1 & if\ \langle x,\nu\rangle \geq 0, \\ -1 & otherwise. \end{cases}$$

Moreover, if $\varphi_{\rm hom}(\omega,\nu)$ denotes the quantity in (5.12), for $\mathbb{P}$-almost every ω the functionals E_ε^ω defined in (5.1) Γ-converge with respect to the convergence in $BV_{\rm loc}(\Omega)$, as ε goes to 0, to the integral functional

$$F^\omega(u;U) = \int_{S(u)\cap U} \varphi_{\rm hom}(\omega,\nu_u)\, d\mathcal{H}^{d-1}$$

for all U with Lipschitz boundary. If in addition $\mathcal{L}$ is ergodic, then $\varphi_{\rm hom}(\cdot,\nu)$ is constant almost surely and is given by

$$\varphi_{\rm hom}(\nu) = \inf_{\delta>0} \limsup_{T\to+\infty} \frac{1}{T^{d-1}} \int_O m_\delta(\omega)(T,\nu)\, d\mathbb{P}(\omega).$$

Proof Given $\omega\in O$ such that $\mathcal{L}(\omega)$ is admissible and a sequence $\varepsilon_n\to 0$, we apply Theorem 4.37 to obtain a subsequence (not relabeled) and a function $\varphi^\omega\colon \Omega\times S^{d-1}\to[0,+\infty)$ such that (5.6) holds. For $u\in BV(\Omega;\{-1,1\})$ and U regular open subset of Ω, we set

$$F_0(u;U) = \int_{S(u)\cap U} \varphi^\omega(x,\nu_u)\, d\mathcal{H}^{d-1}.$$

Given $x_0\in\Omega$, we have that

$$\varphi^\omega(x_0,\nu) = \lim_{\varrho\to 0}\inf_{\delta>0} \frac{1}{\varrho^{d-1}} \min\{F_0(u;Q_\varrho^\nu(x_0)) \colon u = u_{x_0}^\nu \text{ on } \mathbb{R}^d\setminus Q^\nu_{(1-\delta)\varrho}(x_0)\},$$

where

$$u_{x_0}^\nu(x) = \begin{cases} 1 & \text{if } \langle x-x_0,\nu\rangle \geq 0, \\ -1 & \text{otherwise.} \end{cases} \tag{5.14}$$

By the convergence of minima in Proposition 4.31 with $U = Q_\varrho^\nu(x_0)$ we then have

$$\varphi^\omega(x_0,\nu) = \lim_{\varrho\to 0}\inf_{\delta>0}\liminf_{n\to+\infty} \frac{1}{\varrho^{d-1}} \inf\big\{E_{\varepsilon_n}(u;Q_\varrho^\nu(x_0)) \colon \\ u = u_{x_0}^\nu \text{ on } \varepsilon\mathcal{L}(\omega)\setminus Q^\nu_{(1-\delta)\varrho}(x_0)\big\}.$$

Setting $T_n = \frac{\varrho}{\varepsilon_n}$ and through the change of variables $v(x) = u(\varepsilon_n x)$ for $x\in\mathcal{L}(\omega)$, we get

$$\varphi^{\omega}(x_0,\nu)=\lim_{\varrho\to 0}\inf_{\delta>0}\liminf_{n\to+\infty}\frac{1}{T_n^{d-1}}m_\delta(\omega)\Big(\frac{x_0}{\varrho},T_n,\nu\Big),$$

where

$$m_\delta(\omega)(x,T,\nu)=\inf\{E_1^{\omega}(u;Q_T^{\nu}(Tx)),\ u=u_{Tx}^{\nu}\text{ on }\mathcal{L}(\omega)\setminus Q_{(1-\delta)T}^{\nu}(Tx)\}.$$

Note that $m_\delta(\omega)(0,T,\nu)=m_\delta(\omega)(T,\nu)$, where $m_\delta(\omega)(T,\nu)$ is defined in (5.13). Then, the claim follows if we prove that for $\mathbb{P}$-almost every ω and for all $x\in\mathbb{R}^d$ and $\nu\in S^{d-1}$, we have

$$\inf_{\delta>0}\liminf_{T\to+\infty}\frac{1}{T^{d-1}}m_\delta(\omega)(x,T,\nu)=\inf_{\delta>0}\limsup_{T\to+\infty}\frac{1}{T^{d-1}}m_\delta(\omega)(x,T,\nu)\in\mathbb{R},\quad(5.15)$$

and such limit does not depend on x. Indeed, if (5.15) holds, then the density φ^{ω} does not depend on the subsequence nor on x, and the whole family E_ε^{ω} Γ-converges. Moreover, (5.12) holds and $\varphi^{\omega}(x,\nu)=\varphi_{\rm hom}(\omega,\nu)$.

We now show that it suffices to prove (5.15) in the case where the functionals E_ε^{ω} are replaced with the truncated energies

$$E_\varepsilon^{\omega,L}(u;U)=\sum_{\substack{\varepsilon i,\varepsilon j\in\mathcal{L}_\varepsilon^{\omega}(U)\\ \|i-j\|\le L}}\varepsilon^{d-1}a_{ij}^{\omega}(u_j-u_i)^2,$$

with $L>0$. Indeed, if we set

$$m_\delta^{L}(\omega)(x,T,\nu)=\inf\{E_1^{\omega,L}(u;Q_T^{\nu}(Tx)),\ u=u_{Tx}^{\nu}\text{ on }\mathcal{L}(\omega)\setminus Q_{(1-\delta)T}^{\nu}(Tx)\}$$

and assume that for any $L>0$, for $\mathbb{P}$-almost every ω and for all $x\in\mathbb{R}^d$ and $\nu\in S^{d-1}$, there holds

$$\inf_{\delta>0}\liminf_{T\to+\infty}\frac{1}{T^{d-1}}m_\delta^{L}(\omega)(x,T,\nu)=\inf_{\delta>0}\limsup_{T\to+\infty}\frac{1}{T^{d-1}}m_\delta^{L}(\omega)(x,T,\nu)=\varphi_{\rm hom}^{L}(\omega,\nu)\qquad(5.16)$$

and the limit is finite. Then (5.15) is inferred from

$$0\le\frac{m_\delta(\omega)(x,T,\nu)-m_\delta^{L}(\omega)(x,T,\nu)}{T^{d-1}}\le o(1)_{L\to+\infty},\qquad(5.17)$$

where $o(1)_{L\to+\infty}$ does not depend on T and δ. Moreover, from inequality (5.17) we also deduce that $\lim\limits_{L\to+\infty}\varphi_{\rm hom}^{L}(\omega,\nu)=\varphi_{\rm hom}(\omega,\nu)$.

In order to prove (5.17), first observe that, using the interpolation on $\mathcal{L}(\omega)$ of u_{Tx}^{ν} as test functions in the minimum problem defining $m_\delta^{L}(\omega)(x,T,\nu)$ and using ($H_{fs}2$) and ($H_{fs}3$), we get

$$m_\delta^{L}(\omega)(x,T,\nu)\le C\mathcal{H}^{d-1}(S(u_{Tx}^{\nu})\cap Q_T^{\nu}(Tx))\le CT^{d-1}.\qquad(5.18)$$

Let u_T^L be a minimizer of $m_\delta^L(\omega)(x,T,\nu)$; that is, let u_T^L be such that

$$E_1^{\omega,L}(u_T^L; Q_T^\nu(Tx)) = m_\delta^L(\omega)(x,T,\nu).$$

By using the characterization of E_1^ω in (5.8), we get

$$\begin{aligned}
0 &\le \frac{1}{T^{d-1}}(m_\delta(\omega)(x,T,\nu) - m_\delta^L(\omega)(x,T,\nu)) \\
&\le \frac{1}{T^{d-1}}(E_1^\omega(u_T^L; Q_T^\nu(Tx)) - E_1^{\omega,L}(u_T^L; Q_T^\nu(Tx))) \\
&\le \frac{1}{T^{d-1}}\Big(\sum_{\|\xi\|>L-2r}\ \sum_{\alpha\in R_1^\xi(Q_T^\nu(Tx))} a^\omega_{i_\alpha i_{\alpha+\xi}}(u_T^L(x_\alpha) - u_T^L(x_{\alpha+\xi}))^2\Big),
\end{aligned}$$

where R_1^ξ is defined in (5.7). By Lemma 5.3 and using the fixed values of u_T^L near the boundary, we can bound the last expression by

$$\frac{C\mathcal{H}^{d-1}(S(u_T^L)\cap Q_T^\nu(Tx))}{T^{d-1}} \sum_{\|\xi\|>L-2r} J_{\mathrm{lr}}(\|\hat{\xi}\|)\|\xi\|.$$

Assumption ($\mathrm{H}_{fs}2$) yields that

$$\mathcal{H}^{d-1}(S(u_T^L)\cap Q_T^\nu(Tx)) \le C\,E_1^{\omega,L}(u_T^L; Q_T^\nu(Tx)) + C\,T^{d-1},$$

so that, taking into account (5.18), we infer that

$$0 \le \frac{m_\delta(\omega)(x,T,\nu) - m_\delta^L(\omega)(x,T,\nu)}{T^{d-1}} \le C \sum_{\|\xi\|>L-2r} J_{\mathrm{lr}}(\|\hat{\xi}\|)\|\xi\|. \tag{5.19}$$

Then, the validity of (5.17) follows by the integrability assumption on J_{lr}.

The rest of the proof will be devoted to proving (5.16). This will be done in several steps. We will first show that, given the family of auxiliary minimum problems defined as follows,

$$\widetilde{m}^L(\omega)(T,\nu) = \inf\{E_1^{\omega,L}(u; Q_T^\nu),\ u = u^\nu \text{ on } \mathcal{L}(\omega)\setminus Q_{T-L}^\nu\},$$

for $\mathbb{P}$-almost every ω and for all $\nu \in S^{d-1}$ there exists the limit

$$\lim_{T\to+\infty} \frac{1}{T^{d-1}}\widetilde{m}^L(\omega)(T,\nu) = \widetilde{\varphi}^L(\omega,\nu) \in \mathbb{R}. \tag{5.20}$$

Subsequently, we will show that for all $x \in \mathbb{R}^d$ and $\nu \in S^{d-1}$ the limit in (5.16) equals $\widetilde{\varphi}^L(\omega,\nu)$ almost surely.

We first treat the case where ν is a *rational direction*; that is, there exists $t > 0$ such that $t\nu \in \mathbb{Z}^d$.

Step 1 Existence of $\lim_{T\to+\infty} \frac{1}{T^{d-1}} \widetilde{m}^L(\omega)(T,\nu)$ for a rational direction ν.

Let $\nu \in S^{d-1}$ be a rational direction. Then there exists a matrix $A_\nu \in \mathbb{Q}^{d\times d}$ such that $A_\nu e_d = \nu$ and the set $\{A_\nu e_n\}_{n=1}^{d-1}$ is an orthonormal basis for the linear hyperplane Π^ν orthogonal to ν. Moreover, there exists an integer $M = M(\nu) > L$ such that $MA_\nu(z,0) \in \mathbb{Z}^d$ for all $z \in \mathbb{Z}^{d-1}$.

For any rectangle $I = [a,b) \in \mathcal{I}$, we consider the d-dimensional polytope given by

$$I_d = \{MA_\nu(z,t)\colon\ a_n < z_n < b_n \text{ for } n \in \{1,\dots,d-1\} \text{ and } 2|t| < \|b-a\|_\infty\}.$$

Let x_{I_d} denote the barycenter of I_d. We define the stochastic process $\widetilde{\mu}_\nu^L\colon \mathcal{I} \to L^1(\Omega)$ by setting

$$\widetilde{\mu}_\nu^L(I)(\omega) = \inf\left\{E_1^{\omega,L}(u;I_d)\colon\ u = u^\nu \text{ on } \mathcal{L}(\omega)\setminus\left(x_{I_d} + (1-L)(I_d - x_{I_d})\right)\right\} + \alpha\,\mathrm{Per}_{d-1}(I), \tag{5.21}$$

where $\mathrm{Per}_{d-1}(I)$ stands for the perimeter of I in $\mathbb{R}^{d-1}$ and α is a constant to be chosen later. Note that if we consider the square $[-n,n)^{d-1}$ with $n \in \mathbb{N}$, then

$$\widetilde{m}^L(\omega)(2Mn,\nu) = \widetilde{\mu}_\nu^L([-n,n)^{d-1})(\omega) - \alpha\,\mathrm{Per}_{d-1}([-n,n)^{d-1}).$$

Hence, since $\mathrm{Per}_{d-1}([-n,n)^{d-1})$ scales like n^{d-2}, if $\widetilde{\mu}_\nu^L$ satisfies the hypotheses of Theorem 5.7, then the limit

$$\lim_{n\to+\infty} \frac{1}{(2Mn)^{d-1}} \widetilde{m}^L(\omega)(2Mn,\nu) \tag{5.22}$$

exists and it is finite.

In what follows we show that $\widetilde{\mu}_\nu^L$ is indeed a discrete subadditive stochastic process, which is stationary with respect to a measure-preserving group action. We have that, for $I \in \mathcal{I}$, $\widetilde{\mu}_\nu^L(I)$ is a $L^1(O)$-function. Indeed, $\widetilde{\mu}_\nu^L(I)$ is measurable and

$$\widetilde{\mu}_\nu^L(I)(\omega) \le E_1^{\omega,L}(u^\nu;I_d) + \alpha\,\mathrm{Per}_{d-1}(I) \le CM^{d-1}\mathcal{H}^{d-1}(I)$$

for all $I \in \mathcal{I}$ and almost every $\omega \in O$, so that $\widetilde{\mu}_\nu^L(I)$ is essentially bounded.

As for the stationarity of the process, let $z \in \mathbb{Z}^{d-1}$ and note that $(I-z)_d = I_d - z_M^\nu$, where $z_M^\nu = MA_\nu(z,0) \in \Pi^\nu \cap \mathbb{Z}^d$. Hence, the stationarity assumption on the stochastic lattice $\mathcal{L}$ yields that $\widetilde{\mu}_\nu^L(I-z)(\omega) = \widetilde{\mu}_\nu^L(I)(\tau_{z_M^\nu}\omega)$. Setting $\widetilde{\tau}_z = \tau_{-z_M^\nu}$, we obtain a measure-preserving group action on $\mathbb{Z}^{d-1}$ such that

$$\widetilde{\mu}_\nu^L(I)(\widetilde{\tau}_z\omega) = \widetilde{\mu}_\nu^L(I+z)(\omega).$$

We now prove the discrete subadditivity of $\widetilde{\mu}^{\nu}_{L}$. Let $I \in \mathcal{I}$ and let $\{I^k\}_{k=1}^K \subset \mathcal{I}$ be a finite partition of I. For fixed k, let u^k be a function realizing $\widetilde{\mu}^L_\nu(I^k)(\omega)$ and define the function u as

$$u(x) = \begin{cases} u^k(x) & \text{if } x \in I^k_d \text{ for some } k, \\ u^\nu(x) & \text{otherwise.} \end{cases}$$

If x and y belong to the closure of $I^{k_1}_d$ and $I^{k_2}_d$, respectively, with $k_1 \neq k_2$ and give a contribution to $E^{\omega,L}_1(u; I_d)$, then $\|x - y\| \leq L$ and in particular $\operatorname{dist}(x, \partial I^{k_1}_d) \leq L$ and $\operatorname{dist}(y, \partial I^{k_2}_d) \leq L$. Hence, $u(x) = u^\nu(x)$ and $u(y) = u^\nu(y)$, so that x and y lie on different sides of the hyperplane Π^ν orthogonal to ν. We deduce that there exists a constant $C = C(L, M)$ such that

$$E^{\omega,L}_1(u; I_d) \leq \sum_{k=1}^K E^{\omega,L}_1(u^k; I^k_d) + C \sum_{\substack{k,h=1\\k\neq h}}^K \mathrm{Per}_{d-1}(\overline{I^k} \cap \overline{I^h}).$$

Since

$$\mathrm{Per}_{d-1}(I) = \sum_{k=1}^K \mathrm{Per}_{d-1}(I^k) - \sum_{\substack{k,h=1\\k\neq h}}^K \mathrm{Per}_{d-1}(\overline{I^k} \cap \overline{I^h}),$$

we get

$$\begin{aligned}\widetilde{\mu}^L_\nu(I)(\omega) &\leq E^{\omega,L}_1(u; I_d) + \alpha\, \mathrm{Per}_{d-1}(I) \\ &\leq \sum_{k=1}^K \widetilde{\mu}^L_\nu(I^k)(\omega) + (C - \alpha) \sum_{\substack{k,h=1\\k\neq h}}^K \mathrm{Per}_{d-1}(\overline{I^k} \cap \overline{I^h}),\end{aligned}$$

so that we obtain subadditivity if we choose $\alpha > C$. Finally, property (ii) of Definition 5.6 is satisfied, since $\widetilde{\mu}^\nu_L$ is nonnegative. Hence, (5.22) holds.

Consider now an arbitrary sequence $T_k \to +\infty$. By (5.22), we know that there exists

$$\widetilde{\varphi}^L(\omega, \nu) = \lim_{k\to+\infty} \frac{1}{(2M\lfloor T_k \rfloor)^{d-1}} \widetilde{m}^L(\omega)(2M\lfloor T_k \rfloor, \nu)$$

almost surely. In order to shorten the notation, we set $T_{k,1} = 2MT_k$ and $T_{k,2} = 2M\lfloor T_k \rfloor$. Let w^k be a function minimizing $\widetilde{m}^L(\omega)(2M\lfloor T_k \rfloor, \nu)$ and define v^k as

$$v^k(x) = \begin{cases} u^\nu(x) & \text{if } x \in Q^\nu_{T_{k,1}} \setminus Q^\nu_{T_{k,2}}, \\ w^k(x) & \text{otherwise.} \end{cases}$$

Separating the energetic contribution of the interactions in $Q^\nu_{T_{k,2}}$, contributions in $Q^\nu_{T_{k,1}} \backslash Q^\nu_{T_{k,2}}$ and those crossing $\partial Q^\nu_{T_{k,2}}$, and taking into account the boundary conditions, we obtain

$$\begin{aligned} E_1^{\omega,L}(v^k, Q^\nu_{T_{k,1}}) &\le E_1^{\omega,L}(w^k, Q^\nu_{T_{k,2}}) + C_L \mathcal{H}^{d-1}((Q^\nu_{T_{k,1}} \backslash Q^\nu_{T_{k,2}}) \cap \Pi^\nu) \\ &\quad + C_L \mathcal{H}^{d-2}(\partial Q^\nu_{T_{k,2}} \cap \Pi^\nu) \\ &\le \widetilde{m}^L(\omega)(T_{k,2}, \nu) + O(T_{k,1}^{d-2}), \end{aligned}$$

which yields

$$\limsup_{k\to+\infty} \frac{1}{(T_{k,1})^{d-1}} \widetilde{m}^L(\omega)(T_{k,1}, \nu) \le \widetilde{\varphi}^L(\omega, \nu). \tag{5.23}$$

Similarly, we can prove that

$$\widetilde{\varphi}^L(\omega, \nu) \le \liminf_{k\to+\infty} \frac{1}{(T_{k,1})^{d-1}} \widetilde{m}^L(\omega)(T_{k,1}, \nu). \tag{5.24}$$

Combining (5.23) and (5.24), we get the almost-sure existence of the limit for arbitrary sequences and then the proof of Step 1.

Step 2 Invariance of $\widetilde{\varphi}^L$ under the group action τ_z.

Let $O^L = \bigcap_{\nu \in S^{d-1} \cap \mathbb{Q}^d} O^L_\nu$ where O^L_ν is the set of full measure where the limit exists for the rational direction ν. We now prove that the function $\widetilde{\varphi}^L$ is invariant under the group action τ_z. Let $z \in \mathbb{Z}^d$, $\omega \in O^L$, and let $S = S(L, z) = 2\|z\| + L$. Note that

$$Q^\nu_T \subset Q^\nu_{S+T-L}(-z) \tag{5.25}$$

for all $T > 0$. Now, let u be a function realizing $\widetilde{m}^L(\omega)(T, \nu)$ and set

$$v(x) = \begin{cases} u(x) & \text{if } x \in Q^\nu_T, \\ u^\nu_{-z}(x) & \text{if } x \in Q^\nu_{S+T}(-z) \setminus Q^\nu_T. \end{cases}$$

Note that the jump set of v on $Q^\nu_{S+T}(-z) \backslash Q^\nu_T$ is contained in $S(u^\nu_{-z}) \cup \{u^\nu_{-z} \neq u^\nu\}$, and thus its $\mathcal{H}^{d-1}$-measure scales like T^{d-2}. Hence, separating the energetic contribution of the interactions in Q^ν_T, those in $Q^\nu_{S+T}(-z) \setminus Q^\nu_T$ and those crossing ∂Q^ν_T, we get

$$\begin{aligned} E_1^{\omega,L}(v; Q^\nu_{S+T}(-z)) &\le E_1^{\omega,L}(u; Q^\nu_T) + O(T^{d-2}) \\ &= \widetilde{m}^L(\omega)(T, \nu) + O(T^{d-2}). \end{aligned} \tag{5.26}$$

Then, the previous inequality and the stationarity assumption on $\mathcal{L}$ yield that

$$\begin{aligned} \widetilde{\varphi}^L(\tau_z \omega, \nu) &\le \limsup_{T\to+\infty} \frac{1}{(S+T)^{d-1}} E_1^{\omega,L}(v; Q^\nu_{S+T}(-z)) \\ &\le \limsup_{T\to+\infty} \frac{1}{T^{d-1}} \widetilde{m}^L(\omega)(T, \nu) = \widetilde{\varphi}^L(\omega, \nu). \end{aligned}$$

The opposite inequality can be proven similarly, so that the limit indeed exists and, for all $\omega \in O^L$,

$$\widetilde{\varphi}^L(\tau_z\omega, \nu) = \widetilde{\varphi}^L(\omega, \nu). \tag{5.27}$$

In particular, this shows that $\widetilde{\varphi}^L(\cdot\,, \nu)$ is measurable with respect to the σ-algebra $\mathcal{J}$ of invariant sets

$$\mathcal{J} = \{A \in \mathcal{F}\colon\ \mathbb{P}(A\Delta\tau_z A) = 0 \ \text{ for all } \ z \in \mathbb{Z}^d\}.$$

Step 3 We have $\varphi^L_{\rm hom}(\omega, \nu) = \widetilde{\varphi}^L(\omega, \nu)$.

We eventually prove the validity of (5.16) and that $\varphi^L_{\rm hom}(\omega, \nu) = \widetilde{\varphi}^L(\omega, \nu)$ for all $\nu \in S^{d-1}$ and for almost every $\omega \in O$. A crucial tool in the proof will be Birkhoff's ergodic theorem.

Let $\nu \in S^{d-1} \cap \mathbb{Q}^d$ and $x_0 \in \mathbb{Z}^d\backslash\{0\}$. Given $\varepsilon > 0$, we define the events

$$Q_N = \left\{\omega \in O\colon \left|\frac{1}{k^{d-1}}\widetilde{m}^L(\omega)(k,\nu) - \widetilde{\varphi}^L(\omega,\nu)\right| \le \varepsilon \text{ for all } k \ge \frac{N}{2}\right\}.$$

By Step 1 we know that the function χ_{Q_N} converges almost surely to χ_O. Let $\mathcal{J}_{x_0}$ denote the σ-algebra of invariant sets for the measure-preserving map τ_{x_0}. Fatou's lemma for the conditional expectation yields

$$\chi_O = \mathbb{E}[\chi_O|\mathcal{J}_{x_0}] \le \liminf_{N\to+\infty} \mathbb{E}[\chi_{Q_N}|\mathcal{J}_{x_0}]. \tag{5.28}$$

Using (5.28), we know that, given $\eta > 0$, we find $N_0 = N_0(\omega, \eta)$ such that

$$1 \ge \mathbb{E}[\chi_{Q_{N_0}}|\mathcal{J}_{x_0}](\omega) \ge 1 - \eta.$$

Thanks to Birkhoff's ergodic theorem, almost surely for every $\gamma > 0$ there exists $m_0 = m_0(\omega, \gamma)$ such that

$$\left|\frac{1}{m}\sum_{k=1}^{m} \chi_{Q_{N_0}}(\tau_{kx_0}\omega) - \mathbb{E}[\chi_{Q_{N_0}}|\mathcal{J}_{x_0}](\omega)\right| \le \gamma$$

for any $m \ge \frac{1}{2}m_0$. For fixed $m \ge \max\{m_0(\omega,\gamma), N_0(\omega,\eta)\}$, let S denote the maximal integer such that, for all $k = m+1, \dots, m+R$, we have $\tau_{kx_0}(\omega) \notin Q_{N_0}$. In order to estimate S, let $\widetilde{m}$ be the number of unities in the sequence $\{\chi_{Q_{N_0}}(\tau_{kx_0}(\omega))\}_{k=1}^m$. By the definition of S we have

$$\begin{aligned}
\gamma &\ge \left|\frac{\widetilde{m}}{m+S} - \mathbb{E}[\chi_{Q_{N_0}}|\mathcal{J}_{x_0}](\omega)\right| \\
&= \left|1 - \mathbb{E}[\chi_{Q_{N_0}}|\mathcal{J}_{x_0}](\omega) + \frac{\widetilde{m} - m - S}{m+S}\right| \ge \frac{S + m - \widetilde{m}}{m+S} - \eta.
\end{aligned}$$

Since $m - \widetilde{m} \geq 0$ and without loss of generality $\gamma + \eta \leq \frac{1}{2}$, we get that $S \leq 2m(\gamma+\eta)$. Hence, if we choose an arbitrary $m \geq \max\{m_0(\omega,\gamma), N_0(\omega,\eta)\}$ and $\widetilde{S} = 3m(\gamma+\eta)$ we find $l \in [m+1, m+\widetilde{S}]$ such that $\tau_{lx_0}(\omega) \in Q_{N_0}$. Then, by (5.27), we have that

$$\left|\frac{1}{n^{d-1}}\widehat{m}^L(\omega; Q_n^\nu(-lx_0)) - \widetilde{\varphi}^L(\omega,\nu)\right| \leq \varepsilon \tag{5.29}$$

for all $n \geq \frac{N_0}{2}$, where

$$\widehat{m}^L(\omega; Q_n^\nu(-lx_0)) = \inf\left\{E_1^{\omega,L}(u; Q_n^\nu(-lx_0)) \colon u = u_{-lx_0}^\nu \text{ on } \mathcal{L}(\omega)\backslash Q_{n-L}^\nu(-lx_0)\right\}.$$

Setting $l_1 = m + 2\|x_0\|(l-m)$, one can check that $Q_m^\nu(-mx_0) \subset Q_{l_1}^\nu(-lx_0)$ and that each face of the cube $Q_m^\nu(-mx_0)$ has at most the distance $l_1 - m$ from the corresponding face in $Q_{l_1}^\nu(-lx_0)$. Let u be a function minimizing $m_\delta^L(\omega)(-x_0, m, \nu)$ and set

$$v(x) = \begin{cases} u(x) & \text{if } x \in Q_m^\nu(-mx_0), \\ u_{-lx_0}^\nu(x) & \text{otherwise.} \end{cases}$$

Note that the jump set of v on $Q_{l_1}^\nu(-lx_0) \setminus Q_m^\nu(-mx_0)$ is contained in $S(u_{-lx_0}^\nu) \cup \{u_{-lx_0}^\nu \neq u_{-mx_0}^\nu\}$ and thus its $\mathcal{H}^{d-1}$-measure is less than $C\widetilde{S}(l_1)^{d-2}$, for some constant C depending only on x_0. Hence, arguing as in the proof of (5.26), we get

$$E_1^{\omega,L}(v; Q_{l_1}^\nu(-lx_0)) \leq m_\delta^L(\omega)(-x_0, m, \nu) + C\widetilde{S}(l_1)^{d-2}.$$

Dividing the last inequality by $(l_1)^{d-1}$ and since, for m large enough, v is a test function for $\widehat{m}^L(\omega; Q_{l_1}^\nu(-lx_0))$, we deduce

$$\begin{aligned} \frac{\widehat{m}^L(\omega; Q_{l_1}^\nu(-lx_0))}{(l_1)^{d-1}} &\leq \frac{m_\delta^L(\omega)(-x_0, m, \nu)}{(l_1)^{d-1}} + C\frac{\widetilde{S}}{l_1} \\ &\leq \frac{m_\delta^L(\omega)(-x_0, m, \nu)}{m^{d-1}} + 3C(\gamma+\eta). \end{aligned} \tag{5.30}$$

Analogously, given $\delta > 0$ and setting $l_2 = (1-\delta)m - 2\|x_0\|(l-m)$, one can check that $Q_{l_2}^\nu(-lx_0) \subset Q_{(1-\delta)m}^\nu(-mx_0)$. Then, arguing as earlier, we have that

$$\frac{m_\delta^L(\omega)(-x_0, m, \nu)}{m^{d-1}} \leq \frac{\widehat{m}^L(\omega; Q_{l_2}^\nu(-lx_0))}{l_2^{d-1}} + C(3(\gamma+\eta)+\delta). \tag{5.31}$$

Now if γ, η, and δ are small enough (depending only on x_0), we have $l_1 \geq l_2 \geq \frac{m}{2} \geq \frac{N_0}{2}$. Combining (5.30), (5.31), and (5.29), we infer that

$$\limsup_{m \to +\infty} \left| \frac{m_\delta^L(\omega)(-x_0, m, \nu)}{m^{d-1}} - \tilde{\varphi}^L(\omega, \nu) \right| \leq C(3(\gamma + \eta) + \delta) + \varepsilon.$$

Note that the argument also holds if we replace Q^ν with a cube Q_ϱ^ν with $\varrho \in \mathbb{Q}$, where the constants may also depend on ϱ. The extension to arbitrary sequences (and thus to rational points) works similar as for the case of cubes centered at the origin except that now the cubes to be compared are not contained in each other. This is a minor detail that can be fixed by the same arguments already used in the proof of the invariance under the group action. Hence, we infer that, for all points in $\mathbb{Q}^d$ and all rational directions, (5.16) holds almost surely, since the subset of O we exclude is a countable union of null sets. For irrational points and irrational directions $\nu \in S^{d-1}$, the statement follows by continuity arguments.

The claim of the theorem regarding the ergodic case follows from (5.27) since in this case all the functions $\varphi^L_{\mathrm{hom}}(\cdot, \nu)$ are constant and so is the pointwise limit. □

5.2 Poisson Random Sets in the Plane

In this section we consider some random lattices that do not satisfy the regularity assumptions of the stochastic lattices as in the previous section. Namely, in $\mathbb{R}^2$ we consider a *Poisson random set* $\mathcal{L} = \mathcal{L}^\omega$ with *intensity* λ, defined on a probability space $(O, \mathcal{F}, \mathbb{P})$, characterized by the following:

(i) for any bounded Borel set $B \subset \mathbb{R}^2$, the number of points in $B \cap \mathcal{L}$ has a Poisson law with parameter $\lambda|B|$; that is,

$$\mathbb{P}(\{\#(B \cap \mathcal{L}) = n\}) = e^{-\lambda|B|} \frac{(\lambda|B|)^n}{n!};$$

(ii) for any collection of bounded disjoint Borel subsets in $\mathbb{R}^2$, the random variables defined as the number of points of $\mathcal{L}$ in these subsets are independent.

We do not enter into the details of the definition of such a random set. As a technical note, the probability space O is equipped with a dynamical system $T_x : O \mapsto O$, for $x \in \mathbb{R}^2$, such that for any bounded Borel set B and any $x \in \mathbb{R}^2$, we have $\#((B + x) \cap \mathcal{L})(\omega) = \#(B \cap \mathcal{L})(T_x\omega)$. We suppose that T_x is a group of measurable measure-preserving transformations in O and is ergodic.

We simply write $\mathcal{L}$ in the place of $\mathcal{L}(\omega)$ in order to ease the notation and when making almost certain statements.

As usual, we define the ferromagnetic energy of $\mathcal{L}$ as

$$E^{\mathcal{L}}(u) = \sum_{\langle i,j \rangle} (u(i) - u(j))^2,$$

for $u\colon \mathcal{L} \to \{-1, 1\}$, where $\langle i, j\rangle$ denotes nearest-neighbor pairs in the sense of the Delaunay triangulation related to $\mathcal{L}$, and the corresponding scaled energies

$$E^{\mathcal{L}}_{\varepsilon}(u) = \sum_{\langle i,j \rangle} \varepsilon(u_i - u_j)^2, \tag{5.32}$$

for $u\colon \varepsilon\mathcal{L} \to \{-1, 1\}$, where $u_i = u(\varepsilon i)$.

We note that, contrary to a stochastic lattice,

(i) $\mathcal{L}$ is not regular: we have pairs of points of $\mathcal{L}$ arbitrarily close, and squares of arbitrary size not containing points of $\mathcal{L}$;
(ii) $\mathcal{L}$ is isotropic since the properties of Poisson random sets are invariant under (translations and) rotations.

As a consequence of the lack of regularity, given a family of Voronoi cells in $\mathcal{L}$, it is not possible to estimate the perimeter of its union by the number of edges, which is the underlying argument to prove compactness properties for sequences u^{ε} with equibounded energies on stochastic lattices. Nevertheless, compactness properties are possible through the use of the following percolation lemma, which uses a covering of Voronoi cells by *polyominos* (unions of squares). If P is a finite connected union of Voronoi cells of $\mathcal{L}$, we set

$$\mathbf{A}(P) = \{z \in \mathbb{Z}^2 \colon (z + (0,1)^2) \cap P \neq \emptyset\}.$$

Lemma 5.9 (Polyomino lemma) *Let $R > 0$ and $\gamma > 0$. Then there exists a deterministic constant C such that for almost all ω there exists $\varepsilon_0 = \varepsilon_0(\omega) > 0$ such that, if P is a finite connected union of Voronoi cells of $\mathcal{L}$ and $\varepsilon < \varepsilon_0$ satisfy*

$$P \cap \frac{R}{\varepsilon}(-1,1)^2 \neq \emptyset, \qquad \max\Big\{\#\{i : C_i \subset P\}, \#\mathbf{A}(P)\Big\} \geq \varepsilon^{-\gamma},$$

then we have

$$\frac{1}{C}\#\{i : C_i \subset P\} \leq \#\mathbf{A}(P) \leq C\,\#\{i\colon C_i \subset P\}.$$

This lemma states that a large connected family of Voronoi cells C_i can be identified with a union of a comparable number of unit squares, and can be used

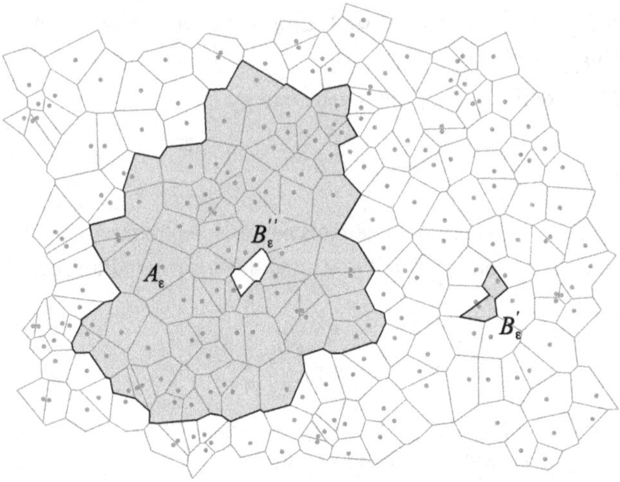

Figure 5.1 Decomposition of a Voronoi set (union of gray cells).

to prove compactness properties for sequences $\{u^\varepsilon\}$ with equibounded $E_\varepsilon^{\mathcal{L}}(u^\varepsilon)$ as follows. Let

$$V_\varepsilon(u^\varepsilon) = \bigcup_{u_i^\varepsilon=1} \varepsilon C_i.$$

Then the argument is that we can write

$$V_\varepsilon(u^\varepsilon) = (A_\varepsilon \cup B'_\varepsilon) \setminus B''_\varepsilon$$

(as pictured in Fig. 5.1), where $|B'_\varepsilon| + |B''_\varepsilon| \to 0$ (using either an isoperimetric argument or the Polyomino Lemma), and $\{A_\varepsilon\}$ is a family of sets with equibounded perimeter.

In the following lemma, ω is fixed such that the polyomino lemma holds. Hence the statement holds almost surely with respect to ω.

Lemma 5.10 (Compactness of Voronoi sets) *Let ω be such that the polyomino lemma holds. Let $\{u^\varepsilon\}$ be such that $\sup_\varepsilon E_\varepsilon(u^\varepsilon) < +\infty$. Then there exists a set of finite perimeter A such that $\chi_{V_\varepsilon(u^\varepsilon)}$ converge to χ_A in $L^1_{\rm loc}(\mathbb{R})$.*

Proof Since we reason locally, we may assume that all $V_\varepsilon(u^\varepsilon)$ are contained in a fixed cube.

We fix $\gamma > 0$ small enough. We subdivide $\partial V_\varepsilon(u^\varepsilon)$ into its connected components. Let $C_\varepsilon^{\gamma,+}$ denote the family of such connected components S with

$$\#\{i \in \mathcal{L} : u_i^\varepsilon = 1, \varepsilon C_i \cap S \neq \emptyset\} \geq \varepsilon^{-\gamma}. \tag{5.33}$$

Note that each such connected component can be identified with the set

$$P = P(S) = \bigcup \Big\{ C_i : u_i^\varepsilon = 1, \varepsilon C_i \cap S \neq \emptyset \Big\}. \tag{5.34}$$

Correspondingly, $\mathcal{C}_\varepsilon^{\gamma,-}$ denotes the family of the remaining connected components.

The first step will be to identify the small sets B'_ε and B''_ε as the "interior" of contours in $\mathcal{C}_\varepsilon^{\gamma,-}$ where the inner trace of $\chi_{V_\varepsilon(u^\varepsilon)}$ is 0 and 1, respectively. In this way the remaining set will have a boundary only composed of "large" components from $\mathcal{C}_\varepsilon^{\gamma,+}$. In order to simplify the argument we may suppose that there are no contours contained in other contours.

By the finiteness of the energy we have

$$\#\mathcal{C}_\varepsilon^{\gamma,-} \leq \frac{C}{\varepsilon}.$$

Note that

$$\#\mathbf{A}\Big(\frac{1}{\varepsilon}S\Big) \leq C\varepsilon^{-\gamma} \text{ for every } S \in \mathcal{C}_\varepsilon^{\gamma,-}.$$

Hence each $S \in \mathcal{C}_\varepsilon^{\gamma,-}$ is contained in a set with boundary at most of length $C\varepsilon^{1-\gamma}$. By an isoperimetric estimate, the measure of the bounded set sorrounded by each $S \in \mathcal{C}_\varepsilon^{\gamma,-}$ is $O(\varepsilon^{2-2\gamma})$. Hence, the total measure of such sets is $O(\varepsilon^{1-2\gamma})$.

Let now $S \in \mathcal{C}_\varepsilon^{\gamma,-}$. For each such S, let P be defined from S by (5.34). We have two cases, whether εP is interior to S or not. Let $\mathcal{C}_{1,\varepsilon}^{\gamma,-}$ denote the first family, $\mathcal{C}_{2,\varepsilon}^{\gamma,-}$ the second one, and define B_ε as the union of the εC_i in the interior of S for some $S \in \mathcal{C}_{1,\varepsilon}^{\gamma,-}$ and such that $u_i^\varepsilon = 1$, and B''_ε as the union of the εC_i in the interior of S for some $S \in \mathcal{C}_{2,\varepsilon}^{\gamma,-}$ and such that $u_i^\varepsilon = -1$. If we set

$$V_\varepsilon = (V_\varepsilon(u^\varepsilon) \setminus B_\varepsilon) \cup B''_\varepsilon$$

then ∂V_ε consists only of components in $\mathcal{C}_\varepsilon^{\gamma,+}$, and $|B_\varepsilon \cup B''_\varepsilon| \leq C\varepsilon^{1-2\gamma}$. In Fig. 5.2 we picture the interior boundary cells of V_ε.

We now write $V_\varepsilon = A_\varepsilon \cup A'_\varepsilon$, where

$$A_\varepsilon = \bigcup_{Q_\varepsilon(z) \subset V_\varepsilon} Q_\varepsilon(z) \quad \text{and} \quad A'_\varepsilon = V_\varepsilon \setminus A_\varepsilon.$$

Note that

$$\partial A_\varepsilon \subset \varepsilon \bigcup_{S \in \mathcal{C}_\varepsilon^{\gamma,+}} \partial \mathbf{A}(P(S)),$$

with $P(S)$ defined in (5.34).

In Fig. 5.3 we picture such union elements of $\mathbf{A}(P(S))$. By Lemma 5.9 we have

$$\mathcal{H}^1(\partial \mathbf{A}(P(S))) \leq C\#\{i \in \mathcal{L} : u_i^\varepsilon = 1, \varepsilon C_i \cap S \neq \emptyset\}.$$

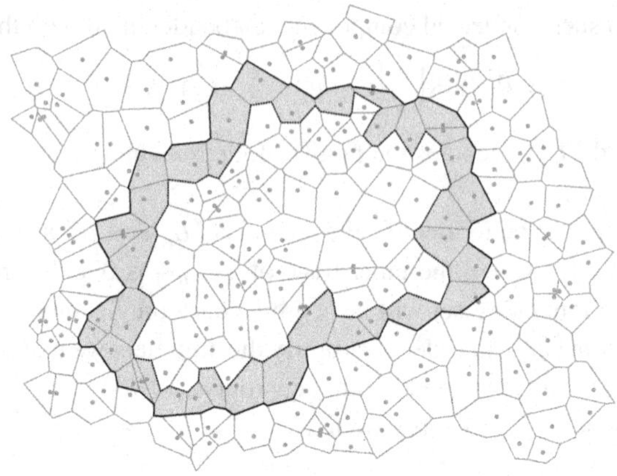

Figure 5.2 Boundary cells of a Voronoi set.

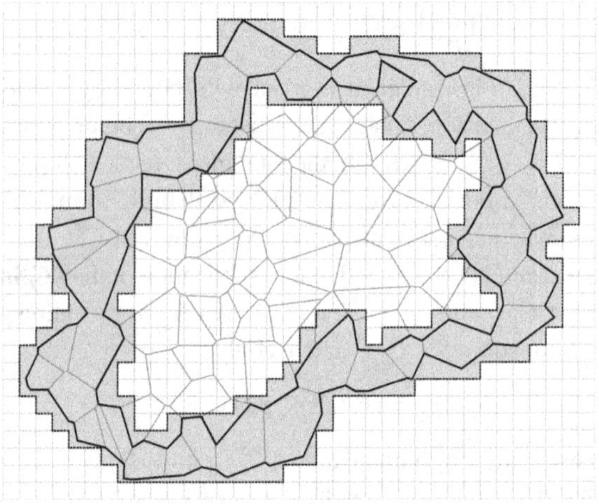

Figure 5.3 Use of the Polyomino Lemma to bound the perimeter of a large Voronoi set.

Summing up over all $S \in C_\varepsilon^{\gamma,+}$, we obtain

$$\mathcal{H}^1(\partial A_\varepsilon) \leq C\, E_\varepsilon(u^\varepsilon).$$

Hence, $\{A_\varepsilon\}$ is a family of sets with equibounded perimeter, and the functions χ_{A_ε} are locally precompact in $L^1(\mathbb{R}^2)$. Again by Lemma 5.9 we have

$$|A'_\varepsilon| \leq C\varepsilon^2 \sum_{S \in C_\varepsilon^{\gamma,+}} \#\mathbf{A}(P(S)) \leq C\varepsilon E_\varepsilon(u^\varepsilon).$$

This shows that $|A'_\varepsilon| \to 0$, and proves the claim, upon setting $B'_\varepsilon = A'_\varepsilon \cup B_\varepsilon$. □

This lemma defines a convergence $u^\varepsilon \to A$; moreover, it gives the first variational ingredient in order to prove the existence of the Γ-limit; the second one regards the possibility to change boundary values in order to prove a homogenization formula. To this end we would like to use a discrete coarea argument. In general, again this is not possible since from an estimate on the length of an interface we cannot deduce an estimate on the energy of such an interface. Indeed, since $\mathcal{L}$ is not regular we can have arbitrarily small edges of Voronoi cells. However, we can prove that we can limit our interfaces to lie on the boundary of regular Voronoi cells defined as follows.

For $\alpha > 0$ we set

$$\mathcal{L}^0_\alpha = \Big\{ i \in \mathcal{L} \colon C_i \text{ contains a ball of radius } \alpha, \\ \operatorname{diam} C_i \leq \frac{1}{\alpha}, \#\{\text{edges of } C_i\} \leq \frac{1}{\alpha} \Big\},$$

the family of *regular Voronoi cells with parameter* α. The following lemma describes some geometrical features of regular Voronoi tessellations.

Lemma 5.11 (A channel property of $\mathcal{L}^0_\alpha$) *Let $\delta > 0$. For every $T \in \mathbb{R}$, $\nu \in S^1$, and $x \in \mathbb{R}^2$, we define*

$$R^\nu_{T,\delta}(x) = \Big\{ y \colon |\langle y - x, \nu \rangle| \leq \delta T, |\langle y - x, \nu^\perp \rangle| \leq \frac{1}{2} T \Big\}.$$

Then there exist $\alpha_0, C_\delta > 0$ such that almost surely there exists $T_0(\omega) > 0$ such that for all $T > T_0(\omega)$ the rectangle $R^\nu_{T,\delta}(x)$ contains at least $C_\delta T$ disjoint paths of Voronoi cells C_i with $i \in \mathcal{L}^0_\alpha$ connecting the two opposite sides of $R^\nu_{T,\delta}(x)$ parallel to ν. This property is uniform as $\frac{x}{T}$ varies on a bounded set of $\mathbb{R}^2$.

This lemma can be proven using combinatoric arguments that allow one to use well-known results for the Bernoulli bond-percolation model. Its use allows one to fix boundary values in the blow-up argument, which, in this two-dimensional setting, involves minimal-path problems. To that end we first let $\mathcal{L}^*$ denote the *dual lattice* to $\mathcal{L}$, that is, the set of all endpoints of edges of Voronoi cells, and let $\pi_0(x)$ denote the point in $\mathcal{L}^*$ closest to x. Note that this is a good definition for almost all x. For the remaining points, we can arbitrarily choose one of the points of $\mathcal{L}^*$ closest to x.

By the isotropy of the Poisson set it is sufficient to characterize the limit surface tension when the normal is, for example, ν_2.

Proposition 5.12 (Surface tension) *Almost surely there exists the limit*

$$\tau = \lim_{t\to+\infty} \frac{1}{t} \min\{\#\{\sigma_i\} : \{\sigma_i\} \text{ path of edges of Voronoi cells with endpoints } \pi_0(0,0) \text{ and } \pi_0(t,0)\}$$

and is a deterministic quantity.

Note that τ depends on λ, and a scaling argument allows one to conclude that $\tau = \sqrt{\lambda}\,\tau_1$, τ_1 being the surface tension for a Poisson random set with intensity 1. The homogenization theorem reads as follows.

Theorem 5.13 (Homogenization on random Poisson sets in the plane) *Let $\mathcal{L}$ be a Poisson random set in $\mathbb{R}^2$ with intensity λ and let $E_\varepsilon^{\mathcal{L}}$ be defined by* (5.32)*. Then almost surely $E_\varepsilon^{\mathcal{L}}$ Γ-converges to $8\sqrt{\lambda}\tau_1\mathcal{H}^1(\partial^* A)$ with τ_1 the deterministic constant given by Proposition* 5.12 *when $\lambda = 1$.*

Remark 5.14 (An extension: finite-range systems) We can use the properties of $\mathcal{L}_\alpha^0$ and the arguments leading to the preceding theorem to prove that, for $R > 0$ large enough (corresponding to α small enough), the energies

$$E_\varepsilon^{\mathcal{L},R}(u) = \sum_{\|i-j\|\leq R} \varepsilon(u_i - u_j)^2$$

almost surely Γ-converge to an isotropic energy $\tau_R\sqrt{\lambda}\mathcal{H}^1(\partial^* A)$.

Note that this energy cannot be directly compared with nearest-neighbor energies, since some nearest neighbors can indeed be at a distance larger than R.

Bibliographical Notes to Chapter 5

Stochastic lattices were considered by Blanc et al. (2006) in the context of limits defined on Sobolev functions. The variational analysis of a general class of discrete energies defined on stochastic lattices with limits in Sobolev spaces was given by Alicandro et al. (2011a). The methods used in their analysis were introduced by Dal Maso and Modica (1986) for the stochastic homogenization of nonlinear integral functionals of the gradient and exploit the Ergodic Theorem 5.7 proved in Akcoglu and Krengel (1981). This is the key ingredient that allows one to elaborate a stochastic equivalent of the subadditive argument used to show the existence of the asymptotic homogenization formula. The homogenization for spin energies on stochastic lattices as illustrated in this chapter has been obtained by Alicandro et al. (2015). A similar analysis for thin films was provided by Braides et al. (2018c).

The results on Poisson clouds discussed in Section 5.2 are proved in Braides and Piatnitski (2022). We refer to, for instance, Daley and Vere-Jones (1988) for equivalent definitions of a Poisson random set and its main properties. The Polyomino Lemma is due to Pimentel (2013). Some of the geometric results on regular Voronoi cells shown previously are also essential in the treatment of discrete energies on Poisson Clouds in the case that the limit energy is finite on Sobolev spaces, as shown by Braides and Caroccia (2022). The treatment of discrete systems on Poisson random point sets is close in spirit to variational models in Machine Learning (García Trillos and Slepčev, 2015, 2016).

6
Extensions

This chapter is devoted to some examples in which, even though the general ideas introduced in the previous chapters can be followed in great part, some additions must be made for a better description of the limit behavior of spin systems. We only treat some model cases, noting that each of these topics is open to a large number and variety of further developments.

In Section 6.1 we consider a ternary system, which is parameterized by $Y = \{-1, 0, 1\}$ and with $Y_0 = \{-1, 1\}$ in the notation of the Compactness Theorem. In this case, a finer description of the limit is obtained by highlighting explicitly the dependence of the limit on the 0-phase. In order to do that, we introduce a dependence on a measure μ describing the concentration of the 0-phase, which may occur at the interface between minimal phases or elsewhere.

In Section 6.2, in the framework of ferromagnetic binary systems, we relax coerciveness conditions on the interactions by allowing some of the coefficients a^{ε}_{ij} to scale as ε, while they still satisfy the hypothesis considered in Section 3.3.5 for perforated domains on N connected components. As a result, the limit is defined on N-tuples of sets of finite perimeter, with an interaction term added to the respective perforated-domain perimeter limit.

Section 6.3 is devoted to the analysis of a system of nearest-neighbor ferromagnetic interactions with the constraint that the 1-phase can be regrouped as a union of sets with given shape (molecules). This constraint enforces incompatibility of ground states differing by translations, so that the limit is parameterized on a larger number of variables. For instance, in the case explicitly treated the number of these variables is four.

In Section 6.4 we relax the decay hypotheses on the coefficients a^{ε}_{ij} of a pairwise ferromagnetic system. In this case, as anticipated in the simple model of Example 4.2, an additional nonlocal term appears in the limit.

6.1 Surfactant Ternary Systems

In this section we consider ternary spin systems; that is, systems for which the order parameter u may take three values, say $Y = \{-1, 0, 1\}$. Comparing this case to the binary one described in Chapter 3, here the presence of more than two competing phases makes it possible to construct richer models where different phenomena occur. In particular, we deal with a special class of ternary systems that are motivated by phase-transition models in the presence of *surfactants* (surface active agents), from which we borrow some terminology.

6.1.1 One-Dimensional Ternary Systems

We start by highlighting a phenomenon due to the presence of competing different "species" of phases. We consider $u\colon \mathcal{L}_\varepsilon \to \{-1, 0, 1\}$, where $\mathcal{L}_\varepsilon = \varepsilon\mathbb{Z} \cap [0, 1]$, and $f\colon \{-1, 0, 1\}^2 \to \mathbb{R}$ satisfying the following properties:

(i) (*symmetry*) $f(u, v) = f(v, u) = f(-u, -v)$;
(ii) (*uniform ground states*) $\operatorname{argmin} f = \{(1, 1), (-1, -1)\}$ and, without loss of generality, $\min f = 0$;
(iii) (*nontrivial optimal phase transition*) $f(-1, 1) > 2 f(0, 1) > 0$.

The energies we consider are

$$E_\varepsilon(u) = \sum_{\varepsilon i \in \mathcal{L}_\varepsilon} f(u_i, u_{i-1}).$$

For simplicity, we suppose that $N = \frac{1}{\varepsilon}$ is an integer and we assume periodic boundary conditions so as to avoid boundary effects. Condition (ii) ensures that ground states are only the constant functions identically equal to 1 and -1. Condition (iii) implies that if we have a transition between 1 and -1, it is energetically favorable to insert a 0 phase. In the framework of phase-transition models, condition (iii) is usually rephrased to say that the 0-phase is of surfactant type, meaning that it lowers the surface tension.

If (u^ε) is a sequence with $\sup E_\varepsilon(u^\varepsilon) = C < +\infty$, then we deduce that $f(u_i^\varepsilon, u_{i-1}^\varepsilon) = 0$ except for a finite set of i. Up to subsequences, this set may be written as

$$I_\varepsilon = \{i_1^\varepsilon, \ldots, i_M^\varepsilon\}$$

for some M. We also define $i_0^\varepsilon = 0$ and $i_{M+1}^\varepsilon = N$. Note that condition (ii) gives

(i) $u_i^\varepsilon = u_{i_k^\varepsilon}^\varepsilon \in \{-1, 1\}$ for $i \in \{i_k^\varepsilon, \ldots, i_{k+1}^\varepsilon - 1\}$ and for all $k \in \{0, \ldots, M\}$;
(ii) if we set $J_\varepsilon = \{\varepsilon i\colon u_i^\varepsilon = 0\}$, then $\#J_\varepsilon \leq M$.

Note the multiscale behavior of the parameters: assuming that u^ε converge, we have

(i) the 0-phase is not detected by L^1-convergence; moreover, the piecewise-constant extensions of u^ε converge to a function $u \in BV((0,1);\{-1;1\})$;
(ii) the 0-phase is concentrated on a finite number of points, which converge, up to subsequences, to points $\{x_1,\dots,x_M\}$. The set of these limits is denoted by J. Note that some of these points may coincide.

It is convenient to consider the measures

$$\mu_\varepsilon = \sum_{i\in J_\varepsilon} \delta_{\varepsilon i}.$$

Up to subsequences, these measures converge weakly* to a measure

$$\sum_{x\in J} N(x)\delta_x,$$

with $N(x) \in \mathbb{N}$ representing the number of sequences in J_ε converging to x.

We may alternatively compute the Γ-limit with respect to the L^1-convergence, thus integrating out the effect of the 0-phase in the limit, or with respect to the convergence in $L^1 \times \mathcal{M}$, where $\mathcal{M}$ is the space of bounded measures equipped with the weak*-convergence. The energies are coercive with respect to both convergences. The second Γ-limit gives more information and is compatible with more constraints, for example, if we fix the total mass of μ_ε or, equivalently, the number of points in J_ε.

Theorem 6.1 (Multiphase limits) *The Γ-limit is finite on pairs of function-measures (u,μ) as earlier, and is of the form*

$$F(u,\mu) = C_0\#(S(u)\setminus J) + \sum_{x\in J}(C_1 + C_2(N(x)-1)),$$

where $C_0 = f(1,-1)$, $C_1 = 2f(0,1)$, and $C_2 = \min\{f(0,0), 2f(0,1)\}$.

Proof If $x \in S(u)\setminus J$, then there exist no sequences of points of J_ε converging to x; hence we have at least one index i with $\varepsilon i \to x$ and $f(u_i^\varepsilon, u_{i-1}^\varepsilon) = f(1,-1)$.

If $x \in J$, then there exist at least two indices i such that $f(u_i^\varepsilon, u_{i-1}^\varepsilon)$ is equal to $f(0,1)$ or to $f(0,-1)$ (which coincide by the hypothesis on f). The other interactions of i where $u_i^\varepsilon = 0$ or $u_{i-1}^\varepsilon = 0$ satisfy

$$f(u_i^\varepsilon, u_{i-1}^\varepsilon) \geq \min\{f(0,1), f(0,0)\},$$

so that, taking into account that $(0,0)$ pairs are counted twice, for any such x we have a contribution not less than $C_1 + C_2(N(x)-1)$.

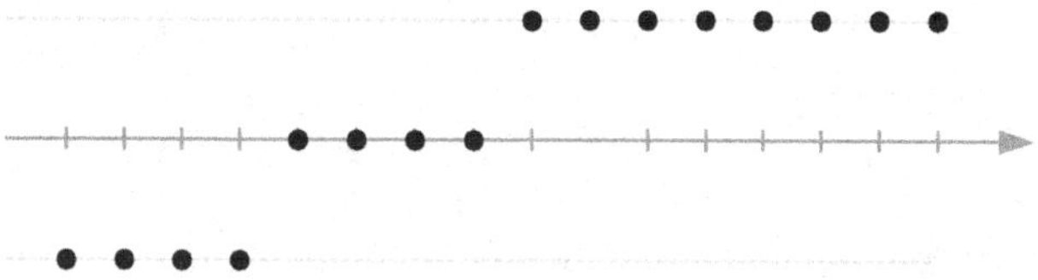

Figure 6.1 Recovery sequence in the case $f(0,0) < 2f(0,1)$ for $N(x) = 4$.

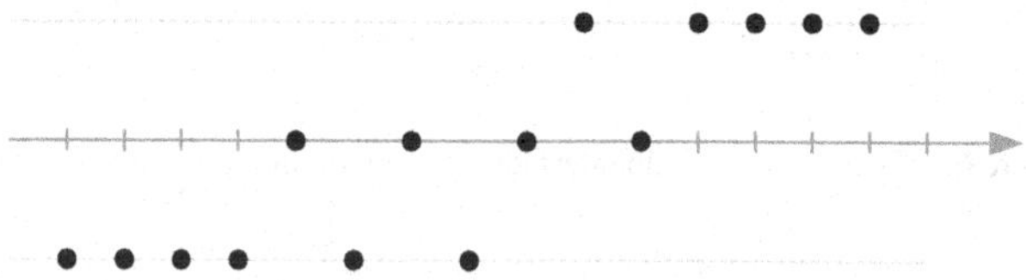

Figure 6.2 Recovery sequence in the case $f(0,0) > 2f(0,1)$ for $N(x) = 4$.

The recovery sequences are constructed by optimizing the preceding process, and are pictured in Figs. 6.1 and 6.2. □

Note that we may rewrite F as

$$F(u,\mu) = \sum_{x\in S(u)} \psi(N(x)) + \sum_{x\in J\setminus S(u)} (C_1 + C_2(N(x)-1)),$$

where ψ is given by

$$\psi(N) = \begin{cases} C_0 & \text{if } N = 0, \\ C_1 - C_2 + C_2 N & \text{if } N \geq 1, \end{cases}$$

and highlights the effect of the 0-variable on the interface (see Fig. 6.3).

Corollary 6.2 *The Γ-limit of E_ε with respect to the L^1-convergence is given by*

$$F(u) = C_1\#(S(u)).$$

Proof The Γ-limit is obtained from the preceding theorem by choosing $J = S(u)$ and $N(x) = 1$; that is, $\mu = \sum_{x\in S(u)} \delta_x$. □

Example 6.3 Let $K > 0$ be a fixed natural number and let $E_\varepsilon(u)$ be defined as earlier with the constraint that $\#\{i\colon u_i = 0\} = K$. The Γ-limit with respect to the L^1-convergence is given by

$$F(u) = \begin{cases} C_1 + C_2(K-1) & \text{if } u \text{ is constant,} \\ (C_1 - C_2)\# S(u) + C_2 K & \text{otherwise.} \end{cases}$$

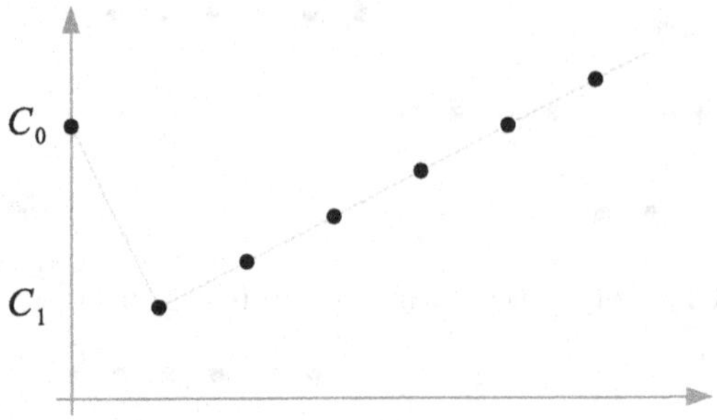

Figure 6.3 Dependence of the interfacial energy on the 0-variable density.

6.1.2 Ternary Systems in Dimension d

We consider $u\colon \mathcal{L}_\varepsilon = \varepsilon\mathbb{Z}^d \cap \Omega \to \{-1, 0, 1\}$, and $f\colon \{-1, 0, 1\}^2 \to \mathbb{R}$ satisfying properties (i)–(ii) of the previous section. The energies we consider are

$$E_\varepsilon(u) = \sum_{\langle i,j\rangle} \varepsilon^{d-1} f(u_i, u_j), \tag{6.1}$$

where as usual $\langle i, j\rangle$ denotes the set of nearest neighbors, with $\varepsilon i, \varepsilon j \in \mathcal{L}_\varepsilon$. Note that the measure of the 0-phase is negligible in the continuum limit. More precisely, since each interaction with a particle i in the 0-phase pays at least a positive energy $\varepsilon^{d-1} \min\{f(0, 1), f(0, 0)\}$, the following estimate holds:

$$\begin{aligned} E_\varepsilon(u) &\geq \#\{\varepsilon i \in \mathcal{L}_\varepsilon\colon u_i = 0\}\, \varepsilon^{d-1} \min\{f(0, 1), f(0, 0)\} \\ &\geq \frac{C}{\varepsilon} |\{x \in \Omega\colon u(x) = 0\}|. \end{aligned} \tag{6.2}$$

This implies that, whenever the energy is bounded, the measure of the 0-phase scales as ε. In addition, as in the Ising model, by identifying u with its piecewise-constant extension, one has

$$E_\varepsilon(u) \geq \min\{f(x, y)\colon x \neq y\}\, \mathcal{H}^{d-1}(S(u)). \tag{6.3}$$

As a result of (6.2) and (6.3), a sequence u^ε such that $\sup_\varepsilon E_\varepsilon(u^\varepsilon) < +\infty$ is in particular L^1-compact in $BV(\Omega; \{-1, 1\})$. Analogously to the previous section, in order to have more information on the 0-phase in the continuum limit, it is convenient to consider the measures

$$\mu_\varepsilon(u) = \sum_{i \in I_0(u)} \varepsilon^{d-1} \delta_{\varepsilon i}, \tag{6.4}$$

where

$$I_0(u) = \{i\colon\ \varepsilon i \in \mathcal{L}_\varepsilon \text{ and } u_i = 0\}. \tag{6.5}$$

By (6.2) we have that $\sup_\varepsilon E_\varepsilon(u^\varepsilon) < +\infty$ implies $\sup_\varepsilon \mu_\varepsilon(u^\varepsilon)(\Omega) < +\infty$ and hence $\mu(u^\varepsilon)$ is compact with respect to the weak*-convergence of measures. Then, by regarding the energy E_ε as a function of $(u, \mu(u)) \in L^1 \times \mathcal{M}$, we may now compute the Γ-limit with respect to the strong convergence in L^1 and the weak*-convergence in $\mathcal{M}$. This second Γ-limit is in particular compatible with the constraint of prescribing the total mass of $\mu(u^\varepsilon)$.

6.1.3 The Blume–Emery–Griffiths Model

In this section we take $d = 2$ and fix

$$f(u, v) = 1 - uv - k(1 - (uv)^2)$$

with $k \in (\frac{1}{3}, 1)$. The condition $k > \frac{1}{3}$ ensures that the 0-phase is relevant at interfaces. To check this, note that, while in dimension 1 it suffices to have $k > 0$ for condition (iii) in the previous section, now we have to take into account that interfaces are lines. In order to estimate the energetic contribution of an interface with the 0-phase we can consider functions of the form

$$u_i = \begin{cases} -1 & \text{if } i_1 < 0, \\ 0 & \text{if } i_1 = 0, \\ 1 & \text{if } i_1 > 0, \end{cases}$$

and compare their energy with that of

$$v_i = \begin{cases} -1 & \text{if } i_1 < 0, \\ 1 & \text{if } i_1 \geq 0. \end{cases}$$

The first ones give an interfacial contribution of $3(1 - k)$ per unit length, while the second ones give an interfacial contribution of 2 per unit length. Hence, the interfacial energy density with surfactant is lower if $k > \frac{1}{3}$, which then implies an analog of assumption (iii) on f in the previous section. The sufficiency of this argument will come from the statements and proofs of the convergence theorem, where terms of the form $1 - 3k$ will appear.

The corresponding energy E_ε, given by

$$E_\varepsilon(u) = \sum_{\langle i,j \rangle} \varepsilon(1 - u_i u_j - k(1 - (u_i u_j)^2)), \tag{6.6}$$

is that of the Blume–Emery–Griffiths model, which describes phase transitions of water and oil in the presence of surfactants.

With a slight abuse of notation, we now extend E_ε to a functional defined on pairs (u, μ) with $u\colon \mathcal{L}_\varepsilon \to \{-1, 0, 1\}$ and $\mu = \mu_\varepsilon(u)$ as in (6.4) and given by

$$E_\varepsilon(u,\mu) = E_\varepsilon(u). \tag{6.7}$$

The following theorem holds, where the dependence on $N(x)$ in the one-dimensional version is replaced with the dependence on $\frac{d\mu}{d\mathcal{H}^1 \llcorner S(u)}$, the density of the measure μ on $S(u)$. We also use the notation

$$\mu^s = \mu - \frac{d\mu}{d\mathcal{H}^1 \llcorner S(u)} \mathcal{H}^1 \llcorner S(u)$$

for the singular part. Note the complex form of energy density φ in dependence of the normal ν. This is the result of the different optimal patterns of the 0-phase at the interface between 1 and -1 in dependence of the orientation.

Theorem 6.4 (Γ-convergence in $L^1 \times \mathcal{M}$) *Let E_ε be the extended functionals defined in* (6.7). *Then, E_ε Γ-converge, with respect to the strong convergence in* $L^1(\Omega)$ *for u and to the weak*-convergence of bounded positive measures in* $\mathbb{R}^2$ *for μ, to a functional F whose domain is $u \in BV(\Omega; \{-1,1\})$ and* $\mathrm{spt}(\mu) \subset \overline{\Omega}$, *where*

$$F(u,\mu) = \int_{S(u)} \varphi\Big(\frac{d\mu}{d\mathcal{H}^1 \llcorner S(u)}, \nu_u\Big) d\mathcal{H}^1 + (2k-2)\mu^s(\overline{\Omega}), \tag{6.8}$$

the function $\varphi\colon [0,+\infty) \times S^1 \to [0,+\infty)$ *being defined by*

$$\varphi(z,\nu) = \max\{\varphi_1(z,\nu), \varphi_2(z,\nu), \varphi_3(z,\nu)\}, \tag{6.9}$$

where

$$\varphi_1(z,\nu) = -4kz + 2(|\nu_1| + |\nu_2|),$$
$$\varphi_2(z,\nu) = (1-3k)z + 2(|\nu_1| \vee |\nu_2|) + (1-k)(|\nu_1| \wedge |\nu_2|),$$
$$\varphi_3(z,\nu) = 2(1-k)z + (1-k)(|\nu_1| + |\nu_2|).$$

Proof We omit the details of the proof of the lim inf inequality, which can be obtained by using the blow-up technique described in Section 3.2 with reference measure λ given by $\lambda(A) = \mathcal{H}^{d-1}(A \cap S(u)) + |A|$. We describe the construction of optimal sequences which highlights the geometry of the 0-phase producing the three types of limit energy densities φ_1, φ_2, and φ_3.

By a density argument, we may reduce to the case when $S(u)$ is a polyhedral set and the measure μ has the form $\mu = h\, d\mathcal{H}^1 \llcorner S(u) + \sum_{k=1}^N c_k \delta_{x_k}$, with h a piecewise-constant function, $N \in \mathbb{N}$ and $\{x_1, x_2, \ldots, x_N\} \subset \Omega$. Since our construction is local, we may further reduce the construction to one of the following two cases:

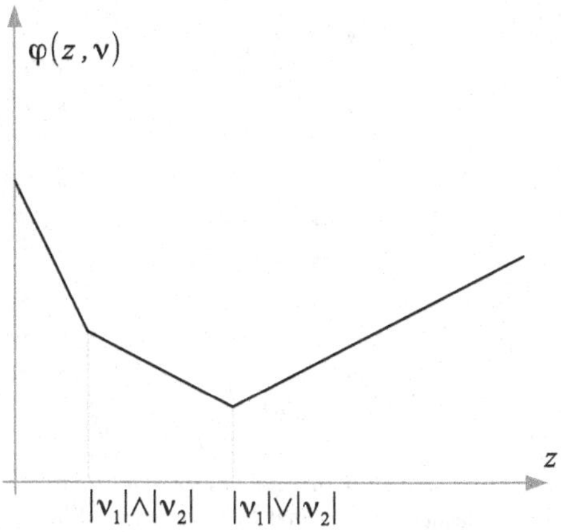

Figure 6.4 The graph of the surface tension density $\varphi(z,\nu)$ as a function of the density z of surfactant at the phase interface.

(i) $(u,\mu) = (u_\nu, \mu_{z,\nu})$, where u_ν is given by

$$u_\nu = \begin{cases} 1 & \text{if } \langle x,\nu\rangle > 0, \\ -1 & \text{if } \langle x,\nu\rangle \le 0, \end{cases} \tag{6.10}$$

and $\mu_{z,\nu} = z\mathcal{H}^1 \llcorner S(u_\nu)$ for some $z \ge 0$;

(ii) $(u,\mu) = (1, c\delta_{x_0})$ for some $c > 0$ and $x_0 \in \Omega$,

To prove the upper estimate in the case (i) we construct u^ε such that $(u^\varepsilon, \mu(u^\varepsilon))$ converge to $(u_\nu, \mu_{z,\nu})$ and

$$\limsup_{\varepsilon\to 0} E_\varepsilon(u^\varepsilon) \le \varphi(z,\nu)\mathcal{H}^1(S(u_\nu)\cap\Omega).$$

We subdivide this construction into three separate cases, in which $\varphi = \varphi_n$ for $n \in \{1,2,3\}$, respectively. Note that, by the definition of φ, we have

$$\varphi(z,\nu) = \begin{cases} \varphi_1(z,\nu) & \text{if } 0 \le z < |\nu_1| \wedge |\nu_2|, \\ \varphi_2(z,\nu) & \text{if } |\nu_1| \wedge |\nu_2| \le z \le |\nu_1| \vee |\nu_2|, \\ \varphi_3(z,\nu) & \text{if } z > |\nu_1| \vee |\nu_2|. \end{cases}$$

Without loss of generality, for simplicity of exposition, we may suppose that $-\nu_1 \ge \nu_2 > 0$. Moreover, by the continuity of $\varphi(z,\cdot)$, by a density argument we may assume that $\frac{\nu_1}{\nu_2} \in \mathbb{Q}$. Let $p, q \in \mathbb{N}$ be such that $\frac{-\nu_1}{\nu_2} = \frac{p}{q}$. Arguing again by density, we may further assume that $z' = z\sqrt{p^2+q^2} \in \mathbb{Q}$. Hence, by possibly replacing (p,q) with (mp,mq) for some $m \in \mathbb{N}$, we may suppose $z' \in \mathbb{N}$.

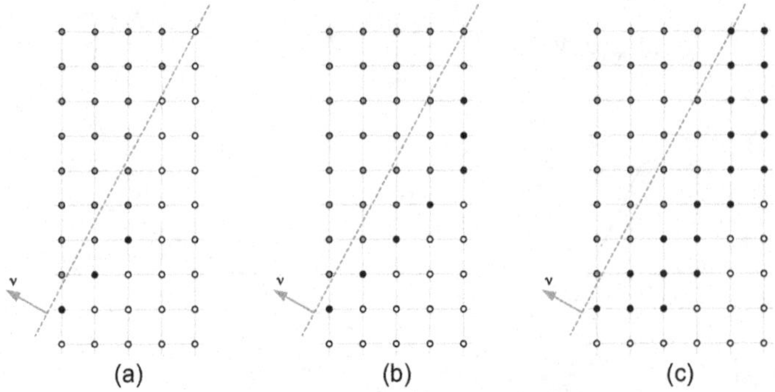

Figure 6.5 The local microstructure of a ground state of the BEG model at a fixed straight interface (the dashed line normal to ν) for three different values of the density of surfactants at the interface. Black, white, and gray dots stand for the 0, 1, and −1 values of the spin field u, respectively. From parts (a) to (c) $u_{z,\nu}$ in the periodicity cell $\{1, 2, \dots, q\} \times \{1, 2, \dots, p\}$ with $q = 5$, $p = 9$ for $z' = 3$, $z' = 7$, and $z' = 20$ corresponding to cases 1, 2, and 3 in the construction of the recovery sequence.

The building block of our constructions is a function u^0 defined on vertical stripes of width q, separating the phases −1 and 1 with a 0-phase placed on the bisector and on the segment between (q, q) and (q, p). More precisely, $u^0 : \{1, 2, \dots, q\} \times \mathbb{Z} \to \{-1, 0, 1\}$ is given by

$$u^0(n_1, n_2) = \begin{cases} 0 & \text{if } n_1 = n_2 \text{ or } n_1 = q,\ q < n_2 \leq p, \\ u_{\overline{\nu}}(n_1, n_2) & \text{otherwise,} \end{cases}$$

where $\overline{\nu} = \frac{e_1 + e_2}{\sqrt{2}}$.

Case 1: $0 < z < |\nu_1| \wedge |\nu_2|$. By the assumptions on ν_1, ν_2 this case corresponds to $z' < q$. Let $u_{z,\nu} : \mathbb{Z}^2 \to \{-1, 0, 1\}$ be such that

$$u_{z,\nu}((n_1, n_2) + (q, p)) = u_{z,\nu}(n_1, n_2), \quad \text{for all } (n_1, n_2) \in \mathbb{Z}^2$$

and, on $\{1, 2, \dots, q\} \times \mathbb{Z} \to \{-1, 0, 1\}$ is defined as (see Fig. 6.5(a))

$$u_{z,\nu}(n_1, n_2) = \begin{cases} u^0(n_1, n_2) & \text{if } 0 < n_1 \leq z', \\ u_\nu(n_1, n_2) & \text{otherwise.} \end{cases}$$

Let then $u^\varepsilon : \varepsilon\mathbb{Z}^2 \to \{-1, 0, 1\}$ be such that $u^\varepsilon(\varepsilon n_1, \varepsilon n_2) = u_{z,\nu}(n_1, n_2)$. It holds that $(u^\varepsilon, \mu(u^\varepsilon)) \to (u_\nu, \mu_{z,\nu})$. In order to estimate the energy of u^ε, we observe that it concentrates on each rectangle of the type

$$R_{\varepsilon,j} = ((0, \varepsilon q] \times (0, \varepsilon p]) + \varepsilon j(q, p) \text{ for } j \in \mathbb{Z},$$

where, by periodicity, it takes the constant value

$$E_\varepsilon(u^\varepsilon; R_{\varepsilon,j}) = \varepsilon\,(4(1-k)z' + 2(p-z') + 2(q-z')) = \varepsilon\,(-4kz' + 2(p+q)).$$

Here we have used the usual notation of localized functionals. Then

$$\begin{aligned} E_\varepsilon(u^\varepsilon) &\le E_\varepsilon(u^\varepsilon; R_{\varepsilon,0})(\#\{j \in \mathbb{Z}\colon\ R_{\varepsilon,j} \subset \Omega\} + 2) \\ &\le \varepsilon\,(-4kz' + 2(p+q))\left(\left\lfloor \frac{\mathcal{H}^1(S(u_\nu)\cap\Omega)}{\varepsilon\sqrt{p^2+q^2}} \right\rfloor + 2\right). \end{aligned}$$

Eventually, letting ε tend to 0, we obtain

$$\limsup_{\varepsilon\to 0} E_\varepsilon(u^\varepsilon) \le \varphi_1(z,\nu)\mathcal{H}^1(S(u_\nu)\cap\Omega).$$

Case 2: $|\nu_1| \wedge |\nu_2| \le z \le |\nu_1| \vee |\nu_2|$. This case corresponds to $q \le z' \le p$. Let $v_{z,\nu}\colon \mathbb{Z}^2 \to \{-1,0,1\}$ be such that

$$v_{z,\nu}((n_1,n_2) + (q,p)) = v_{z,\nu}(n_1,n_2), \quad \text{for all } (n_1,n_2) \in \mathbb{Z}^2$$

and, on $\{1,2,\dots,q\} \times \mathbb{Z} \to \{-1,0,1\}$, is defined as (see Fig. 6.5(b))

$$v_{z,\nu}(n_1,n_2) = \begin{cases} -1 & \text{if } n_1 = q \text{ and } n_2 > z', \\ u^0(n_1,n_2) & \text{otherwise.} \end{cases} \tag{6.11}$$

Let then $v^\varepsilon\colon \varepsilon\mathbb{Z}^2 \to \{-1,0,1\}$ be such that $v^\varepsilon(\varepsilon n_1, \varepsilon n_2) = v_{z,\nu}(n_1,n_2)$. It holds that $(v^\varepsilon, \mu(v^\varepsilon)) \to (u_\nu, \mu_{z,\nu})$. By arguing as in Case 1, taking into account that

$$\begin{aligned} E_\varepsilon(v^\varepsilon; R_{\varepsilon,j}) &= \varepsilon\,(4(1-k)q + 3(1-k)(z'-q) + 2(p-z')) \\ &= \varepsilon\,((1-3k)z' + (1-k)q + 2p), \end{aligned}$$

we get

$$E_\varepsilon(v^\varepsilon) \le \varepsilon\,((1-3k)z' + (1-k)q + 2p)\left(\left\lfloor \frac{\mathcal{H}^1(S(u_\nu)\cap\Omega)}{\varepsilon\sqrt{p^2+q^2}} \right\rfloor + 2\right).$$

Eventually, letting ε tend to 0, we obtain

$$\limsup_{\varepsilon\to 0} E_\varepsilon(v^\varepsilon) \le \varphi_2(z,\nu)\mathcal{H}^1(S(u_\nu)\cap\Omega). \tag{6.12}$$

Case 3: $z > |\nu_1| \vee |\nu_2|$. This case corresponds to $z' > p$. We extend the function u^0 to $\mathbb{Z}^2$ in such a way that

$$u^0((n_1, n_2) + (q, p)) = u^0(n_1, n_2), \quad \text{for all } (n_1, n_2) \in \mathbb{Z}^2. \tag{6.13}$$

We now construct $w_{z,\nu}: \mathbb{Z}^2 \to \{-1, 0, 1\}$ by modifying the function u^0, suitably increasing the numbers of its zeros in order to match the density constraint on the surfactant phase. More precisely, we set $z'' = z' - \lfloor \frac{z'}{p} \rfloor p$ and, recalling the definition of I_0 in (6.5),

$$\tilde{I}_0 = \bigcup_{m=0}^{\lfloor \frac{z'}{p} \rfloor} (I_0(u^0) + me_1) \cap \bigcup_{j \in \mathbb{Z}} \{(n_1, n_2) \in \mathbb{Z}^2 : \; jp + 1 \le n_2 \le jp + z''\},$$

$$\hat{I}_0 = \bigcup_{m=0}^{\lfloor \frac{z'}{p} \rfloor - 1} (I_0(u^0) + me_1) \cap \bigcup_{j \in \mathbb{Z}} \{(n_1, n_2) \in \mathbb{Z}^2 : \; jp + z'' + 1 \le n_2 \le (j+1)p\},$$

we define (see Fig. 6.5(c))

$$w_{z,\nu}(n_1, n_2) = \begin{cases} 0 & \text{if } (n_1, n_2) \in \tilde{I}_0 \cup \hat{I}_0, \\ u^0(n_1, n_2) & \text{otherwise.} \end{cases}$$

Let then $w^\varepsilon : \varepsilon\mathbb{Z}^2 \to \{-1, 0, 1\}$ be such that $w^\varepsilon(\varepsilon n_1, \varepsilon n_2) = w_{z,\nu}(n_1, n_2)$. It holds that $(w^\varepsilon, \mu(w^\varepsilon)) \to (u_\nu, \mu_{z,\nu})$. Then, the energy of each stripe $S_{\varepsilon,j} = \mathbb{R} \times (jp, (j+1)p]$ with $j \in \mathbb{Z}$ is

$$\begin{aligned} E_\varepsilon(w^\varepsilon; S_{\varepsilon,j}) &= \varepsilon\left((1-3k)p + (1-k)q + 2p + 2(1-k)(z'-p)\right) \\ &= \varepsilon\left((1-k)(p+q) + (1-k)2z'\right). \end{aligned}$$

It follows that

$$\begin{aligned} E_\varepsilon(w^\varepsilon) &\le E_\varepsilon(w^\varepsilon; S_{\varepsilon,0})(\#\{j \in \mathbb{Z} : \; S_{\varepsilon,j} \subset \Omega\} + 2) \\ &\le \varepsilon\left((1-k)(p+q) + (1-k)2z'\right)\left(\left\lfloor \frac{\mathcal{H}^1(S(u_\nu) \cap \Omega)}{\varepsilon\sqrt{p^2+q^2}} \right\rfloor + 2\right). \end{aligned}$$

Eventually, letting ε tend to 0, we obtain

$$\limsup_{\varepsilon \to 0} E_\varepsilon(w^\varepsilon) \le \varphi_3(z, \nu)\mathcal{H}^1(S(u_\nu) \cap \Omega).$$

We now prove upper estimate (ii). To this end, we set

$$u^\varepsilon(n_1, n_2) = \begin{cases} 0 & \text{if } (\varepsilon n_1, \varepsilon n_2) \in \left(x_0 + \left(-\frac{\sqrt{c\varepsilon}}{2}, \frac{\sqrt{c\varepsilon}}{2}\right)^2\right) \cap \varepsilon\mathbb{Z}^2, \\ 1 & \text{otherwise.} \end{cases}$$

We observe that $\varepsilon \# I_0(u^\varepsilon) = c + o(1)_{\varepsilon\to 0}$ and that each surfactant particle whose interactions do not cross the boundary of $x_0 + \big(-\frac{\sqrt{c\varepsilon}}{2}, \frac{\sqrt{c\varepsilon}}{2}\big)^2$ gives a contribution to the energy that is equal to $2(1-k)$. Moreover, since the number of the remaining surfactants scales as $\frac{1}{\sqrt{\varepsilon}}$, we have

$$E_\varepsilon(u^\varepsilon) = \varepsilon\, 2(1-k)\# I_0(u^\varepsilon) + o(1)_{\varepsilon\to 0} = c\, 2(1-k) + o(1)_{\varepsilon\to 0}.$$

Letting ε tend to 0, we get the conclusion. □

Corollary 6.5 (Γ-convergence in L^1) *Let $\frac{1}{3} < k < 1$ and let E_ε be defined as in (6.6). Then the functionals E_ε Γ-converge with respect to the $L^1(\Omega)$-topology to the functional F whose domain is $BV(\Omega; \{-1,1\})$ given by*

$$F(u) = \int_{S(u)} \psi(\nu_u) d\mathcal{H}^1, \tag{6.14}$$

where the function $\psi : S^1 \to [0,+\infty)$ is given by

$$\psi(\nu) = (1-k)(3|\nu_1| \vee |\nu_2| + |\nu_1| \wedge |\nu_2|).$$

The proof of this result follows by optimizing μ and by the previous theorem.

Remark 6.6 (Threshold phenomenon) A threshold phenomenon occurs at the phase interface, as can be deduced from the graph of φ pictured in Fig. 6.4. For a given $\nu \in S^1$ the surface tension $\varphi(z,\nu)$ of the system decreases only up to a certain value of the density z of the surfactant, namely $z = |\nu_1| \vee |\nu_2|$. Increasing further the amount of surfactant in the system, the energy increases in two different ways. Either the density of the surfactant on the interfaces increases (the surfactant is absorbed onto the interface) and thus the surface tension grows, or the singular part of the surfactant measure increases (the surfactant is not absorbed onto the interface) and the energy grows proportionally to its mass.

6.2 High-Contrast Media

We now analyze some effects of the loss of the coercivity condition on (part of the) nearest neighbors. We only treat two model cases, stating the corresponding general results at the end of the section.

6.2.1 Weak Interactions: Multiphase Limits

We consider a model with weak interactions between two different connected components of the underlying lattice.

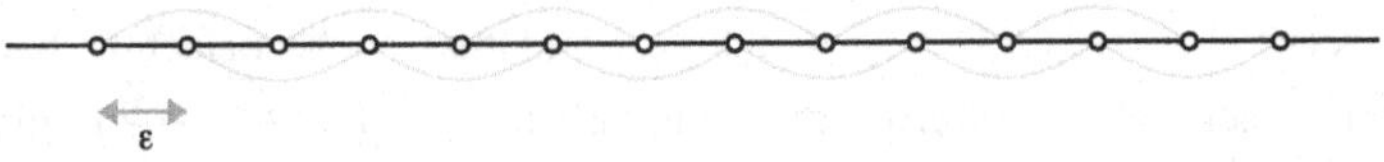

Figure 6.6 Decoupled lattices.

Example 6.7 (One-dimensional multiphase limits) (i) (*decoupled media*) We preliminarily remark that, if the coefficients a_{ij} vanish for some set of nearest neighbors, then the limit may be described by independent variables. As an example, let $\Omega = \mathbb{R}$, and set $a_{ij} = 1$ if $|i-j| = 2$, and 0 otherwise; that is, the points of the lattice $\varepsilon\mathbb{Z}$ interact only with second neighbors. As pictured in Fig. 6.6, we have nearest-neighbor interactions on the lattice $2\varepsilon\mathbb{Z}$ (even lattice) and on the translated lattice $\varepsilon + 2\varepsilon\mathbb{Z}$ (odd lattice), and they are independent. Applying the compactness argument for nearest-neighbor interactions separately on both lattices, we obtain that if a sequence $\{u_\varepsilon\}$ has equibounded energy, then u_ε converge to a pair of piecewise-constant functions $(u_{\mathrm{e}}, u_{\mathrm{o}})$, in the sense that the restrictions of u_ε to the even and odd lattice converge to u_{e} and u_{o}, respectively. The functionals E_ε Γ-converge to

$$F(u_{\mathrm{e}}, u_{\mathrm{o}}) = 8\big(\#S(u_{\mathrm{e}}) + \#S(u_{\mathrm{o}})\big),$$

with the two functions completely independent. Alternatively, we can interpret the energies as depending on a vector variable with values in $\{-1,1\}^2$ (see Example 4.9).

(ii) (*weakly interacting media*) We modify the coefficients of the previous example by introducing a weak interaction between nearest neighbors as follows:

$$a_{ij}^\varepsilon = \begin{cases} \varepsilon & \text{if } |i-j| = 1,\\ 1 & \text{if } |i-j| = 2,\\ 0 & \text{otherwise.} \end{cases}$$

Note that the coerciveness hypothesis (H_f2) is still not satisfied, since the nearest-neighbor coefficients are strictly positive, but not uniformly strictly positive. The energies can be written as

$$\begin{aligned} E_\varepsilon(u) &= \sum_{|i-j|=2} (u_i - u_j)^2 + 2\sum_i \varepsilon(u_i - u_{i-1})^2 \\ &= \sum_{|i-j|=2} (u_i - u_j)^2 + 2\int_{\mathbb{R}} (u_{\mathrm{e}}^\varepsilon - u_{\mathrm{o}}^\varepsilon)^2\,dx, \end{aligned}$$

where $u^\varepsilon_{\rm e}$ and $u^\varepsilon_{\rm o}$ are the discretizations of u on the even and odd lattice, respectively. The integral term continuously converges to $2\int_{\mathbb{R}}(u_{\rm e}-u_{\rm o})^2\,dx$; hence, the Γ-limit is

$$F(u_{\rm e},u_{\rm o}) = 8(\#S(u_{\rm e})+\#S(u_{\rm o})) + 2\int_{\mathbb{R}}(u_{\rm e}-u_{\rm o})^2\,dx.$$

Note that in these examples the limit depends on a vector, the pair $(u_{\rm e},u_{\rm o})$, and not on a single function. The second example describes mixtures of media with highly different properties, the so-called *high-contrast media.* We note that similar problems can be seen also in the continuous framework, but in that case we can mimic the role of the two disjoint lattices only in dimension greater than 3, taking two unbounded disjoint connected components as a continuous equivalent of the lattices.

We extend the model case considered in the preceding example to a general d-dimensional setting. We consider two complementary K-periodic sets C_1 and $C_2=\mathbb{Z}^d\setminus C_1$ such that both the sets are connected in $\mathbb{Z}^d$ (see Definition 3.19). Note that, for topological reasons, $d\geq 3$. This restriction is not necessary if we consider long-range interactions in place of nearest neighbors, as in Example 6.7.

The model energy of such a system is

$$E_\varepsilon(u) = \sum_{\substack{\langle i,j\rangle_1\\ \varepsilon i,\varepsilon j\in\Omega}} \varepsilon^{d-1}(u_i-u_j)^2 + \sum_{\substack{\langle i,j\rangle_2\\ \varepsilon i,\varepsilon j\in\Omega}} \varepsilon^{d-1}(u_i-u_j)^2 + 2\sum_{\substack{\langle i,j\rangle_{1,2}\\ \varepsilon i,\varepsilon j\in\Omega}} \varepsilon^{d}(u_i-u_j)^2, \tag{6.15}$$

where $\langle i,j\rangle_k=\{(i,j): i,j\in C_k,\ \|i-j\|=1\}$ for $k=1,2$ and

$$\langle i,j\rangle_{1,2}=\{(i,j): i\in C_1, j\in C_2,\ \|i-j\|=1\}.$$

Note that the last normalization factor is 2 since we do not have double counting of pairs (i,j). The last term is a weak interaction between the sets εC_1 and εC_2 through nearest neighbors, with coefficient ε.

These energies are not coercive, but we can apply the arguments for perforated domains to each of the two restrictions

$$E^k_\varepsilon(u) = \sum_{\langle i,j\rangle_k} \varepsilon^{d-1}(u_i-u_j)^2. \tag{6.16}$$

Definition 6.8 (Multiphase convergence) We say that a sequence $\{u^\varepsilon\}$ *converges to a pair* (A_1,A_2) of sets of finite perimeter in Ω if the extensions of the restrictions of u^ε to each εC_k converge to A_k in the sense of Proposition 3.20.

Remark 6.9 (Compactness) Given a sequence $\{u^\varepsilon\}$ with $\sup_\varepsilon E_\varepsilon(u^\varepsilon)<+\infty$, there exists a subsequence converging to some (A_1,A_2) as above. Indeed, in this case it is sufficient to remark that both $\sup_\varepsilon E^1_\varepsilon(u^\varepsilon) < +\infty$ and $\sup_\varepsilon E^2_\varepsilon(u^\varepsilon) < +\infty$ and apply Proposition 3.20.

We can then describe the limit of E_ε with respect to the compact convergence in Definition 6.8.

Theorem 6.10 *The family E_ε Γ-converges to the functional*

$$F(A_1, A_2) = \int_{\Omega\cap\partial^* A_1} \varphi^1_{\rm hom}(\nu)d\mathcal{H}^{d-1} + \int_{\Omega\cap\partial^* A_2} \varphi^2_{\rm hom}(\nu)d\mathcal{H}^{d-1} + m|\Omega \cap (A_1 \triangle A_2)|,$$

where for $k \in \{1,2\}$ the function $\varphi^k_{\rm hom}$ is the homogenized energy density of the perforated domain C_k*; that is, the density defined in* (3.12) *for the limit of E^k_ε as*

$$\varphi^k_{\rm hom}(\nu) = \lim_{T\to+\infty} \frac{1}{T^{d-1}} \min\Big\{ \sum_{\substack{\langle i,j\rangle_k \\ i,j\in Q^\nu_T}} (v_i - v_j)^2 : v\colon \mathbb{Z}^d \cap Q^\nu_T \to \{-1,1\}, \quad v = 2\chi_{H^\nu} - 1 \text{ in } Q^\nu_T \setminus Q^\nu_{T-2}\Big\} \tag{6.17}$$

and m is given by

$$m = \frac{2}{K^d}\#\{(i,j) : i \in \{1,\dots,K\}^d \cap C_1, j \in C_2, \|i-j\| = 1\}. \tag{6.18}$$

Proof If u^ε converge to (A_1, A_2), considering separately $E^1_\varepsilon(u^\varepsilon)$ and $E^2_\varepsilon(u^\varepsilon)$ provides a lower bound with the two surface energies. The bulk term is obtained by noting that, up to an asymptotically negligible set of $l \in \mathbb{Z}^d$, for both $k = 1,2$ the restrictions of u^ε to $\varepsilon\big(lK + (\{1,\dots,K\}^d \cap C_k)\big)$ must be constant, which gives the constant m. Note that the interaction term can also be written as

$$m\int_\Omega (\chi_{A_1} - \chi_{A_2})^2\, dx,$$

in accordance with Example 6.7.

The recovery sequences are obtained simply by taking separately a recovery sequence for each A_k on $\Omega \cap \varepsilon C_k$. □

6.2.2 Weak Inclusions: Lower-Order Effects

We now examine the case of nearest-neighbor systems in which some nodes are only weakly connected to the others, as in the following one-dimensional example.

Example 6.11 Let $\Omega = (0,1)$ and let E_ε be defined by

$$E_\varepsilon(u) = \sum_{i\in I^*_\varepsilon}(u_{2i} - u_{2i-2})^2 + \varepsilon \sum_{i\in I_\varepsilon}\Big((u_i - u_{i-1})^2 + g(u_i)\Big),$$

where $I_\varepsilon = \{i \in \mathbb{Z}\colon \varepsilon i, \varepsilon(i-1) \in (0,1)\}$, $I^*_\varepsilon = \{i \in \mathbb{Z}\colon 2\varepsilon i, 2\varepsilon(i-1) \in (0,1)\}$, and $g(1), g(-1)$ are given numbers.

In this system, even nodes are strongly connected, while nearest neighbors are weakly connected so that odd nodes possess only weak connections. In addition, there is a forcing bulk term given by g.

As in the examples of the previous section, for every sequence $\{u^\varepsilon\}$ with equibounded energy we can consider the restriction to the even lattice $2\varepsilon\mathbb{Z}$ and let $u = u_e$ denote the corresponding limit, which exists up to subsequences. Note that if $g = 0$, then the odd nodes are negligible in the limit with respect to this convergence, which is simply $E(u) = 4\#S(u)$. In the general case, we can optimize the values on the odd nodes, obtaining as a limit

$$E(u) = 4\#S(u) + m_1|\{u = 1\}| + m_{-1}|\{u = -1\}|,$$

where, for $k \in \{-1, 1\}$,

$$m_k = \frac{1}{2}\Big(\min\{g(k), 8 + g(-k)\} + g(k)\Big).$$

Now we formalize this kind of interaction in a general setting, for a system of nearest-neighbor interactions in $\mathbb{Z}^d$ with $d \geq 2$. We consider a set $\mathcal{D} \subset \{1,\dots,K-1\}^d$, and we suppose that

$$C = \mathbb{Z}^d \setminus (\mathcal{D} + K\mathbb{Z}^d)$$

is connected in $\mathbb{Z}^d$. Note that $\mathcal{D} + K\mathbb{Z}^d$ is composed of disconnected components.

We define the energies

$$E_\varepsilon(u) = \sum_{\substack{\langle i,j\rangle_1\\ \varepsilon i, \varepsilon j\in\Omega}} \varepsilon^{d-1}(u_i - u_j)^2 + \sum_{\substack{\langle i,j\rangle_0\\ \varepsilon i, \varepsilon j\in\Omega}} \varepsilon^{d}(u_i - u_j)^2, \qquad (6.19)$$

where $\langle i,j\rangle_1 = \{(i,j)\colon i,j \in C,\ \|i-j\| = 1\}$ and $\langle i,j\rangle_0 = \{(i,j)\colon i \text{ or } j \notin C,\ \|i-j\| = 1\}$. Note that the nearest-neighbor interaction coefficient between points in εC is 1 and is ε (*weak interactions*) between points not in εC (either both or one of them in $\varepsilon(\mathcal{D} + K\mathbb{Z}^d)$). The energy can be considered as that of a perforated medium when its complement is not empty, but interacts with the medium itself with weak interactions.

The effect of the weak interaction is not felt directly in the Γ-limit, as remarked here below.

Remark 6.12 Note that the Γ-limit of E_ε with respect to the usual convergence is the same as that of the perforated-medium energy

$$E^1_\varepsilon(u) = \sum_{\substack{\langle i,j\rangle_1 \\ \varepsilon i, \varepsilon j\in\Omega}} \varepsilon^{d-1}(u_i - u_j)^2, \tag{6.20}$$

when the second term is not present. Indeed, clearly that energy is a lower bound since $E^1_\varepsilon \leq E_\varepsilon$. Conversely, since we have recovery sequences for the perforated domain case that are converging to a set of finite perimeter, using those sequences gives a negligible contribution for the weak interactions. Note, however, that the sequence is not coercive.

Weak interactions may give a bulk contribution when perturbations are considered; for example, of the simple form

$$G_\varepsilon(u) = E_\varepsilon(u) + \sum_{\varepsilon i\in\Omega\cap\varepsilon\mathbb{Z}^d} \varepsilon^d g(u_i). \tag{6.21}$$

Note that the last term is continuous with respect to the L^1-convergence, giving an integral term of the form $\int_\Omega g(u)\,dx$. However, in this case our energies are not coercive with respect to the strong convergence in $L^1(\Omega)$. We will use as the notion of convergence the L^1-convergence of the extension of functions in εC as in the study of perforated domains, which has been proven to be compact in Proposition 3.20.

Theorem 6.13 *The functionals G_ε Γ-converge to the functional G defined by*

$$G(A) = \int_{\Omega\cap\partial^* A} \varphi^1_{\rm hom}(\nu)d\mathcal{H}^{d-1} + m_1|\{u=1\}\cap\Omega| + m_{-1}|\{u=-1\}\cap\Omega|, \tag{6.22}$$

where $\varphi^1_{\rm hom}$ is the homogenized energy density of the perforated domain *C given in* (6.17), *and m_k is given by*

$$m_k = \frac{1}{K^d}\min_v\Big\{\sum_{i\in\mathcal{D}} g(v_i) + \sum_{i,j\in\{0,\dots,K\}^d} (v_i - v_j)^2 : v_i = k \text{ on } \{0,\dots,K\}^d\setminus\mathcal{D}\Big\}$$
$$+\Big(1 - \frac{\#\mathcal{D}}{K^d}\Big)g(k). \tag{6.23}$$

Proof We first consider the lower bound. We can split the lower bound into an estimate on E^1_ε (which gives the surface term in the limit) and the remaining part

$$\sum_{\substack{\langle i,j\rangle_0 \\ \varepsilon i,\varepsilon j\in\Omega}} \varepsilon^d(u^\varepsilon_i - u^\varepsilon_j)^2 + \sum_{\varepsilon i\in\Omega\cap\varepsilon\mathbb{Z}^d} \varepsilon^d g(u^\varepsilon_i).$$

Except for a negligible number of $l \in \mathbb{Z}^d$, on each cube $\varepsilon(Kl + \{0,\dots,K\}^d)$, the value of u^ε on εC is constant so that we can estimate

$$\begin{aligned}
&\sum_{\substack{\langle i,j\rangle_0\\ i,j\in Kl+\{0,\dots,K\}^d}} \varepsilon^d (u_i^\varepsilon - u_j^\varepsilon)^2 + \sum_{i\in Kl+\{0,\dots,K\}^d} \varepsilon^d g(u_i^\varepsilon)\\
&\qquad \geq \sum_{\substack{\langle i,j\rangle_0\\ i,j\in Kl+\{0,\dots,K\}^d}} \varepsilon^d (u_i^\varepsilon - u_j^\varepsilon)^2 + \sum_{i\in Kl+\mathcal{D}} \varepsilon^d g(u_i^\varepsilon)\\
&\qquad + \sum_{i\in Kl+\{0,\dots,K\}^d\setminus\mathcal{D}} \varepsilon^d g(u_i^\varepsilon) \geq \varepsilon^d K^d m_{u^{\varepsilon,l}},
\end{aligned}$$

where $u^{\varepsilon,l}$ is the constant value of u^ε. Since $\sum_l u^{\varepsilon,l}\chi_{\varepsilon(lK+[0,K]^d)}$ has the same limit as u^ε, we obtain the lower bound.

It is sufficient to show the upper bound for u constant, for which we take the minimizer v in (6.23) and extend it K-periodically. □

6.2.3 General Double-Porosity Limits

We briefly outline how the preceding results can be generalized. We may have at the same time two types of geometrical issues.

(i) (*long-range interactions*) The perforated domains C_k can be taken as in Section 3.3.5; in particular, we may have longer range of interaction than only nearest neighbors. We assume that for $k \neq k'$ there is no strong interaction between points of C_k and of $C_{k'}$.

(ii) (*more components*) We may have more than two strong components; that is, we may have N disjoint periodic sets $C_1,\dots,C_N$, and the complement C_0. By definition interactions involving points in C_0 or between different strong components are weak. Note that in this case C_0 may not be composed of disjoint bounded inclusions.

The resulting energies have the form

$$\begin{aligned}
G_\varepsilon(u) &= \sum_{k=1}^N \sum_{\varepsilon i,\varepsilon j\in\Omega\cap\varepsilon C_k} \varepsilon^{d-1} a_{ij}(u_i - u_j)^2 + \sum_{k=0}^N \sum_{\substack{\varepsilon i\in\Omega\cap\varepsilon C_k\\ \varepsilon j\in\Omega\setminus\varepsilon C_k}} \varepsilon^d a_{ij}(u_i - u_j)^2\\
&\quad + \sum_{\varepsilon i,\varepsilon j\in\Omega\cap\varepsilon C_0} \varepsilon^d a_{ij}(u_i - u_j)^2 + \sum_{\varepsilon i\in\Omega\cap\varepsilon\mathbb{Z}^d} \varepsilon^d g(u_i). \qquad (6.24)
\end{aligned}$$

We can define a convergence $u^\varepsilon \to (A_1,\dots,A_N)$ as the convergence to A_k of each restriction of u^ε to $\varepsilon\, C_k$. With respect to that convergence we have a *double-porosity* limit energy of the form

$$G(A_1,\dots,A_N)=\sum_{k=1}^{N}\int_{\Omega\cap\partial^* A_k}\varphi^k_{\rm hom}(\nu)d\mathcal{H}^{d-1}+\int_\Omega \widetilde{g}(u^1,\dots,u^N)\,dx, \quad (6.25)$$

where $\varphi^k_{\rm hom}$ is defined in (6.17), $u^k=2\chi_{A_k}-1$, and the interaction term $\widetilde{g}$ takes into account both weak interactions and the forcing term g. It can be described by the formula

$$\widetilde{g}(z_1,\dots,z_N)=\lim_{T\to+\infty}\frac{1}{T^d}\min\Big\{\sum_{k=0}^{N}\sum_{\substack{i\in[0,T]^d\cap C_k\\ j\in[0,T]^d\setminus C_k}} a_{ij}(v_i-v_j)^2$$
$$+\sum_{i,j\in[0,T]^d\cap C_0} a_{ij}(v_i-v_j)^2+\sum_{i\in[0,T]^d\cap\mathbb{Z}^d} g(v_i)\colon v=z_k \text{ on } C_k\Big\},$$

which sums up the arguments in the two previous sections.

6.3 Ferromagnetic Systems with Modulated Phases

We consider some constrained systems, which are not directly included in the Compactness Theorem but may be treated with a little adaptation. The simplest one is a ferromagnetic interaction on a subset of spin functions that can be characterized in geometrical terms.

We consider all functions $u\colon \varepsilon\mathbb{Z}^2\to\{-1,1\}$ such that the set $\{i\in\mathbb{Z}^2\colon u_i=1\}$ can be decomposed into disjoint triples $\{i,i+e_1,i+e_2\}$ (see Fig. 6.7). Any such triple will be called a *molecule*, which is *chiral* since its symmetric analogs are not considered as the same molecule. In the simplest model, which we are going to examine, we only have one kind of chiral molecule. We consider the ferromagnetic energies

$$E^\chi_\varepsilon(u)=\sum_{\langle i,j\rangle}\varepsilon(u_i-u_j)^2$$

restricted to such functions (χ for chiral). For simplicity here we consider functions defined on the whole $\varepsilon\mathbb{Z}^2$ in order to avoid boundary effects.

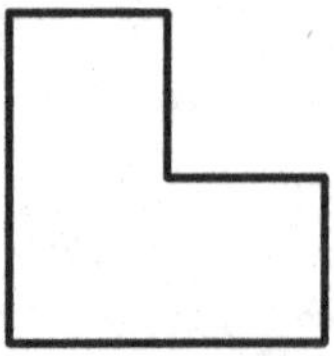

Figure 6.7 A "molecule."

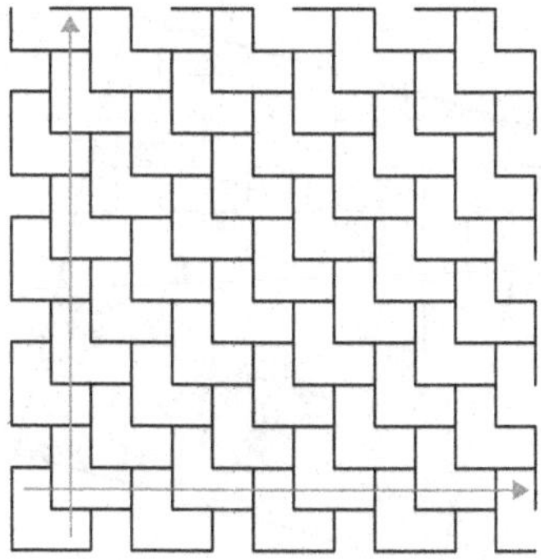

Figure 6.8 A ground state set of molecules.

Even though we are treating ferromagnetic energies, the limit will be defined on functions taking values in a larger class of parameters than only 1 and -1. To understand why, we note that we may interpret the function $u = 1$ as a disjoint union of molecules $\{i, i+e_1, i+e_2\}$. Using this standpoint, we have four ground states; namely, the emply set, corresponding to the constant state -1, and the three functions $u_k^{\rm gs}$, $k \in \{1,2,3\}$ whose sets of molecules, corresponding to tilings of $\mathbb{R}^2$, are $\{\{i, i+e_1, i+e_2\} : i \in (1,1)\mathbb{Z} + (0,3)\mathbb{Z} + (0,k)\}$. Note that all three functions $u_k^{\rm gs}$ correspond to the constant 1. In Fig. 6.8 we have pictured the molecules corresponding to $u_3^{\rm gs}$.

If a function is not identically 1, then the corresponding set of molecules is univocally determined. In this case, for each 2×2 square we have that either u is identically -1, u takes both values 1 and -1, or u is identically 1, in which case it can be identified univocally with one of the three functions $u_k^{\rm gs}$. Moreover, u cannot be equal to two different $u_k^{\rm gs}$ on two neighboring such squares. This implies, as in Remark 4.23, that we can define a compact convergence of a family u^ε after identifying the union of ε-squares corresponding to $\{i \in \mathbb{Z}^2 : u^\varepsilon = u_k^{\rm gs}\}$ with a set A_k^ε and the union of ε-squares where $u^\varepsilon = -1$ with A_0^ε. Then the convergence of u^ε is understood as the convergence of the corresponding A_j^ε to $\{A_0, A_1, A_2, A_3\}$ for all $j \in \{0,1,2,3\}$. These sets are a partition of $\mathbb{R}^2$ into sets of finite perimeter. Let u denote the function with value j on A_j, which is a function in $BV_{\rm loc}(\mathbb{R}^2; \{0,1,2,3\})$. The homogenization result is the following, where for uniformity of notation we let $u_0^{\rm gs}$ denote the constant -1.

Theorem 6.14 (Homogenization of chiral molecules) *Energies E_ε^χ Γ-converge as $\varepsilon \to 0$ to the functional*

$$F^\chi(u) = \int_{S(u)} \varphi^\chi(u^+, u^-, \nu_u)\, d\mathcal{H}^1,$$

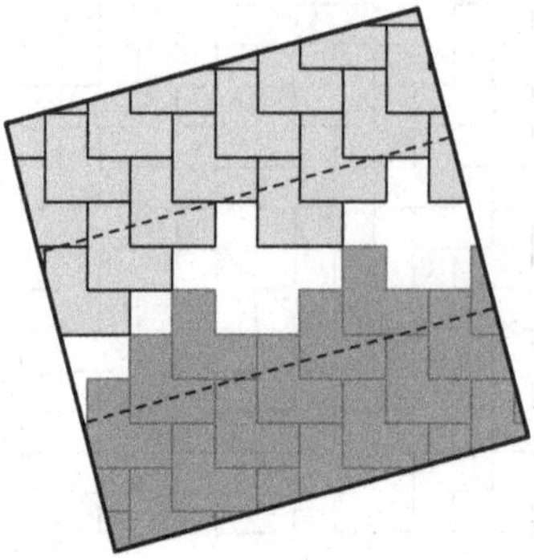

Figure 6.9 A competitor for the minimum in the homogenization formula.

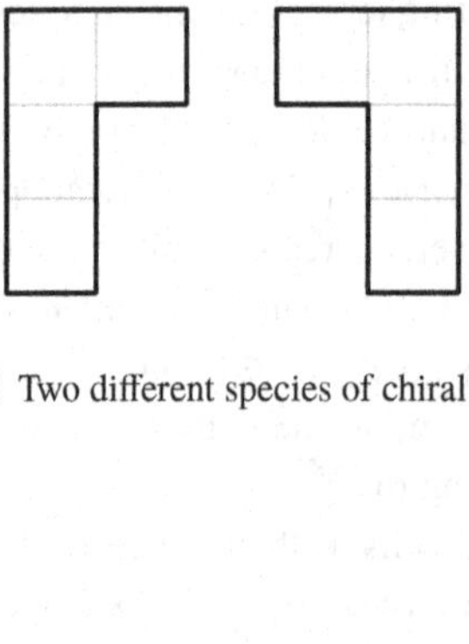

Figure 6.10 Two different species of chiral molecules.

where φ^χ satisfies

$$\varphi^\chi(k,l,\nu) = \lim_{T\to+\infty} \frac{1}{T} \min\Big\{ \sum_{\substack{\langle i,j\rangle \\ i,j\in Q^\nu_T}} \varepsilon(v_i - v_j)^2 : v \in \mathcal{A}(k,l,Q^\nu_T)\Big\},$$

where $\mathcal{A}(k,l,Q^\nu_T)$ is the set of admissible "molecular" v such that $v = u^{\rm gs}_k$ on $\partial Q^\nu_T \cap \{x\colon \langle x,\nu\rangle \geq 2\}$ and $v = u^{\rm gs}_l$ on $\partial Q^\nu_T \cap \{x\colon \langle x,\nu\rangle \leq -2\}$.

A pictorial description of the minimum problem in the definition of φ^χ is given in Fig. 6.9.

Remark 6.15 (Systems of more chiral molecules) The preceding result can be extended to more than one type of chiral molecule. An example is given by the two types in Fig. 6.10, which give in total nine parameters (four for each type of molecule plus one for the −1 phase).

6.4 Nonlocal Limits

In this section we approximate nonlocal functionals of the form

$$\int_{S(u)} \varphi(x,\nu)\, d\mathcal{H}^{d-1} + \int_\Omega \int_\Omega K(x,y)|u(x)-u(y)|\, dxdy \tag{6.26}$$

defined on $BV(\Omega;\{-1,1\})$ with energies for spin systems of the form

$$E_\varepsilon(u) = \sum_{\varepsilon i,\varepsilon j \in \mathcal{L}_\varepsilon(\Omega)} \varepsilon^{d-1} a^\varepsilon_{ij}(u_i-u_j)^2 \tag{6.27}$$

as in (4.74). In this case, hypothesis (H_f4) on the coefficients a^ε_{ij} in Theorem 4.37 is removed, while (H_f3) still holds, guaranteeing that the limit energies are finite on $BV(\Omega;\{-1,1\})$. Such energies include two relevant examples: fractional perimeters and Ohta–Kawasaki-type energies.

As for the first term, we assume that φ is such that there exist $\tilde{a}^\varepsilon_{ij}$ satisfying (H_f1)–(H_f4) and the corresponding energies

$$\tilde{E}_\varepsilon(u) = \sum_{\varepsilon i,\varepsilon j \in \mathcal{L}_\varepsilon(\Omega)} \varepsilon^{d-1} \tilde{a}^\varepsilon_{ij}(u_i-u_j)^2 \tag{6.28}$$

Γ-converge to the first term in (6.26). The key observation is that the nonlocal term in (6.26) can be approximated continuously by a sequence of energies $E^{\rm nl}_\varepsilon$, so that the two terms can be dealt with separately. In order to define $E^{\rm nl}_\varepsilon$, we simply discretize the double integral in (6.26). If $K \in C^1(\Omega\times\Omega\setminus\{x=y\})$, for εi and $\varepsilon i + \varepsilon\xi \in \mathcal{L}_\varepsilon(\Omega)$, we set

$$k^\varepsilon_{i,\xi} = \varepsilon^{d+1} K(\varepsilon i, \varepsilon i + \varepsilon \xi) \tag{6.29}$$

and we define

$$E^{\rm nl}_\varepsilon(u) = \frac{1}{4}\sum_{\xi\in\mathbb{Z}^d} \sum_{\varepsilon i,\varepsilon i+\varepsilon\xi \in \mathcal{L}_\varepsilon(\Omega)} \varepsilon^{d-1} k^\varepsilon_{i,\xi}(u_{i+\xi}-u_i)^2. \tag{6.30}$$

Note that we may also write

$$E^{\rm nl}_\varepsilon(u) = \sum_{\xi\in\mathbb{Z}^d} \sum_{\varepsilon i,\varepsilon i+\varepsilon\xi \in \mathcal{L}_\varepsilon(\Omega)} \varepsilon^{d-1} k^\varepsilon_{i,\xi}|u_{i+\xi}-u_i|$$

and, setting $Q(\varepsilon,i) = \varepsilon i + [0,\varepsilon)^d$ for $i \in \mathbb{Z}^d$, we have

$$\begin{aligned} E^{\rm nl}_\varepsilon(u) &= \sum_{\varepsilon i,\varepsilon j \in \mathcal{L}_\varepsilon(\Omega)} \varepsilon^{2d} K(\varepsilon i,\varepsilon j)|u_i-u_j| \\ &= \sum_{i,j\in\mathbb{Z}^d} \int_{\Omega\cap Q(\varepsilon,i)} \int_{\Omega\cap Q(\varepsilon,j)} K(x,y)|u(x)-u(y)|\, dx\, dy + o(1)_{\varepsilon\to 0} \\ &= \int_\Omega\int_\Omega K(x,y)|u(x)-u(y)|\, dx\, dy + o(1)_{\varepsilon\to 0}, \end{aligned}$$

where u is the piecewise-constant interpolation of u_i.

Proposition 6.16 (Continuity of the nonlocal term) *Let $\Omega \subset \mathbb{R}^d$ be a bounded open set with Lipschitz boundary and let $K\colon \Omega \times \Omega \to [0, +\infty)$ be such that $K \in C^1(\Omega \times \Omega \setminus \{x = y\})$ and $K(x, y) \le C\|y - x\|^{-d-s}$ in a neighborhood of $\{x = y\}$ for some $s \in (0, 1)$. Let $u^\varepsilon\colon \mathcal{L}_\varepsilon(\Omega) \to \{-1, 1\}$ be such that $u^\varepsilon \to u$ in $L^1(\Omega)$ and $\sup_\varepsilon \mathcal{H}^{d-1}(S(u^\varepsilon)) < +\infty$. Then $u \in BV(\Omega; \{-1, 1\})$ and*

$$\lim_{\varepsilon\to 0} E^{\mathrm{nl}}_\varepsilon(u^\varepsilon) = \int_\Omega \int_\Omega K(x, y)|u(x) - u(y)|\, dx\, dy. \tag{6.31}$$

Proof We extend u^ε outside $\mathcal{L}_\varepsilon(\Omega)$ by setting $u^\varepsilon_i = 1$ if $\varepsilon i \in \varepsilon\mathbb{Z}^d \setminus \Omega$. Since $\partial\Omega$ is Lipschitz, we still have that $\sup_\varepsilon \mathcal{H}^{d-1}(S(u^\varepsilon)) < +\infty$, which implies that for any $h \in \mathbb{R}^d$ it holds

$$\int_{\mathbb{R}^d} |u^\varepsilon(x + h) - u^\varepsilon(x)|\, dx \le C\|h\|, \tag{6.32}$$

with $C > 0$ independent of ε and h. Given $\delta > 0$, we split $E^{\mathrm{nl}}_\varepsilon(u^\varepsilon)$ into two terms, accounting for the interactions at distance greater and less than δ, respectively; that is,

$$\begin{aligned} E^{\mathrm{nl}}_\varepsilon(u^\varepsilon) = &\sum_{\xi\in\mathbb{Z}^d\cap B_{\frac{\delta}{\varepsilon}}} \sum_{\varepsilon i, \varepsilon i+\varepsilon\xi \in \mathcal{L}_\varepsilon(\Omega)} \varepsilon^{d-1} k^\varepsilon_{i,\xi} |u^\varepsilon_{i+\xi} - u^\varepsilon_i| \\ &+ \sum_{\xi\in\mathbb{Z}^d\setminus B_{\frac{\delta}{\varepsilon}}} \sum_{\varepsilon i, \varepsilon i+\varepsilon\xi \in \mathcal{L}_\varepsilon(\Omega)} \varepsilon^{d-1} k^\varepsilon_{i,\xi} |u^\varepsilon_{i+\xi} - u^\varepsilon_i| = I^1_{\varepsilon,\delta} + I^2_{\varepsilon,\delta}, \end{aligned} \tag{6.33}$$

where B_r denotes the open ball with center 0 and radius r. By the continuity assumption on K we obtain

$$\lim_{\varepsilon\to 0} I^2_{\varepsilon,\delta} = \int_{(\Omega\times\Omega)\cap\{\|y-x\|>\delta\}} K(x, y)|u(x) - u(y)|\, dx\, dy. \tag{6.34}$$

We now proceed by estimating $I^1_{\varepsilon,\delta}$. The decay assumption on K implies

$$I^1_{\varepsilon,\delta} \le \sum_{\xi\in\mathbb{Z}^d\cap B_{\frac{\delta}{\varepsilon}}} \frac{\varepsilon^{-s}}{\|\xi\|^{d+s}} \int_{\mathbb{R}^d} |u_\varepsilon(x + \varepsilon\xi) - u_\varepsilon(x)|\, dx;$$

hence, by (6.32),

$$I^1_{\varepsilon,\delta} \le C\varepsilon^{1-s} \sum_{\xi\in\mathbb{Z}^d\cap B_{\frac{\delta}{\varepsilon}}} \frac{1}{\|\xi\|^{d+s-1}}. \tag{6.35}$$

Noting that

$$\sum_{\xi\in\mathbb{Z}^d\cap B_{\frac{\delta}{\varepsilon}}}\frac{1}{\|\xi\|^{d+s-1}}=\sum_{k=1}^{\lfloor\frac{\delta}{\varepsilon}\rfloor-1}\sum_{\xi\in\mathbb{Z}^d\cap(B_{k+1}\setminus B_k)}\frac{1}{\|\xi\|^{d+s-1}}$$
$$\leq\sum_{k=1}^{\lfloor\frac{\delta}{\varepsilon}\rfloor-1}\frac{\#(\mathbb{Z}^d\cap(B_{k+1}\setminus B_k))}{k^{d+s-1}}\leq C\sum_{k=1}^{\lfloor\frac{\delta}{\varepsilon}\rfloor-1}\frac{1}{k^s}\leq C\left(\frac{\delta}{\varepsilon}\right)^{1-s},$$

from (6.35) we deduce that

$$I^1_{\varepsilon,\delta}\leq C\delta^{1-s}. \tag{6.36}$$

Eventually, by (6.33), (6.34), and (6.36), we infer that

$$\lim_{\varepsilon\to0}E^{\rm nl}_\varepsilon(u_\varepsilon)=\int_{\Omega\times\Omega\cap\{\|y-x\|>\delta\}}K(x,y)|u(x)-u(y)|\,dx\,dy+r_\delta,$$

with

$$\limsup_{\delta\to0}\frac{r_\delta}{\delta^{1-s}}<+\infty,$$

and the conclusion follows by the arbitrariness of $\delta>0$. □

By using Proposition 6.16, we derive the following Γ-convergence result.

Theorem 6.17 *Let $E_\varepsilon=\widetilde{E}_\varepsilon+E^{\rm nl}_\varepsilon$, with $\widetilde{E}_\varepsilon$ and $E^{\rm nl}_\varepsilon$ as in* (6.28) *and* (6.30), *respectively. Then, if K satisfies the assumption of Proposition* 6.16, *functionals E_ε Γ-converge with respect to the convergence in $BV_{\rm loc}(\Omega)$ to the functional*

$$E(u)=\int_{S(u)}\varphi(x,\nu)\mathcal{H}^{d-1}+\int_\Omega\int_\Omega K(x,y)|u(x)-u(y)|\,dx\,dy \tag{6.37}$$

with domain $BV(\Omega;\{-1,1\})$.

Example 6.18 (Fractional perimeters) The previous approximation includes the case in which the nonlocal term in the limit functional is a fractional perimeter of the set $\{u=1\}$ inside Ω. The latter can be defined, for a given $s\in(0,1)$ and for any set $A\subset\Omega$, as

$$\mathrm{Per}_s(A;\Omega)=\int_A\int_{\Omega\setminus A}\frac{1}{\|y-x\|^{d+s}}\,dx\,dy. \tag{6.38}$$

Note that $\mathrm{Per}_s(A;\Omega)$ coincides with half of the fractional Sobolev seminorm $|\chi_A|_{W^{s,1}(\Omega)}$. Moreover, if $u=\chi_A-\chi_{\Omega\setminus A}$, then

$$\mathrm{Per}_s(A;\Omega)=\frac{1}{4}|u|_{W^{s,1}(\Omega)}=\int_\Omega\int_\Omega\frac{|u(y)-u(x)|}{\|y-x\|^{d+s}}\,dx\,dy, \tag{6.39}$$

which corresponds to choosing $K(x,y) = \|x-y\|^{-d-s}$ in (6.37) and $k^{\varepsilon}_{i,\xi} = \frac{\varepsilon^{1-s}}{\|\xi\|^{d+s}}$ in (6.30). A direct computation shows that, for any $R > 0$,

$$\limsup_{\varepsilon\to 0} \sum_{\xi\in\mathbb{Z}^d\cap B_{\frac{R}{\varepsilon}}} \frac{\varepsilon^{1-s}}{\|\xi\|^{d+s}} < +\infty;$$

$$\limsup_{\varepsilon\to 0} \sum_{\xi\in\mathbb{Z}^d\cap(B_{\frac{R}{\varepsilon}}\setminus B_{\frac{R}{2\varepsilon}})} \frac{\varepsilon^{1-s}}{\|\xi\|^{d+s}} \geq C > 0.$$

Note that the assumption (H_f3) is satisfied by the functionals E_ε, while (H_f4) is violated.

Example 6.19 (Ohta–Kawasaki-type energies) A canonical mathematical model in the study of energy-driven pattern formation is based on the following energy

$$\begin{aligned}\mathcal{E}_\varepsilon(u) = \varepsilon \int_\Omega |\nabla u|^2\,dx + \frac{1}{\varepsilon}\int_\Omega (1-u^2)^2\,dx \\ + \gamma_0 \int_\Omega\int_\Omega G(x,y)u(x)u(y)\,dxdy.\end{aligned} \tag{6.40}$$

Here u is an $H^1(\Omega)$ phase parameter and G is the *Green's function of* $-\Delta$. The functionals $\mathcal{E}_\varepsilon$ Γ-converge to the functional $\mathcal{E}$ given by

$$\mathcal{E}(u) = \frac{8}{3}\mathcal{H}^{d-1}(S(u)) + \gamma_0\int_\Omega\int_\Omega G(x,y)u(x)u(y)\,dxdy \tag{6.41}$$

with domain $BV(\Omega;\{-1,1\})$. Note that the integral in the second term of (6.41) can be rewritten as

$$\begin{aligned}\int_\Omega\int_\Omega G(x,y)u(x)u(y)\,dxdy = \int_\Omega\int_\Omega G(x,y)\,dxdy \\ -2\int_\Omega\int_\Omega G(x,y)|u(x)-u(y)|\,dxdy,\end{aligned}$$

so that, thanks to the growth and regularity properties of the Green's function G, we can apply Theorem 6.17 to obtain a discrete variational approximation of anisotropic versions of the functional in (6.41). More precisely, let

$$E^{\mathrm{OK}}_\varepsilon(u) = \widetilde{E}_\varepsilon(u) + \gamma_0 \sum_{\varepsilon i,\varepsilon j\in\mathcal{L}_\varepsilon(\Omega)} \varepsilon^{2d} G(\varepsilon i,\varepsilon j)u_i u_j, \tag{6.42}$$

with $\widetilde{E}_\varepsilon$ as in (6.28). Then, the energies E_ε Γ-converge with respect to the $BV_{\mathrm{loc}}(\Omega)$ convergence to the functional

$$E^{\mathrm{OK}}(u) = \int_{S(u)} \varphi(x, \nu)\, d\mathcal{H}^{d-1} + \gamma_0 \int_{\Omega} \int_{\Omega} G(x, y) u(x) u(y)\, dx\, dy$$

with domain $BV(\Omega; \{-1, 1\})$.

Bibliographical Notes to Chapter 6

Surfactant energies from ternary models as presented here have been considered by Alicandro et al. (2012). A variational description of phase separation phenomena in the presence of surfactants has been introduced in the context of continuum models by Llaradji et al. (1991), and several generalizations have also been considered by Gompper and Schick (1994). We refer to Fonseca et al. (2007) for a variational treatment of a surfactant model as a generalization of the Modica–Mortola functional. Most of the ternary discrete models considered in the literature are variants of that introduced by Blume et al. (1971) and presented here. The modeling of more general surfactant models might require the presence of more species (Kumar and Mittal, 1999). The corresponding variational analysis of n-ary surfactant models can be carried out as for ternary systems; the details can be found in the paper by Alicandro et al. (2012).

Surface energies in discrete high-contrast systems have been studied by Braides et al. (2016a). There is a vast mathematical literature on continuum high-contrast media, especially for elliptic energies, since the seminal paper by Arbogast et al. (1990).

Systems modeling chiral molecules as presented here have been studied by Braides et al. (2016b), and are closely related to the problem of self-assembly of molecules (Bowden, 1997; Whitesides, 2002).

The two examples of nonlocal energies arising from discrete systems have been examined by Alicandro and Gelli (2016). For the Ohta–Kawasaki model we refer to Ohta and Kawasaki (1986), and for the fractional perimeter to, for example, Ludwig (2014).

7
Frustrated Systems

In this chapter we focus on spin systems with pairwise interactions for which coefficients may also take negative values as a paradigmatic case of interactions favoring oscillations at the lattice level. The prototypical case is that of *antiferromagnetic* nearest-neighbor energies of the form

$$-\sum_{\langle i,j\rangle}(u_i-u_j)^2.$$

While ground states for these energies on the cubic lattice $\mathbb{Z}^d$ exist with $u_i \neq u_j$ for all nearest neighbors, which then minimize separately each interaction, such ground states do not exist for other lattices (e.g. the triangular lattice). In general, for arbitrary lattices and energies mixing positive and negative coefficients, ground states, if they do exist, may not minimize separately each interaction. Such types of ground states are termed as *frustrated*. As a result of frustration, we may have nonconstant periodic ground states, which we call *patterns*. Ground states differing by a translation are referred to as *modulated phases*.

In this chapter we will first give some examples highlighting different behaviors of frustrated systems, and subsequently define a class of such systems whose Γ-limits are energies defined on partitions into sets of finite perimeter.

The main issues can be summarized as

(i) suitably define ground states so that functions with bounded energy are locally described by them;
(ii) define a notion of convergence using, as a set of parameters, the ground states themselves in such a way that they are equicoercive;
(iii) find conditions in order to apply the Compactness Theorem to derive an energy defined on partitions into sets of finite perimeter;
(iv) give examples of spin systems where such a description with a surface energy is not possible.

7.1 Model Cases

We first show how we may have limits where the relevant parameter in the final energy must be carefully chosen so as to distinguish between different ground states. A very simple case is that of antiferromagnetic spin systems where ground states are two two-periodic states oscillating between 1 and -1 as in Example 4.8. We now see how we can modify that example to obtain more complex ground states.

Example 7.1 (Modulated phases) We consider $N \in 2\mathbb{N}$ and $\varepsilon = \frac{1}{N}$, and functions $u\colon \varepsilon\mathbb{Z} \to \mathbb{R}$ satisfying the periodicity constraint $u_k = u_{k+N}$ for any $k \in \mathbb{Z}$. We define the energies

$$E_\varepsilon(u) = \sum_{i=0}^{N}(u_i u_{i+1} + u_{i-1}u_{i+1} + 1).$$

Ground states for these energies are four-periodic functions with alternating second neighbors; that is, they are translations of the four-periodic function $\overline{v}$ with $\overline{v}_i = -1$ if $i \in \{0,1\}$ and $\overline{v}_i = 1$ if $i \in \{2,3\}$. Note that $\overline{v}$ separately minimizes antiferromagnetic interactions on even and odd lattices. When the odd and even lattices are not decoupled, as in this case, it is convenient to introduce a *pattern variable*.

Note that if $\sup_\varepsilon E_\varepsilon(u^\varepsilon) < +\infty$, then, up to subsequences, there exist $K \in \mathbb{N}$ and a finite number of indices i_j^ε with $j \in \{1,\dots,K\}$ such that $\varepsilon i_j^\varepsilon$ converge to $x_j \in [0,1]$ as $\varepsilon \to 0$ and, having set $x_0 = 0$ and $x_{K+1} = 1$, if $x_j \neq x_{j-1}$, then there exists $\phi_j \in \{0,1,2,3\}$ such that

$$u_i^\varepsilon = \overline{v}_{i+\phi_j} \text{ definitively locally in } (x_{j-1},x_j). \tag{7.1}$$

Then, if $\phi\colon [0,1] \to \{0,1,2,3\}$ is a piecewise-constant function, we say that $u^\varepsilon \to \phi$ if there exists a finite set of points $\{x_j\}_{j=0}^M$ with $x_0 = 0$, $x_M = 1$, and $x_{j-1} < x_j$ for $j \in \{1,\dots,M\}$ such that $\phi = \phi_j$ on (x_{j-1},x_j) and (7.1) holds. Such a ϕ is called a *phase parameter* (see Fig. 7.1).

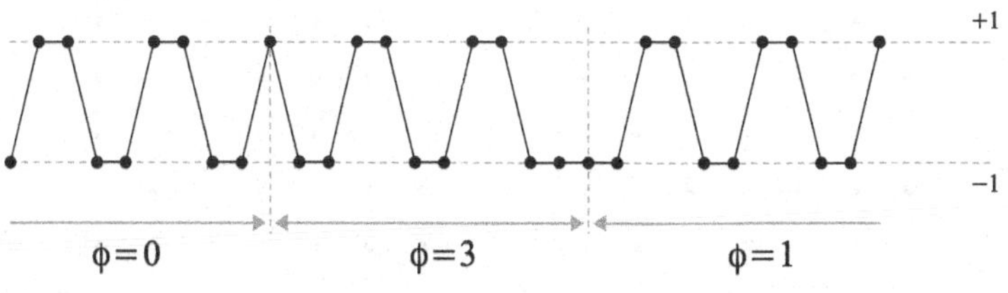

Figure 7.1 A function u and its corresponding *phase parameter* ϕ.

The Γ-limit has the form

$$F(\phi) = \sum_{t\in S(\phi)\cap[0,1)} \psi(|\phi(t^+) - \phi(t^-)|),$$

where ϕ is extended periodically outside $(0, 1)$ and the value of ψ is obtained by minimizing the transition problem between two values of the phase, for which we obtain

$$\psi(l) = \begin{cases} 3 & \text{if } l = 1, \\ 2 & \text{if } l = 2, \\ 1 & \text{if } l = 3. \end{cases}$$

In the following example, we consider some discrete energies in the lattice $\mathbb{Z}^2$ with negative coefficients and look at their ground states.

Example 7.2 (Ground states for antiferromagnetic energies in dimension 2) Since we are interested in the shape of the minimizers, we can consider the nonscaled functionals with $\varepsilon = 1$.

(i) *(nearest neighbors)* We set $a_{ij} = -1$ for any i, j nearest neighbors, and 0 otherwise; that is, we consider energies $E = E_1$ given by

$$E(u) = -\sum_{\langle i,j\rangle}(u_i - u_j)^2.$$

Minimizers alternate the values 1 and −1 for any pair of points at distance 1, so that we have two possible ground states given by $u_i = (-1)^{i_1+i_2}$ (where $i = (i_1, i_2)$) and by the translated checkerboard $u_i = -(-1)^{i_1+i_2}$, as pictured in Fig. 7.2.

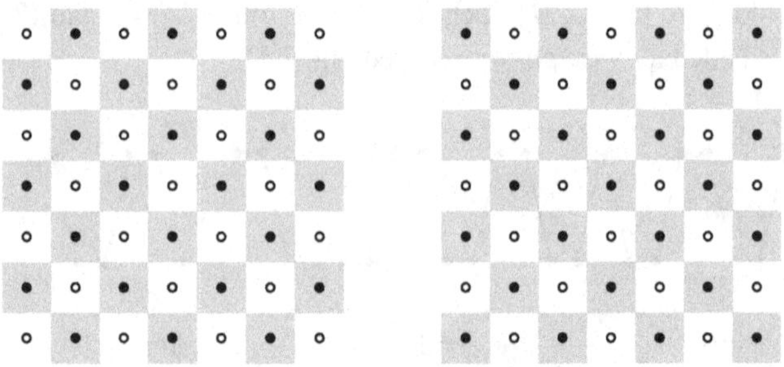

Figure 7.2 Checkerboards with different parity.

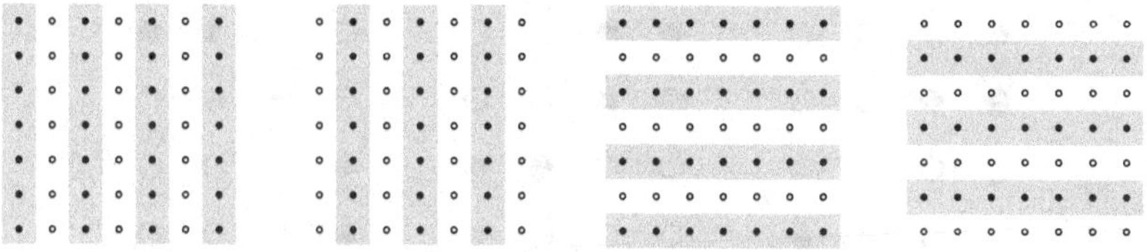

Figure 7.3 Stripes with different orientations and parities.

(ii) *(nearest and next-to-nearest neighbors)* Given $\alpha > 0$, we set $a_{ij} = -1$ for any i, j nearest neighbors, $a_{ij} = -\alpha$ if $\|i - j\| = \sqrt{2}$ and 0 otherwise; that is, we consider energies given by

$$E(u) = -\sum_{\langle i,j\rangle}(u_i - u_j)^2 - \sum_{\|i-j\|=\sqrt{2}} \alpha(u_i - u_j)^2.$$

If α is large enough, the cross bonds are stronger, and the minimizers alternate the values 1 and -1 for any pair (i, j) such that $\|i-j\| = \sqrt{2}$; hence, the corresponding minimizing sets are horizontal and vertical stripes, with two different parities for each one (see Fig. 7.3), and we have four different ground states.

We will see that in the cases of the preceding example, the Γ-limit can be described using the ground states themselves as parameters, with a representation in terms of partitions into sets of finite perimeter.

Now, we consider homogeneous long-range antiferromagnetic pair interactions. It is convenient to write energies as depending on many-body potentials adopting the notation

$$E_\varepsilon(u) = \sum_{i\in\mathbb{Z}^d} \varepsilon^{d-1}\phi(\{u_{i+j}\}_j)$$

of Chapter 4.

Example 7.3 (Multiple-phase and antiphase boundaries for two-dimensional long-range antiferromagnetic interactions) We now extend Example 7.1 to dimension 2 (in the special case of next-to-nearest neighbor interactions only) by considering

$$\phi(\{u_j\}_j) = (u_{(1,0)} + u_{(-1,0)})^2 + (u_{(0,1)} + u_{(0,-1)})^2, \tag{7.2}$$

where we have an antiferromagnetic interaction between all points at distance 2 on the lattice $\mathbb{Z}^2$.

The resulting energy can be decoupled as the contributions on horizontal and vertical directions, noting that on each vertical or horizontal line we have

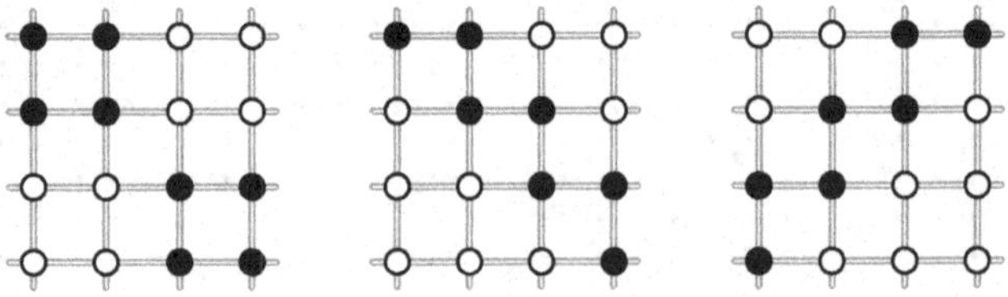

Figure 7.4 Three patterns of ground states. From left to right: (a) the checkerboard, (b) the 45-degrees, and (c) the minus 45-degrees stripes.

interactions as in Example 7.1. As a result, minimal states are four-periodic functions with alternating pairs of 1 and −1 in each direction. Such functions are determined by their values on a square of period two; hence, we have 16 different ground states, subdivided into three families of modulated phases with different patterns: four-periodic checkerboards, with 8 elements, and stripes at plus or minus 45-degrees, respectively, with 4 elements each. In Fig. 7.4 we picture the configurations on a period for three different elements, one for each family. Note that the patterns of the ground states stay the same if we add nearest- (and also next-to-nearest-) neighbor interactions with small enough coefficients.

In the following examples we will consider the triangular lattice in $\mathbb{R}^2$. Note that interactions on the triangular lattice can be parameterized on $\mathbb{Z}^2$ by suitably identifying first neighbors in the triangular lattice with first and second neighbors in the square lattice. For the convenience of pictorial representation, we will use a parameterization on a unit triangular lattice.

Example 7.4 ("Stabilization" for antiferromagnetic interactions on the triangular lattice) Antiferromagnetic nearest-neighbor systems on a triangular lattice possess infinitely many periodic and nonperiodic ground states, as we will see in Example 7.14. Here we show that the addition of a (even small) ferromagnetic or antiferromagnetic next-to-nearest-neighbor interaction produces the existence of only a finite number of ground states.

We first add to the antiferromagnetic nearest-neighbor energy a ferromagnetic next-to-nearest-neighbor interaction, choosing

$$\begin{aligned}\phi(\{u_j\}_j) &= u_{(0,0)}u_{(1,0)} + u_{(1,0)}u_{(1/2,\sqrt{3}/2)} + u_{(1/2,\sqrt{3}/2)}u_{(0,0)} + 1\\ &\quad + c\Big((u_{(0,0)} - u_{(0,\sqrt{3})})^2 + (u_{(0,0)} - u_{(3/2,-\sqrt{3}/2)})^2\\ &\quad + (u_{(0,0)} - u_{(-3/2,-\sqrt{3}/2)})^2\Big).\end{aligned} \tag{7.3}$$

Due to the presence of the ferromagnetic interactions, the ground states (corresponding here to the zero energy states) have u constant on the three triangular

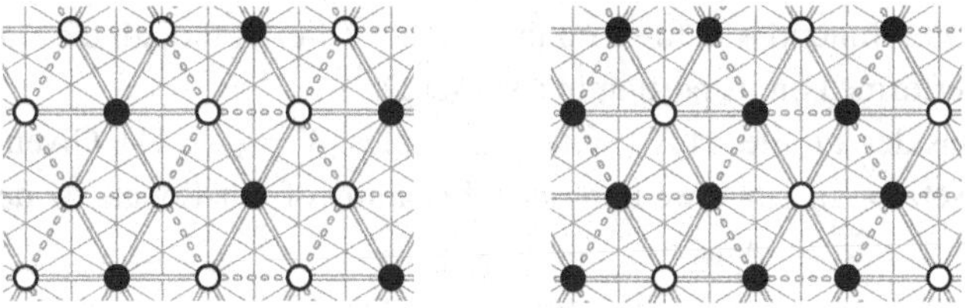

Figure 7.5 Hexagonal patterns in the triangular lattice.

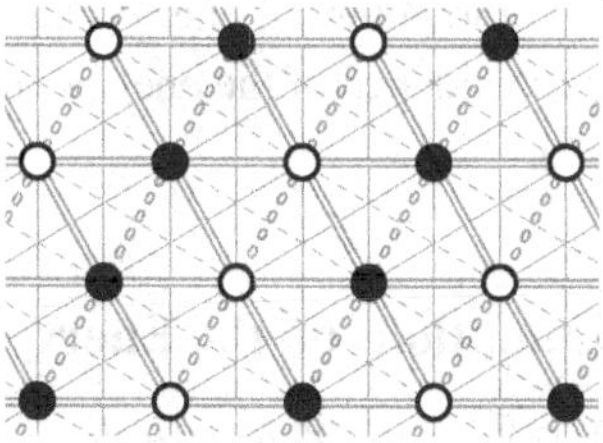

Figure 7.6 Striped patterns in the triangular lattice.

sublattices with triangles of side length $\sqrt{3}$. As a result, the ground states are the six nonconstant functions which are constant on each of the three sublattices. They form two families of modulated phases with hexagonal symmetries, two elements of which are pictured in Fig. 7.5 (the other ones differing by a translation). We adopt the convention that antiferromagnetic interactions are indicated by double lines. Dashed lines highlight frustrated interactions.

A similar argument can be carried out if, instead of ferromagnetic, antiferromagnetic next-to-nearest-neighbor interactions are chosen. In that case we have

$$\phi(\{u_j\}_j) = u_{(0,0)}u_{(1,0)} + u_{(1,0)}u_{(1/2,\sqrt{3}/2)} + u_{(1/2,\sqrt{3}/2)}u_{(0,0)} + 1$$
$$+ c\Big(u_{(0,0)}u_{(0,\sqrt{3})} + u_{(0,0)}u_{(3/2,-\sqrt{3}/2)} + u_{(0,0)}u_{(-3/2,-\sqrt{3}/2)} + 1\Big).$$

Now the six ground states have a striped pattern and are determined by their values on two adjacent triangles (see Fig. 7.6).

7.2 An Integral-Representation Result on Patterns

The compactness theorem in Chapter 4 does not directly apply to describe the preceding systems since we do not have constant minimizers. Nevertheless, the fact that we have a finite number of ground states will allow to use those ground states themselves as the set of parameters Y_0 in the Compactness Theorem. We will consider functions defined on the whole $\varepsilon\mathbb{Z}^d$ even though the summation is only performed on its portion in a subset Ω.

We consider a finite set $\mathcal{X}$ (in most examples, $\mathcal{X} = \{-1, 1\}$). Given $\Omega \subseteq \mathbb{R}^d$ an open set, for all $\varepsilon > 0$ we introduce the family of functionals E_ε defined on functions $u \colon \varepsilon\mathbb{Z}^d \to \mathcal{X}$ as

$$E_\varepsilon(u; \Omega) = \sum_{\varepsilon i \in \mathcal{L}_\varepsilon(\Omega)} \varepsilon^{d-1} \phi(\{u_{i+k}\}_{k \in \mathbb{Z}^d}), \tag{7.4}$$

where $\phi \colon (\mathcal{X})^{\mathbb{Z}^d} \to [0, L]$ is the potential energy density of the system. In the case $\Omega = \mathbb{R}^d$ we drop the dependence of the energy on the set and simply write $E_\varepsilon(u)$ in place of $E_\varepsilon(u; \mathbb{R}^d)$.

7.2.1 Assumptions on the Energy Density

Given $h \in \mathbb{N}$, we say that $v \colon \mathbb{Z}^d \to \mathcal{X}$ is *h-periodic* if, for all $i \in \{1, \ldots, d\}$,

$$v_{j+he_i} = v_j$$

for all $j \in \mathbb{Z}^d$. We assume that there exist K functions $v^1, \ldots, v^K$ h-periodic such that the following conditions are satisfied.

(H1) *(existence of ground states)* We have $\phi(\{z_{j+i}\}_j) = 0$ for all $i \in \mathbb{Z}^d$ if and only if there exists $l \in \{1, \ldots, K\}$ such that $z_j = v^l_j$ for all $j \in \mathbb{Z}^d$.

(H2) *(coerciveness)* There exist $M, M' \in \mathbb{N}$ with $M' \geq M \geq 2$ and $C_M > 0$ such that if $u \colon \mathbb{Z}^d \to \mathcal{X}$ and $i \in \mathbb{Z}^d$ are such that for all $l \in \{1, \ldots, K\}$, there exists $j \in Q_{Mh}(i)$ such that $u_j \neq v^l_j$, then there exists $i_M \in Q_{M'h}(i)$ such that

$$\phi(\{u_{i_M+j}\}_{j \in \mathbb{Z}^d}) \geq C_M.$$

(H3) *(mild nonlocality)* Given $z, w \colon \mathbb{Z}^d \to \mathcal{X}$ and $m \in \mathbb{N}$ such that $z_j = w_j$ for all $j \in Q_m \cap \mathbb{Z}^d$, then

$$|\phi(\{w_j\}_{j \in \mathbb{Z}^d}) - \phi(\{z_j\}_{j \in \mathbb{Z}^d})| \leq c_m,$$

where the constants c_m are such that

$$\sum_{m=1}^{+\infty} c_m m^{d-1} < +\infty$$

if Ω is not bounded and such that

$$\sum_{m=1}^{+\infty} c_m m < +\infty$$

if Ω is bounded.

Remark 7.5 (General lattices) For simplicity of notation we parameterize our functions on the lattice $\mathbb{Z}^d$, but the result is independent from the choice of the lattice, with another lattice $\mathcal{L}$ replacing $\mathbb{Z}^d$ in (H1)–(H3). If $\mathcal{L}$ is invariant under translations by its elements, then the proof remains unchanged since we may still use the notation $\{u_{i+j}\}_j$ as defining a function on $\mathcal{L}$. Otherwise, one has to slightly extend the proof to inhomogeneous energy densities, as we will see in Section 7.3.

Remark 7.6 (Comments on the hypotheses) Hypothesis (H1) and the positiveness of ϕ entail that the energy density be chosen and normalized in such a way that it is not minimized pointwise on a function u which is not a periodic minimizer.

Note that the same energy may be rewritten using different functions ϕ. For example, a system of nearest and next-to-nearest-neighbor interactions in the square lattice for spin systems $\mathcal{X} = \{-1, 1\}$ can be described by the energy density

$$\phi(\{u_j\}_j) = \sigma_{\rm n}\Big(u_{(1,0)}u_{(0,0)} + u_{(0,1)}u_{(0,0)} + u_{(-1,0)}u_{(0,0)} + u_{(0,-1)}u_{(0,0)}\Big)$$
$$+ \sigma_{\rm nn}\Big(u_{(1,1)}u_{(0,0)} + u_{(-1,1)}u_{(0,0)} + u_{(-1,1)}u_{(0,0)} + u_{(-1,-1)}u_{(0,0)} + 4\Big)$$

with $\sigma_{\rm n}$, $\sigma_{\rm nn}$ positive. In this case $\phi(\{u_j\}_j) = -4\sigma_{\rm n} < 0$ if $u_{(0,0)} = 1$ and $u_i = -1$ if $i \neq (0,0)$, so that the positiveness assumption is not satisfied. We can anyhow regroup the interactions in such a way that the energy is described by a new positive ϕ satisfying (H1)–(H3).

Since the energy is invariant under translation, by (H1) the functions w^l defined as $w^l_j = v^l_{j+j'}$ for some $j' \in \mathbb{Z}^d$ are ground states of the system; that is, every pattern gives a family of modulated phases indexed by the possible translations.

7.2.2 Discrete-to-Continuum Analysis

In what follows, to each discrete function we associate a K-ple of sets and define a notion of convergence accordingly.

Given $u\colon \varepsilon\mathbb{Z}^d \to \mathcal{X}$ and $l \in \{1, 2, \ldots, K\}$, we define

$$I_l(u) = \{j \in \mathbb{Z}^d\colon\ u_i = v^l_i \text{ for all } i \in Q_h(jh) \cap \mathbb{Z}^d\}, \tag{7.5}$$

where we recall that h is the period of v^l, and define the *phase* l as the set

$$A_{\varepsilon,l}(u) = \varepsilon \bigcup_{j \in I_l(u)} Q_h(jh).$$

Definition 7.7 (Discrete-to-continuum convergence) Let $\{u_\varepsilon\}$ be a sequence of discrete functions and let $\mathbb{A} = (A_1, \dots, A_K) \subseteq (\mathbb{R}^d)^K$. We say that $\{u^\varepsilon\}$ *converges to* $\mathbb{A}$ and we write $u^\varepsilon \to \mathbb{A}$ if $|(A_{\varepsilon,l}(u^\varepsilon)\triangle A_l) \cap Q_R| \to 0$ as $\varepsilon \to 0$ for all $l \in \{1, \dots, K\}$ and for all $R > 0$.

Remark 7.8 To any K-ple of sets $\mathbb{A} = (A_1, \dots, A_K) \subseteq (\mathbb{R}^d)^K$ we associate the function $p_\mathbb{A} : \mathbb{R}^d \to \{0, 1, \dots, K\}$ defined as

$$p_\mathbb{A}(x) = \sum_{l=1}^{K} l\chi_{A_l}(x).$$

Note that whenever $\mathbb{A} = (A_1, \dots, A_K)$ is a partition of $\mathbb{R}^d$ into sets of finite perimeter, then the convergence $u^\varepsilon \to \mathbb{A}$ is equivalent to the convergence of the functions $p_{\mathbb{A}_\varepsilon(u^\varepsilon)}$ to $p_\mathbb{A}$ in $L^1_{\rm loc}(\mathbb{R}^d)$, where $\mathbb{A}_\varepsilon(u^\varepsilon) = (A_{\varepsilon,1}, A_{\varepsilon,2}, \dots, A_{\varepsilon,K})$.

The following result states that, up to subsequences, as $\varepsilon \to 0$ the phases of a discrete system with equibounded energy form a partition of $\mathbb{R}^d$ into sets of finite perimeter.

Theorem 7.9 (Compactness) *Let ϕ satisfy hypothesis* (H2), *and let $\{u^\varepsilon\}$ be such that $\sup_\varepsilon E_\varepsilon(u^\varepsilon) < +\infty$. Then, up to subsequences, $\{u^\varepsilon\}$ converges to $\mathbb{A} = (A_1, A_2, \dots, A_K)$, where, for all $l \in \{1, 2, \dots, K\}$, A_l is a set of finite perimeter in $\mathbb{R}^d$ and the family $\{A_1, \dots, A_K\}$ forms a partition of $\mathbb{R}^d$.*

With respect to this convergence we have the following Γ-convergence result.

Theorem 7.10 (Homogenization) *Let the energy functionals E_ε satisfy hypotheses* (H1)–(H3) *and Ω be a Lipschitz set. Then, there exists the Γ-limit of $E_\varepsilon(\cdot; \Omega)$ as $\varepsilon \to 0$ with respect to the convergence in Definition* 7.7, *and we have*

$$\Gamma\text{-}\lim_{\varepsilon\to 0} E_\varepsilon(u; \Omega) = F(\mathbb{A}; \Omega) = \sum_{l,l'=1}^{K} \int_{\Omega\cap\partial^* A_l \cap\partial^* A_{l'}} \varphi(l, l', \nu_l)\, d\mathcal{H}^{d-1} \tag{7.6}$$

for all $\mathbb{A} = \{A_1, \dots, A_K\}$ partitions of $\mathbb{R}^d$ into sets of finite perimeter, where ν_l is the inner normal to $\partial^* A_l$, $\varphi \colon \{1, \dots, K\}^2 \times S^{d-1} \to [0, +\infty)$ satisfies

$$\begin{aligned}\varphi(l, l', \nu) &= \lim_{\delta\to 0} \liminf_{T\to+\infty} \frac{1}{T^{d-1}} \inf\{E_1(u, Q^\nu_T),\ u = u^{l,l',\nu} \text{ on } \mathbb{Z}^d \setminus Q^\nu_{(1-\delta)T}\} \\ &= \lim_{\delta\to 0} \limsup_{T\to+\infty} \frac{1}{T^{d-1}} \inf\{E_1(u, Q^\nu_T),\ u = u^{l,l',\nu} \text{ on } \mathbb{Z}^d \setminus Q^\nu_{(1-\delta)T}\},\end{aligned} \tag{7.7}$$

and $u^{l,l',\nu}$ is defined by

$$u_i^{l,l',\nu} = \begin{cases} v_i^l & \text{if } i \in \mathbb{Z}^d \cap \Pi^+_\nu, \\ v_i^{l'} & \text{otherwise,} \end{cases} \tag{7.8}$$

where

$$\Pi_\nu^+ = \bigcup_{\langle j,\nu\rangle \geq 0} Q_h(hj) \tag{7.9}$$

and h is the common period of ground states.

The proof of both theorems will reduce to the results proved in Chapter 4 once we rewrite the energies E_ε with respect to functions of a larger set of variables. This will be done in the next section.

7.2.3 Simplified Variables and Compactness

Suppose that ϕ satisfies assumptions (H1)–(H3) in Section 7.2.1. We will construct an equivalent energy by a suitable family of energy densities ϕ_i^ε satisfying the hypotheses of the Compactness Theorem. It is not restrictive to suppose that $M \in h\mathbb{Z}$ and that $M' \leq 2M$.

By a coarse-graining argument on the lattice $M\mathbb{Z}^d$, we may reduce to the assumption of having constant ground states by considering the values of a function $u\colon \mathbb{Z}^d \to \mathcal{X}$ on a periodicity cell $Q_M(iM) \cap \mathbb{Z}^d$ as a vector in $\mathcal{X}^{M^d}$. We now show that this identification still describes the same system and can be studied in terms of the Compactness Theorem.

Let

$$Y = \mathcal{X}^{Q_M} = \{v\colon Q_M \to \mathcal{X}\}$$

and define the bijection $\Psi\colon Y^{\mathbb{Z}^d} \to \mathcal{X}^{\mathbb{Z}^d}$ as follows. Given $z\colon \mathbb{Z}^d \to Y$, we set

$$\Psi(z)_i = z_{\tilde{i}}(i - \tilde{i}M), \qquad \text{for all } i \in Q_M(\tilde{i}M). \tag{7.10}$$

We then define $\tilde{\phi}\colon Y^{\mathbb{Z}^d} \to [0, L]$ as

$$\tilde{\phi}(\{z_{\tilde{j}}\}_{\tilde{j}}) = \frac{1}{M^d} \sum_{i \in Q_M} \phi(\{\Psi(z)_{i+j}\}_j), \tag{7.11}$$

and $\phi_i^\varepsilon = \phi^\varepsilon = \tilde{\phi}$ independent of ε.

The hypotheses (H1)–(H3) of the Compactness Theorem are satisfied: (H1) is satisfied with $\tilde{v}^l = \Psi^{-1}(v^l)$ in place of v^l, (H2) holds with $C = C_M M^{-d}$ and (H3) with c_m replaced by $\tilde{c}_m = c_{(m-1)M}$ (taking $c_0 = 2L$). Hence, we can apply the Compactness Theorem to the functionals

$$\tilde{E}_\varepsilon(w; U) = \sum_{\varepsilon M j \in U \cap \varepsilon M \mathbb{Z}^d} \varepsilon^{d-1} \tilde{\phi}(\{w_j\}_j)$$

defined for $w\colon \varepsilon M\mathbb{Z}^d \to Y$ with the notation $w_j = w(\varepsilon M j)$. We note that $E_\varepsilon(u; U) = \tilde{E}_\varepsilon(\Psi^{-1}(u); U)$.

Given $u^\varepsilon : \varepsilon\mathbb{Z}^d \to \mathcal{X}$, we may consider the piecewise-constant interpolations of $w^\varepsilon = \Psi^{-1}(u^\varepsilon)$ on $\varepsilon M\mathbb{Z}^d$. By the results in Chapter 4, such a sequence with equibounded energy is precompact in $BV_{\rm loc}(\mathbb{R}^d;Y)$ and its limits belong to $BV_{\rm loc}(\mathbb{R}^d;Y_0)$, where Y_0 is the set of ground states. This convergence corresponds to the convergence of the sets

$$\tilde{A}_{\varepsilon,l}(w) = \bigcup_{i\in\tilde{I}_l(w)} Q_{\varepsilon M}(i\varepsilon M),$$

where $\tilde{I}_l(w) = \{j \in \mathbb{Z}^d : w_i = v^l_i, \text{ for all } i \in Q_{Mh}(jMh) \cap \mathbb{Z}^d\}$. Since by construction $\tilde{A}_{\varepsilon,l}(\Psi^{-1}(u^\varepsilon)) \subseteq A_{\varepsilon,l}(u^\varepsilon)$, the corresponding inclusions hold in the limit. Since the limit sets form a partition we deduce that the limits of u^ε and $\Psi^{-1}(u^\varepsilon)$ are the same.

Finally we note that the Γ-convergence of E_ε with respect to the convergence $u^\varepsilon \to \mathbb{A}$ is equivalent to the Γ-convergence of $\tilde{E}_\varepsilon$ with respect to the convergence $w^\varepsilon \to \mathbb{A}$.

Remark 7.11 Note that, if Y is identified with a subset of a Euclidean space, then the convergence $w^\varepsilon \to \mathbb{A}$ is equivalent to the $L^1_{\rm loc}$-convergence of their piecewise-constant interpolations to $\sum_{l=1}^K v^l \chi_{A_l}$. This shows that the Γ-limit can be set in the framework of separable metric spaces, and that it enjoys lower-semicontinuity properties with respect to the $L^1_{\rm loc}$-convergence of the elements of partitions.

7.2.4 Examples of Homogenized Energy Densities

We now briefly examine the examples in Section 7.1 and describe the energy densities of the corresponding continuum limits.

Example 7.12 (Two-dimensional long-range antiferromagnetic interaction) In order to compute the homogenized energy density in Example 7.3, it is convenient to introduce a vector parameter, after noting that the energy can be rewritten as the sum of four decoupled antiferromagnetic energies on the lattices $2\mathbb{Z}^2 + i$ with

$$i \in \{(0,0), (1,0), (0,1), (1,1)\}.$$

A ground state v is then described by the vector

$$y = (v_{(0,0)}, v_{(1,0)}, v_{(0,1)}, v_{(1,1)})$$

in $Y_0 = \{-1,1\}^4$, which plays the role of $l \in \{1,\dots,K\}$ in Section 7.2.1. Using such a parameter, we obtain the energy density simply by

$$\varphi(y, y', \nu) = \|y - y'\|^2 \, \|\nu\|_1.$$

This formula highlights that we have a superposition of the separate energy densities $(y_k - y_k')^2\|\nu\|_1$; that is, $4\|\nu\|_1$ whenever we have an interface for the kth component; the factor 4 in place of 8 is due to the fact that each such interaction is considered on the lattice $2\mathbb{Z}$.

Example 7.13 ("Stabilized" antiferromagnetic triangular lattice) Similarly to the case of Example 7.3 treated earlier, in order to compute the homogenized energy density in Example 7.4, it is convenient to label a ground state v with $(v_{(0,0)}, v_{(1,0)}, v_{(1/2,\sqrt{3}/2)}) \in Y_0 = \{-1,1\}^3 \setminus \{(1,1,1),(-1,-1,-1)\}$. By decoupling the interactions on the three ferromagnetic triangular lattices, one obtains

$$\varphi(y, y', \nu) = c\|y - y'\|^2 \, \varphi_{\rm hex}(\nu),$$

where $\varphi_{\rm hex}$ denotes the homogenized energy for the ferromagnetic interactions on one of the three lattices as in Section 3.3.4.

7.2.5 Total Frustration

We now check the well-known fact that antiferromagnetic interactions in the triangular lattice are not described as in Theorem 7.10.

Example 7.14 (Total frustration for the triangular lattice for antiferromagnetic interactions) We may consider the normalized antiferromagnetic energy density ϕ as

$$\phi(\{u_j\}_j) = u_{(0,0)}u_{(1,0)} + u_{(1,0)}u_{(1/2,\sqrt{3}/2)} + u_{(1/2,\sqrt{3}/2)}u_{(0,0)} + 1. \qquad (7.12)$$

The corresponding energy is minimized by functions u not taking a constant value on each unit triangle of the lattice.

In order to check that a parameterization with a finite number of ground states is not possible, we show that there are infinitely many distinct periodic ground states. To that end it suffices to exhibit a compact-support perturbation of a periodic ground state. Indeed, we can choose as a periodic ground state the one with alternate horizontal lines of 1 and -1, and we remark that we can exchange the values on those lines on a rhombus as pictured in Fig. 7.7. By placing periodically copies of such a rhombus we can construct minimal states with arbitrary period.

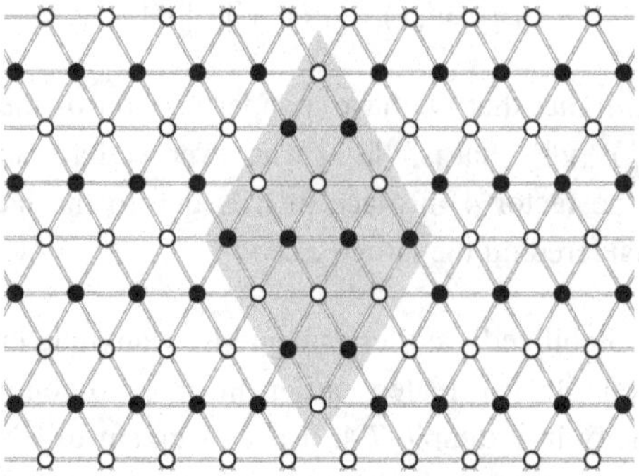

Figure 7.7 Compact-support perturbation of a periodic ground state.

7.3 Analysis of Inhomogeneous Energies

Theorem 7.10 can be applied to treat inhomogeneous energies, when the energy densities themselves depend on the index i; namely, we have

$$E(u;\Omega) = \sum_{i\in\mathbb{Z}^d\cap\Omega} \phi_i(\{u_{i+j}\}_{j\in\mathbb{Z}^d}), \tag{7.13}$$

and $i \mapsto \phi_i : \mathcal{X}^{\mathbb{Z}^d} \to [0, L]$ is itself h-periodic; moreover, the functionals E_ε are defined accordingly as in (7.4). With the change of parameters in Section 7.2.3 such energies can be rewritten as homogeneous energies on a coarser lattice. Indeed, in the notation of that section, it suffices to define

$$\phi(\{u_j\}_j) = \frac{1}{h^d} \sum_{i\in\{1,\dots,h\}^d} \phi_i(\{\Psi(u)_{i+j}\}_{j\in\mathbb{Z}^d}). \tag{7.14}$$

Under the assumption that, up to addition of a constant in order to normalize the minimum to 0, if necessary, ϕ satisfies (H1)–(H3) in the Compactness Theorem, we can then apply that result. Note that the homogenized energy density φ is described by the same formulas (7.7) with E as in (7.13).

Remark 7.15 (General lattices) The extension to inhomogeneous interactions allows one to consider also more general non-Bravais lattices, which are not invariant under translations by their elements, for example, the hexagonal lattice, upon interpreting $\phi_i(\{u_{i+j}\}_{j\in\mathcal{L}-i})$ as defined on the lattice $\mathcal{L} - i$.

7.3.1 Ferromagnetic-Antiferromagnetic Mixtures

The extension to inhomogeneous energies allows us to consider some homogenization problems for mixtures of ferromagnetic and antiferromagnetic interactions. In this section we consider nearest-neighbor systems with periodic coefficients themselves taking the values -1 and 1. The overall behavior may be from nonfrustrated to totally frustrated with various types of degeneracies.

Example 7.16 (Nonfrustrated systems) Let $a_{ij} \in \{-1, 1\}$ be periodic nearest-neighbor coefficients on a periodic lattice $\mathcal{L}$. A simple criterion that ensures that such a system of *mixtures of ferromagnetic and antiferromagnetic nearest-neighbor interactions*

$$\sum_{\langle i,j \rangle} a_{ij}(u_i - u_j)^2$$

is nonfrustrated is that every cycle in the lattice, that is, every finite sequence of coefficients $\{a_{i_n i_{n-1}} : n = 1, \dots, N\}$ with $i_N = i_0$, possesses an even number of negative a_{ij}. In the square lattice this condition is equivalent to requiring that each square has an even number of ferromagnetic interactions. In this case, if T denotes the period of the system, the value of $u_{(0,0)}$ determines the whole nonfrustrated u, which is T-periodic if $u_{(T,0)} = u_{(0,T)} = u_{(0,0)}$, and it is $2T$ periodic otherwise. Hence, (H1)–(H3) are satisfied with ϕ_i given simply by the sum of the interactions with one vertex in i, with $(u_j - u_0)^2$ if the interaction between i and $i + j$ is ferromagnetic, and $(u_j + u_0)^2$ if it is antiferromagnetic, $K = 2$, $h = 2T$, and $M' = 2M = 2h$. A nonfrustrated system is equivalent to a ferromagnetic system up to a suitable change of sign, for example, of all sites that take the value -1 at a ground state. Hence, the homogenized energy density is always $8\|\nu\|_1$.

A suitable choice of the geometry allows us to exhibit mixtures for which ground states are nonfrustrated. We now exhibit three two-dimensional examples of nearest-neighbor systems, with different patterns for the minimizers. We describe the energies through the energy densities ϕ_i as in (7.13).

(i) (ferromagnetic vertical interaction, and alternating columns of antiferromagnetic and ferromagnetic horizontal interactions) If

$$\phi_i(\{u_j\}_j) = \begin{cases} (u_{(1,0)} + u_{(0,0)})^2 + (u_{(0,1)} - u_{(0,0)})^2 & \text{if } i_1 \text{is even,} \\ (u_{(1,0)} - u_{(0,0)})^2 + (u_{(0,1)} - u_{(0,0)})^2 & \text{if } i_1 \text{is odd,} \end{cases}$$

then the two ground states are alternating vertical pairs of lines of 1 and -1 spins (see Fig. 7.8).

(ii) (alternating horizontal and vertical lines of ferromagnetic and antiferromagnetic interactions) If

$$\phi_i(\{u_j\}_j) = (u_{(1,0)} + (-1)^{i_2}\, u_{(0,0)})^2 + (u_{(0,1)} + (-1)^{i_1}\, u_{(0,0)})^2,$$

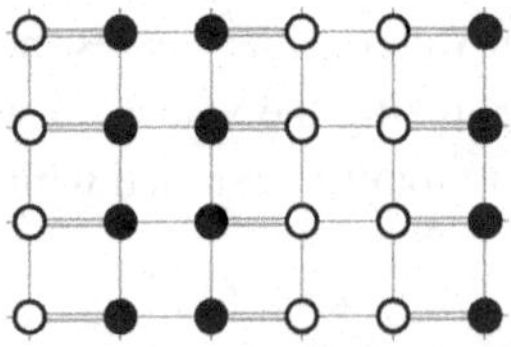

Figure 7.8 Nonfrustrated lines.

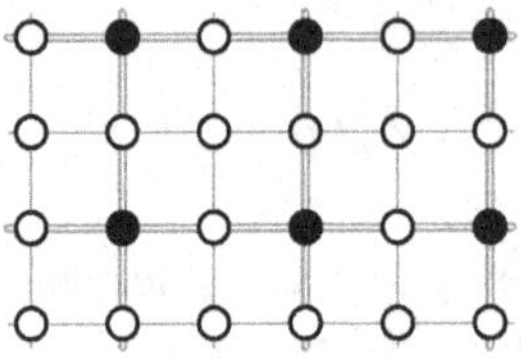

Figure 7.9 Nonfrustrated inclusions.

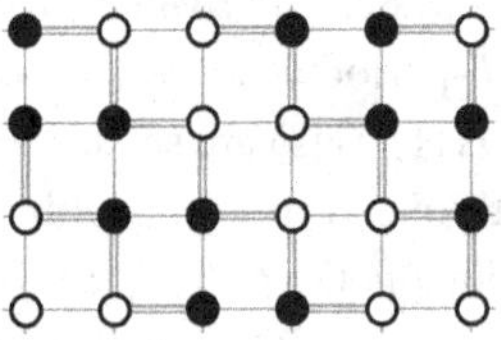

Figure 7.10 Nonfrustrated stripes.

then the two ground states are uniform states except on the lattice $2\mathbb{Z}^2$, where they have the opposite value (see Fig. 7.9).

(iii) (a zig-zag pattern of alternating ferromagnetic and antiferromagnetic interactions) If

$$\phi_i(\{u_j\}_j) = (u_{(1,0)} + (-1)^{i_1+i_2}\, u_{(0,0)})^2 + (u_{(0,1)} + (-1)^{i_1+i_2}\, u_{(0,0)})^2,$$

then the two ground states are stripes at –45 degrees (see Fig. 7.10).

Remark 7.17 (Systems parameterized by the "majority phase") We may consider a h-periodic system of nearest-neighbor interactions where the antiferromagnetic interactions form a periodic set composed of isolated components. If the distance between such connected components is large enough, then ground states are nonconstant only on isolated sets, so that we may take this majority value 1 or -1 as parameterizing the ground states, ensuring a relevant continuum parameter of ferromagnetic type. The Γ-convergence can be given in terms of that parameter as for the case of discrete perforated domains. An interesting remark is that in the two-dimensional case, even though the limit is a perimeter functional defined on one-dimensional interfaces, the homogenization formula giving its integrand φ cannot in general be translated in a least-path formula.

Note that Theorem 7.10 may not directly be applied if we had more than one pattern minimizing the isolated antiferromagnetic contributions, since in that case (H2) would be violated.

The preceding observation suggests that Theorem 7.10 may be extended to the case when (H2) does not hold, provided we can define an equivalence class among ground states that can be mixed with zero interfacial energy. The following example clarifies this observation.

Example 7.18 (Equivalent ground states and interfacial microstructure) We consider the 3-periodic system in dimension 2 with

$$\phi_i(\{u_j\}_j) = \begin{cases} (u_{(1,0)} + u_{(0,0)})^2 + (u_{(0,1)} + u_{(0,0)})^2 & \text{if } i = (0,0), \\ (u_{(1,0)} - u_{(0,0)})^2 + (u_{(0,1)} - u_{(0,0)})^2 & \text{otherwise.} \end{cases}$$

Note that the system is frustrated; hence, no function satisfies $E(u) = 0$. In this case ϕ directly defined as in (7.14) does not satisfy the hypotheses of Theorem 7.10 since its pointwise minimization does not correspond to a ground state for E. We then have to regroup the terms in the sum giving E and renormalize the energy. We can choose ϕ defined on functions defined on $3\mathbb{Z}^2$ by

$$\begin{aligned}\phi(\{w_j\}_j) &= 4u_{(1,0)}u_{(0,0)} + 4u_{(-1,0)}u_{(0,0)} - 8 \\ &\quad + \frac{1}{2}\sum_{i\in\{-1,0,1,2\}^2}\Big((u_{i+e_1} - u_i)^2 + (u_{i+e_1+e_2} - u_{i+e_1})^2 \\ &\quad + (u_{i+e_2} - u_{i+e_1+e_2})^2 + (u_i - u_{i+e_2})^2\Big),\end{aligned}$$

where $u = \Psi(w)$. Minimizing this energy density, we obtain two pairs of functions defined on $\{-1,0,1,2\}^2$ that have a constant value on all points except at $(0,0)$ where they can take equivalently the value 1 or -1. In particular, two functions with the same constant value on most points give two equivalent ground states, with no interfacial energy between them (see Fig. 7.11).

Note that the optimal microstructure for the interfacial energy is obtained by maximizing nonfrustrated interactions along the interface (see e.g. Fig. 7.12 for an optimal microstructure for a 45-degree interface).

Example 7.19 (A noncoercive case) We show an example of application of Theorem 7.10 to a case when (H1) and (H2) are not satisfied. As a consequence, the resulting Γ-limit is degenerate and noncoercive on partitions of sets of finite perimeter.

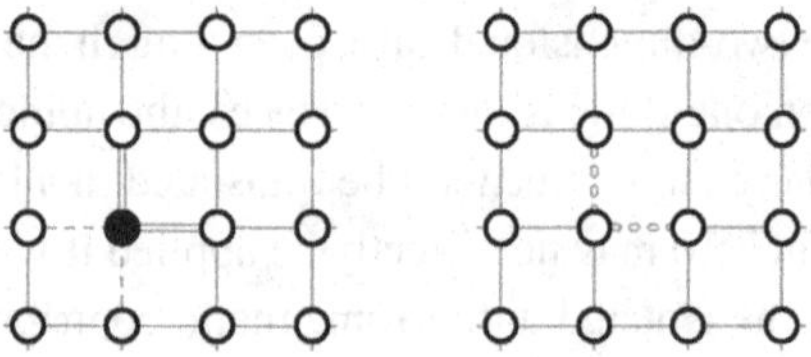

Figure 7.11 Two equivalent ground states.

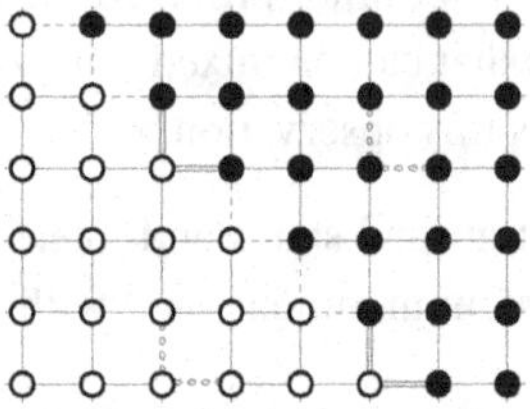

Figure 7.12 Optimal interfacial microstructure.

We consider the inhomogeneous system on $\mathbb{Z}^2$ with

$$\phi_i(\{u_j\}_j) = \begin{cases} (u_{(1,0)} - u_{(0,0)})^2 + (u_{(1,1)} - u_{(1,0)})^2 \\ \quad +(u_{(0,1)} - u_{(1,1)})^2 + (u_{(0,0)} - u_{(0,1)})^2 & \text{if } i_1 = 0 \text{ modulo } 4, \\ (u_{(1,0)} - u_{(0,0)})^2 + (u_{(1,1)} + u_{(1,0)})^2 \\ \quad +(u_{(0,1)} - u_{(1,1)})^2 + (u_{(0,0)} - u_{(0,1)})^2 - 1 & \text{if } i_1 = 1 \text{ modulo } 4, \\ (u_{(1,0)} + u_{(0,0)})^2 + (u_{(1,1)} + u_{(1,0)})^2 \\ \quad +(u_{(0,1)} + u_{(1,1)})^2 + (u_{(0,0)} + u_{(0,1)})^2 & \text{if } i_1 = 2 \text{ modulo } 4, \\ (u_{(1,0)} + u_{(0,0)})^2 + (u_{(1,1)} - u_{(1,0)})^2 \\ \quad +(u_{(0,1)} + u_{(1,1)})^2 + (u_{(0,0)} + u_{(0,1)})^2 - 1 & \text{if } i_1 = 3 \text{ modulo } 4. \end{cases}$$

These interactions are four periodic in the horizontal direction, are ferromagnetic on squares with the left-hand side vertices i with $i_1 = 0$ modulo 4, and are antiferromagnetic on squares with the left-hand side vertices i with $i_1 = 2$ modulo 4. On the other squares the horizontal interactions are the same as the vertical interaction on the left edge, so that we have three ferromagnetic interactions and an antiferromagnetic one or the converse, which are normalized accordingly.

We note that functions with zero energy restricted to four-by-two cells have one of the four patterns in Fig. 7.13. Note that each of these functions u can be viewed as a pair of a uniform ferromagnetic state and an alternating antiferro-

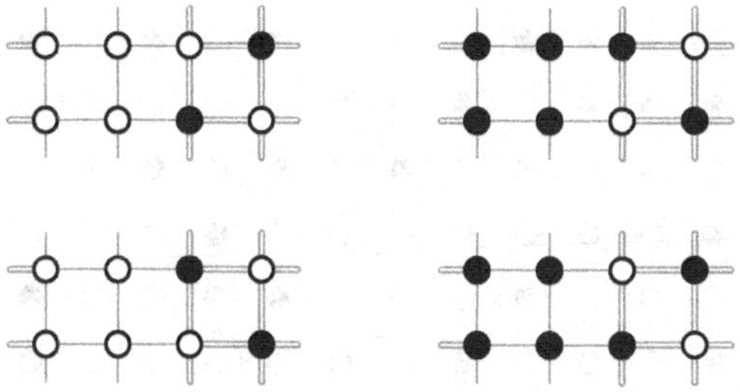

Figure 7.13 Minimal patterns.

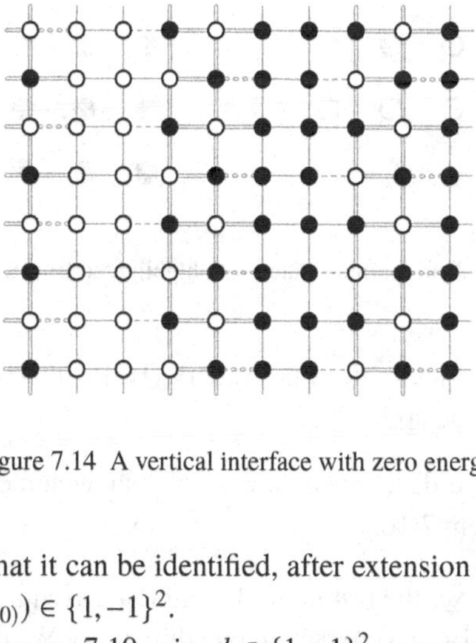

Figure 7.14 A vertical interface with zero energy.

magnetic state, so that it can be identified, after extension by periodicity, with the value $(u_{(0,0)}, u_{(2,0)}) \in \{1,-1\}^2$.

We can apply Theorem 7.10 using $l \in \{1,-1\}^2$ as parameters, with a slight abuse of notation. Note that the energy density φ is degenerate on vertical interfaces; that is,

$$\varphi(l,l',e_1) = 0 \text{ for all } l,l' \in \{1,-1\}^2$$

(see Fig. 7.14 for a vertical interface between $(1,1)$ and $(-1,1)$ with zero energy), while for horizontal interfaces we have

$$\varphi(l,l',e_2) = \frac{1}{2}|l-l'|^2.$$

An example of minimal horizontal interface between $(1,1)$ and $(-1,1)$ is given in Fig. 7.15. By approximation of an arbitrary interface with interfaces in the coordinate directions we obtain

$$\varphi(l,l',\nu) = \frac{1}{2}|l-l'|^2|\nu_1|.$$

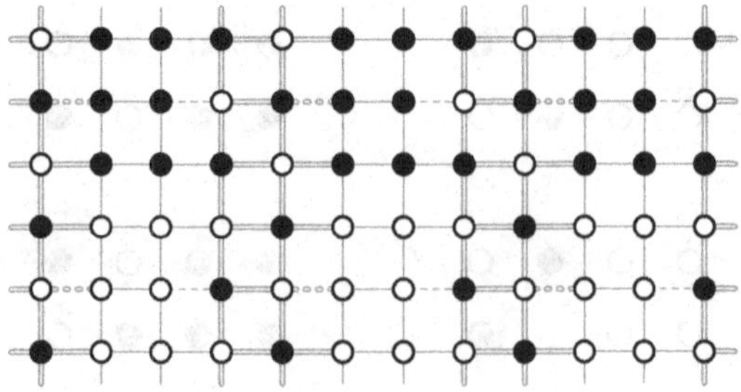

Figure 7.15 A horizontal minimal interface.

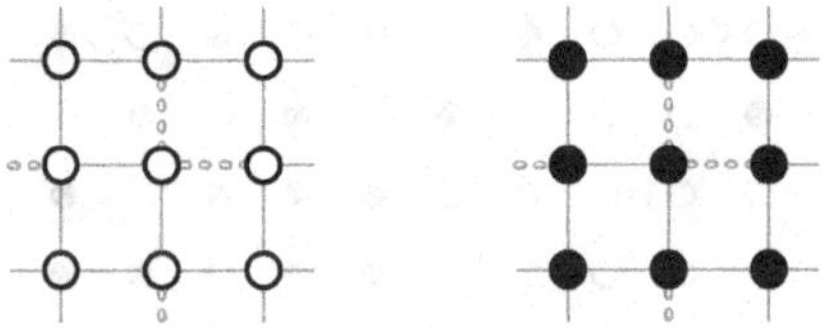

Figure 7.16 Uniform minimal states with highlighted frustrated interactions.

Note that, since our system does not satisfy (H1) and (H2), the compactness Theorem 7.9 cannot be applied.

Example 7.20 (Some degenerate cases) We show some cases to which we cannot apply Theorem 7.10.

(i) This example mixes the features of Example 7.18 (equivalent ground states) and Example 7.19 (degeneracy in one direction). We consider a system as in Example 7.18 but with a shorter periodicity; namely, with ϕ defined on functions defined on $2\mathbb{Z}^2$ by

$$\begin{aligned}\phi(\{w_j\}_j) = {} & (u_{(1,0)} + u_{(0,0)})^2 + (u_{(0,1)} + u_{(0,0)})^2 - 8 + (u_{(-1,0)} - u_{(0,0)})^2 \\ & + (u_{(0,-1)} - u_{(0,0)})^2 + \frac{1}{2}\Big((u_{(1,0)} - u_{(1,1)})^2 + (u_{(1,1)} - u_{(0,1)})^2 \\ & + (u_{(0,1)} - u_{(-1,1)})^2 + (u_{(-1,1)} - u_{(-1,0)})^2 + (u_{(-1,0)} - u_{(-1,-1)})^2 \\ & + (u_{(-1,-1)} - u_{(0,-1)})^2 + (u_{(0,-1)} - u_{(1,-1)})^2 + (u_{(1,-1)} - u_{(0,0)})^2\Big),\end{aligned}$$

where $u = \Psi(w)$.

In this case we have nonperiodic minimal states, since we can construct 45-degree interfaces with zero interfacial energy between two uniform minimal periodic states as those in Fig. 7.16. Hence, we can construct alternating layers of those two states of arbitrary (nonperiodic) thickness

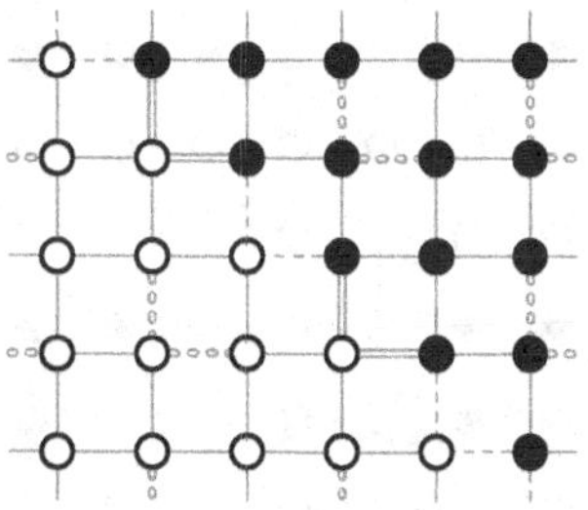

Figure 7.17 Interface with zero energy.

(see Fig. 7.17). Note that in this example the minimum of ϕ is negative, even though the limit energy is nonnegative.

(ii) (a totally frustrated example) We now give a two-dimensional example on the square lattice with a behavior similar to antiferromagnetic interactions on a triangular lattice. In this example the interactions corresponding to the sides of a unit lattice square are three ferromagnetic interactions and one antiferromagnetic interaction, or the converse. To that end, we may choose ϕ_i as follows:

$$\phi_i(\{u_j\}_j) = \begin{cases} (u_{(1,0)} + u_{(0,0)})^2 + (u_{(1,1)} - u_{(1,0)})^2, \\ \quad +(u_{(0,1)} - u_{(1,1)})^2 + (u_{(0,0)} - u_{(0,1)})^2 - 4 & \text{if } i_2 \text{ is even,} \\ (u_{(1,0)} - u_{(0,0)})^2 + (u_{(1,1)} + u_{(1,0)})^2, \\ \quad +(u_{(0,1)} + u_{(1,1)})^2 + (u_{(0,0)} + u_{(0,1)})^2 - 4 & \text{if } i_2 \text{ is odd.} \end{cases}$$

The interactions are chosen so that all squares in a horizontal line are of the same type.

A ground state of this energy is the function v, given by $v^i = (-1)^{\lfloor \frac{i_2}{2} \rfloor}$ (double alternate horizontal stripes). In order to show that the energy is not minimized only on periodic ground states it suffices to exhibit a function with $E(u) = 0$ consisting of a compact-support perturbation of the function v defined here. Such a function u is described in Fig. 7.18.

7.4 Some Remarks on Boundary Conditions

The Compactness Theorem in Chapter 4 allows one to treat Neumann boundary conditions; that is, when no boundary condition is required, on a bounded domain, or Dirichlet boundary conditions as defined on a neighborhood of the boundary of the set (see Section 4.7.1). We have seen that it is possible to use that theorem to treat problems with oscillating ground states for energies on

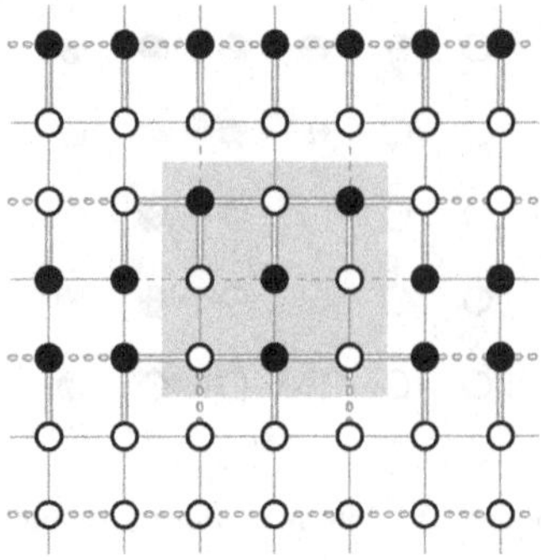

Figure 7.18 Inclusion with no interfacial energy.

functions u defined on the whole $\varepsilon\mathbb{Z}^d$. Unfortunately, if we want to consider functions defined only on some $\Omega \cap \varepsilon\mathbb{Z}^d$, the resulting energy densities ϕ_i^ε will depend necessarily on i, and the hypothesis that all such energy densities have the same ground states is a too restrictive assumption. In general, even for energy densities with finite range, interactions with i close to the boundary of Ω will tend to be minimized by patterns different from those minimizing the energy densities in the interior (which are those described by Theorem 7.10). As a result, we expect to still describe the behavior in (the interior of) Ω as in Theorem 7.10, but additional conditions will derive from boundary relaxation. Note, however, that if ϕ_i^ε satisfy the hypotheses of Theorem 7.10 in compact subsets of Ω, then the functional defined in (7.6) is still a lower bound for the limit energies, so that the Gamma-limit is finite only on ground-state parameterized partitions.

7.4.1 Nonlocal Effects of Boundary Conditions

In this section we examine some effects of the presence of a boundary.

Example 7.21 (One-dimensional examples: parity effects) A simple example illustrating parity effects due to oscillating ground states is the one-dimensional nearest-neighbor antiferromagnetic system with Dirichlet boundary conditions

$$\min\Big\{\sum_{i=1}^{N}(u_i + u_{i-1})^2 : u_0 = u_N = 1\Big\}.$$

The minimum is 0, achieved on $v_i = (-1)^i$ if N is even, while it is 4, achieved by any function v^j of the form

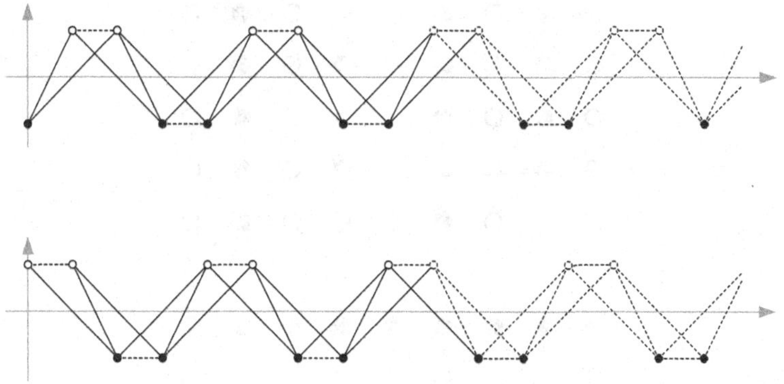

Figure 7.19 Two types of boundary pattern.

$$v_i^j = \begin{cases} (-1)^i & \text{if } i < j, \\ (-1)^{i+1} & \text{if } i \geq j \end{cases}$$

for some $j \in \{1, \ldots, N\}$.

Example 7.22 (One-dimensional examples: boundary layers) We consider the functionals of Example 7.1 on a bounded interval; that is, at fixed ε, on functions $u\colon \{0, \ldots, N\} \to \{-1, 1\}$ for some N, and take the form

$$E_\varepsilon(u) = \sum_{i=1}^{N} u_i u_{i-1} + \sum_{i=1}^{N-1} (u_{i-1} u_{i+1} + 1).$$

We can regroup the terms as follows,

$$E_\varepsilon(u) = \sum_{i=1}^{N-1} \Big(\frac{1}{2}(u_i u_{i-1} + u_{i+1} u_i) + u_{i+1} u_{i-1} + 1 \Big) + \frac{1}{2}(u_1 u_0 + u_N u_{N-1}),$$

to highlight the fact that the first and last u_i are not balanced, and have a *boundary-layer* effect.

The functional is minimized if the first parenthesis is zero for all i; that is, u coincides with one of the four ground states in Example 7.1, and the last term is minimal. The latter condition depends on the parity of N: if N is even, then the minimum is -1 and is achieved on the two states with alternating first two and last two values (as in the first one in Fig. 7.19 at 0), while if N is odd, the value is 0 at all ground states and we have four minimizers.

Example 7.23 (Minimization of striped antiferromagnetic states in a rectangle) We consider a system of nearest and next-to-nearest antiferromagnetic

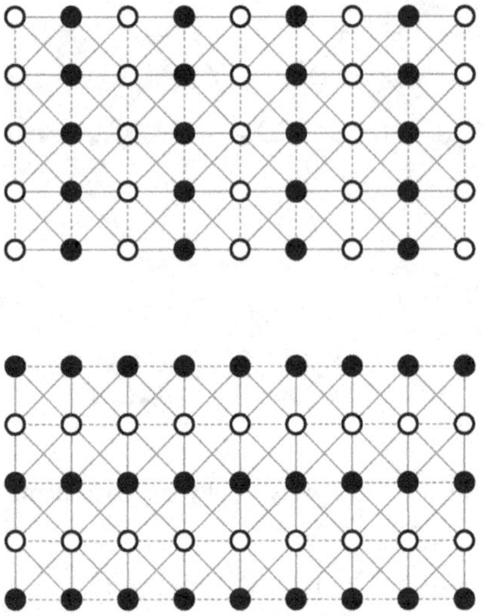

Figure 7.20 Striped patterns in a coordinate rectangle.

interactions in a rectangle. At fixed ε the rectangle can be parameterized on $\{0,\ldots,N_1\} \times \{0,\ldots,N_2\}$. We suppose that $N_1 > N_2$, and in addition that $c_2 > 2c_1 > 0$. In this case a candidate minimizer is the vertical striped state in Fig. 7.20, which optimizes the energy on each cell and, compared with the horizontal striped state in the figure, minimizes the number of frustrated connections on the boundary, which give a total contribution $c_1(N_2 - N_1)$. Note that in this case the energy of the upper and lower sides of the rectangle is negative. The minimality of the vertical striped configuration can be proved by remarking that if a square in the interior of the rectangle has a striped pattern, then the same pattern minimizes the interactions in the same whole vertical column or horizontal line since $c_2 > 2c_1$.

If we are allowed variations of the boundary, provided it tends to the rectangle as $\varepsilon \to 0$, then we may have a lower value, still keeping a connected set of points, for example with the boundary in Fig. 7.21, which gives a total contribution of $-c_1N_1$.

7.4.2 Frustrated Thin Films

For thin films the boundary effects are particularly relevant and can be highlighted by the variation of the number of ground states necessary to describe the limit in dependence of the thin-film thickness. In the following examples we refer to energies defined in previous sections without renormalizing them

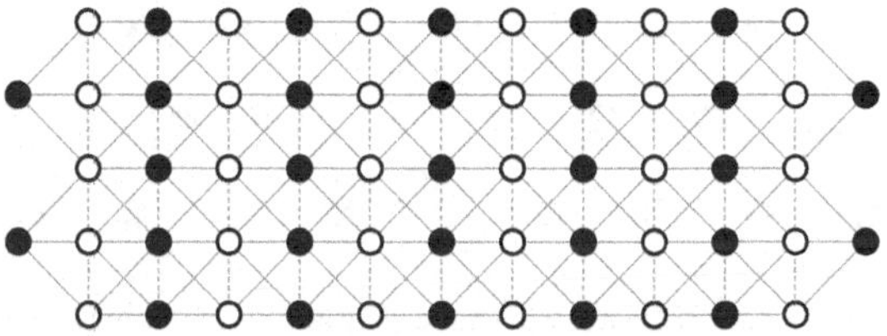

Figure 7.21 Variations of a rectangular boundary.

in order to have ground states with zero energy, since the actual computation of the interfacial energies is not relevant to our arguments.

Example 7.24 (Striped antiferromagnetic states: reduction of the number of parameters) The same reasoning of Example 7.23 shows that, in dimension 2, coordinate thin films of nearest and next-to-nearest antiferromagnetic interactions in the square lattice can be described by only two parameters representing the two variants of vertical stripes (same as in the first picture of Fig. 7.20). We then have a reduction with respect to the number of parameters of the bulk case (four).

Example 7.25 (Triangular nearest-neighbor antiferromagnetic systems: rigidity) The triangular antiferromagnetic system is totally frustrated. Its thin films instead have a finite number of ground states, increasing with the number N of layers.

If $N = 1$, we have a one-dimensional nearest-neighbor interaction on $\mathbb{Z}$ with two parameters.

If $N = 2$, then ground states have no frustrated interactions on the boundary, so that they are alternating states on both copies of $\mathbb{Z}$ with resulting four parameters.

If $N = 3$, then ground states have no frustrated interactions on the boundary and only one frustrated interaction per triangle. We then have eight ground states that are two-periodic in the horizontal direction, corresponding to alternating states on the three copies of $\mathbb{Z}$. The only possible locally minimal configuration that is not periodic is reproduced in Fig. 7.22. Note that if some frustrated connections form a convex angle, then they determine all subsequent connections in that angle (in particular, they determine all connections if the angle is of π; that is, if they are aligned). The minimal-energy configuration in Fig. 7.22 is then determined and equals a periodic ground state outside the central transition layer.

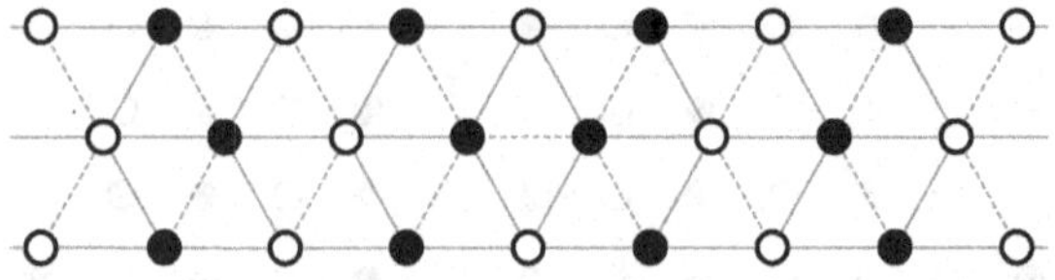

Figure 7.22 Zero-energy asymmetric transition for three layers.

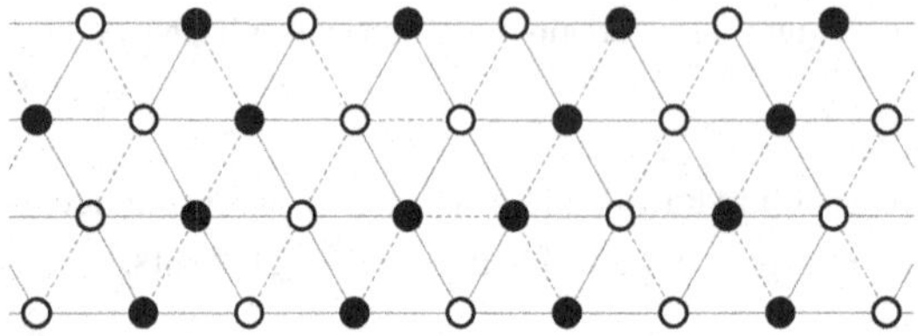

Figure 7.23 Zero-energy asymmetric transition for four layers.

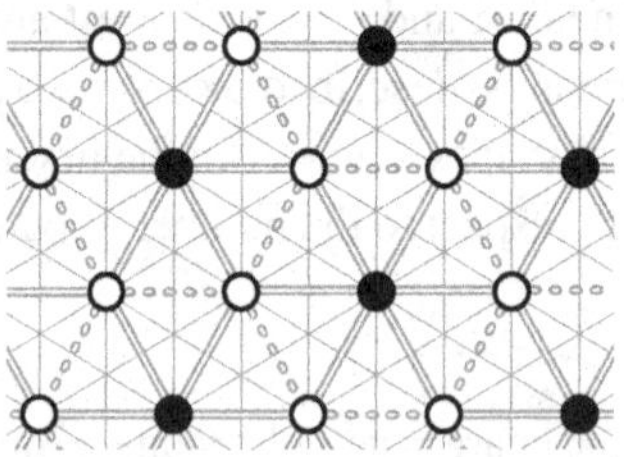

Figure 7.24 Ground state with a hexagonal pattern.

It is interesting to note that the configuration in Fig. 7.22 gives a zero-energy interfacial term between two states, and highlights that the final one-dimensional energy density φ is not symmetric; that is, $\varphi(u, v) \neq \varphi(v, u)$.

If $N > 3$, then it is likely that the number of parameters is 2^N, even though the classification of the geometries of possible zero-energy transitions seems to get increasingly complex, as in Fig. 7.23 for $N = 4$.

Example 7.26 ("Stabilized" antiferromagnetic triangular systems: increase of the number of parameters) We consider a thin film in a triangular lattice with antiferromagnetic nearest-neighbor interactions and with ferromagnetic next-to-nearest-neighbor interactions as in Example 7.4, where it is shown to have six ground states with a hexagonal pattern, as pictured in Fig. 7.24. We slightly change the notation of Example 7.4 by denoting by $\sigma_{\rm n}$ the coefficient of the nearest-neighbor interactions and by $\sigma_{\rm nn}$ that of next-to-nearest-neighbor interactions. In that example those coefficients are 1 and c respectively, so that

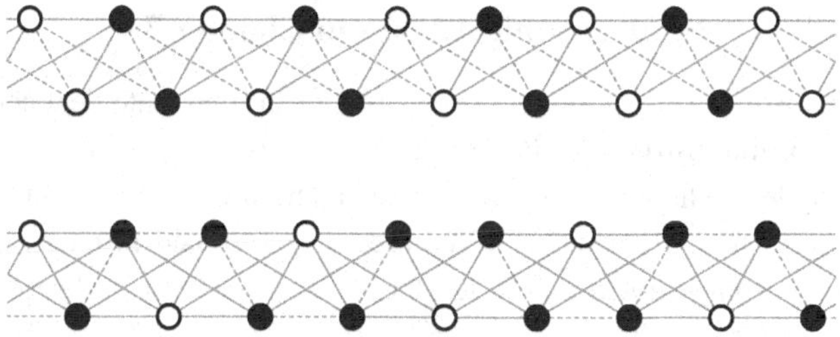

Figure 7.25 Competing patterns for ground states.

the energy is the same, up to a multiplicative constant, if we take $c = \frac{\sigma_{\rm nn}}{\sigma_{\rm n}}$. Note that both $\sigma_{\rm n}$ and $\sigma_{\rm nn}$ are positive.

If $N = 1$ we have a simple nearest-neighbor antiferromagnetic interaction system on $\mathbb{Z}$ described by two parameters. If $N = 2$, then we have two possible competing types of patterns, shown in Fig. 7.25. Note that the upper one (with four ground states) is the same as in Example 7.25; that is, when $\sigma_{\rm nn} = 0$, the other one (with six ground states) is the trace of the hexagonal pattern in Example 7.4. The first pattern is minimal if $\sigma_{\rm n} > 3\sigma_{\rm nn}$.

If $N \geq 3$, then we note that if $\sigma_{\rm n}$ is small enough, then the effect of the nearest neighbors is negligible and the minimization of second neighbor interactions gives the hexagonal pattern (six parameters), while if $\sigma_{\rm nn}$ is small enough, then the system behaves as for $\sigma_{\rm nn} = 0$. A comparison between the energies of the two patterns suggests the conjecture that indeed this happens when

$$\sigma_{\rm nn} < \frac{1}{3(N-1)}\sigma_{\rm n}. \tag{7.15}$$

Since we have shown that for $N = 3$ we have eight parameters, for that case we have a *thin film with a larger number of parameters than in the bulk case.*

Remark 7.27 (Nonmonotonicity of the number of parameters in dependence of the thickness) Note that the conjectured condition (7.15) implies that if we take coefficients satisfying

$$0 < \sigma_{\rm nn} < \frac{1}{6}\sigma_{\rm n},$$

then the related number of parameters is 2^N for $N \leq 3$, while it is 6 for N so large that (7.15) is violated. Hence, we have a nonmonotone dependence of the number of parameters on N.

Bibliographical Notes to Chapter 7

The mechanism of frustration as responsible for ground-state degeneracy and emergence of microstructures in discrete systems is a key ingredient of many physical models. The choice of a suitable function of the relevant variables that describes the system with the required accuracy and in terms of which the system becomes not (or only less) frustrated, hence simpler to be analyzed, is often very subtle. A list of examples of frustrated spin models and of possible simpler descriptions can be found in the book by Deep (2013). The study of frustrated systems by Γ-convergence has been first performed by Alicandro et al. (2006). The compactness and representation theorem using the parameterization by ground states follow the approach introduced by Braides and Cicalese (2017). A homogenization theorem for mixtures of ferromagnetic and antiferromegnetic interactions with a low percentage of the latter has been given by Braides and Piatnitski (2013), where the representation is given with respect to the majority phase.

Frustration appears also in vector spin systems. Even though in that case it may be enhanced by geometric constraints, the examples given in Section 7.1 may still be an illustration of phenomena of frustration-producing chirality transitions as in the work by Cicalese and Solombrino (2015) and Cicalese et al. (2019). The examples presented in the antiferromagnetic case illustrate the basic mechanism of the frustration phenomenon considered in Bach et al. (2021) for a nearest-neighbor vectorial antiferromagnetic spin model on the triangular lattice.

The problem of mixtures of ferromagnetic and antiferromagnetic interactions is complex and very intriguing. In Section 7.3.1 we have included some examples, mainly taken from the work by Braides and Cicalese (2017), when we have only nearest-neighbor interactions and the coefficients a_{ij} only take the values 1 and -1, so that they are somewhat balanced. The cases of long-range interactions and unbalanced coefficients are widely open and delicate.

The concept of ground state in the homogenization theorem is related to minimization problems in the whole space. An analysis of the symmetries of the ground states in the whole space of discrete models with competing ferromagnetic short-range and antiferromagnetic long-range interactions was performed by Giuliani et al. (2006, 2007, 2014) and more recently by Daneri and Runa (2019) exploiting, among others, the reflection-positivity method systematized in the context of statistical mechanics by Frohlich et al. (1978). The adaptation of those techniques to variational problems on general domains is a challenging issue, within the general problem of adapting methods valid in the whole space to bounded sets, which involve a delicate handling of boundary conditions. In particular, an interesting issue is whether such ground states in the whole space satisfy the hypotheses of the homogenization theorem.

The study of random mixtures of ferromagnetic and antiferromagnetic interactions is a wide and fascinating field, almost completely unexplored from the variational standpoint. We note that a nearest-neighbor model of i.i.d. coefficients in a square lattice with probability p of having an antiferromagnetic interaction and $1 - p$ of having a ferromagnetic interaction highlights a change of parameter in the description of the system as p varies. Indeed for $p = 0$ we have ferromagnetic interactions and the ground states are the constant states, while for $p = 1$ we have antiferromagnetic interactions and the ground states are oscillating states. We then deduce that a change in the parameters must occur at a certain value of p. In a paper by Braides et al. (2018b) it is shown that there exists a positive value p_0 that can be explicitly estimated, such that for $p < p_0$ (dilute case) in two dimensions ground states possess a majority phase. This has been proved by the use of the classical Peierls argument of Statistical Physics (Peierls, 1936; Lebowitz and Mazel, 1998). Even in this dilute regime, the description of the Γ-limit and the general d-dimensional case are open. As p moves toward $\frac{1}{2}$ the argument cannot be applied since interactions act in a nonlocal fashion and therefore their effect becomes very sensitive to the geometry of the infinite clusters. It is very suggestive to think that the description of such systems at $p = \frac{1}{2}$ is linked to the theory of *spin glasses* (Edwards and Anderson, 1975; Parisi, 1980; Mezard et al., 1987; Guerra and Toninelli, 2002).

In Braides et al. (2018b) also an interesting deterministic counterpart of the dilute case of mixtures of ferromagnetic and antiferromagnetic interactions with $a_{ij} \in \{-1, 1\}$ has been studied for periodic systems, showing that, for p_0 as above, the percentage of N-periodic nearest-neighbor systems with a percentage of antiferromagnetic interactions less than p_0 that cannot be described by the majority phase tends to 0 as N tends to $+\infty$. This is a new problem of a combinatorial nature that seems open to many developments. The deterministic statement was inspired by a lecture by M. Bhargava on proportions of elliptic curves (see e.g. Bhargava and Shankar, 2015).

The content of Section 7.4 is original; since only a few geometries are considered, here also there is room for generalizations and further applications, for example, to the effect of boundary conditions in general domains or to thin films in three dimensions.

8
Perspectives toward Dense Graphs

In this chapter we consider systems with a large number of connections. In this case the parameterization on a lattice may lose relevance, but some results can be still obtained using refined combinatorical notions on graphs.

The definition of ferromagnetic energies can be extended to sequences of graphs $G_k = (\mathcal{N}_k, \mathcal{E}_k)$, where $\mathcal{N}_k$ denotes the set of nodes and $\mathcal{E}_k$ is the set of the connections; that is, a symmetric subset of $\mathcal{N}_k \times \mathcal{N}_k$. Such energies are defined as

$$E_k(u) = \alpha_k \sum_{(i,j)\in\mathcal{E}_k} (u(i) - u(j))^2,$$

where the scaling factor α_k depends on the cardinality and structure of $\mathcal{E}_k$. In the case of ferromagnetic energies on cubic lattices, assuming for simplicity that the active coefficients a_{ij} are all equal to 1, we have $\mathcal{N}_k = \varepsilon_k\mathbb{Z}^d \cap \Omega$ for some ε_k and $\mathcal{E}_k$ giving the interacting nodes, and $\alpha_k = \varepsilon_k^{d-1}$.

In general, in order to quantify the density of a sequence of graphs with a diverging number of nodes, we will use as a parameter the ratio of active connections with respect to the total number of possible connections.

We note that the ferromagnetic energies we have considered until now satisfy

$$\lim_{k\to+\infty} \frac{\#\mathcal{E}_k}{(\#\mathcal{N}_k)^2} = 0.$$

The corresponding sequence of graphs are said to be *sparse*, while if this condition does not hold the sequence is said to be *dense*. Dense graph sequences will be examined in Section 8.2.

8.1 Locally Dense Graphs: Coarse Graining

In view of an analysis of convergence results applied to dense graphs in the next section, we examine ferromagnetic systems beyond the summability conditions

(H_f3) and (H_f4) in Section 4.9. Those assumptions ensure that essentially (up to asymptotically negligible terms) the range of the interactions is finite. We show that for some families of energies with long-range interactions a compactness result still holds and we can obtain in the limit a perimeter functional. In the following sections we consider an example of ferromagnetic energies that are defined on sparse graphs, but do not satisfy the abovementioned growth conditions. Such graphs are *locally dense*, in the sense that, at a scale that is diverging in the reference lattice (but small after scaling), all nodes are interacting.

8.1.1 Coarse Graining in Dimension 1

Let $\Omega = \mathbb{R}$ and consider an energy of the form

$$G_\varepsilon(u) = \sum_{|i-j|\leq R_\varepsilon} (u_i - u_j)^2,$$

where the sum is over $i, j \in \mathbb{Z}$ and R_ε satisfies the asymptotic properties

$$R_\varepsilon \to +\infty, \ \ \varepsilon R_\varepsilon \to 0 \text{ as } \varepsilon \to 0.$$

Note that (H_f4) holds by the second condition, while (H_f3) does not hold since R_ε diverges. Now, we look for a scaling of G_ε such that in the limit we may have an interface with finite energy. If we consider a discretization u^ε of the sign function, the energy $G_\varepsilon(u^\varepsilon)$ is of order R_ε^2, since each point interacts with all points at a distance less than R_ε, and we have to take into account an $\varepsilon R_\varepsilon$-neighbor of 0; that is, a number of interacting points of order R_ε (see Fig. 8.1). Hence, to obtain equibounded energies for sequences approximating a jump, we scale G_ε by a factor $\frac{1}{R_\varepsilon^2}$ and define

$$E_\varepsilon(u) = \sum_{|i-j|\leq R_\varepsilon} \frac{1}{R_\varepsilon^2}(u_i - u_j)^2.$$

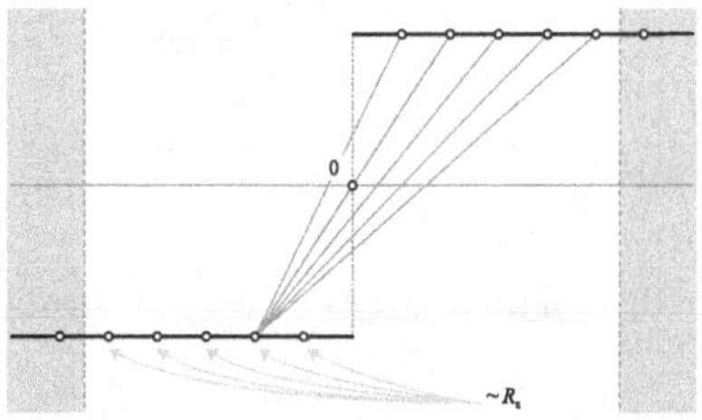

Figure 8.1 Estimate of the energy of an interface.

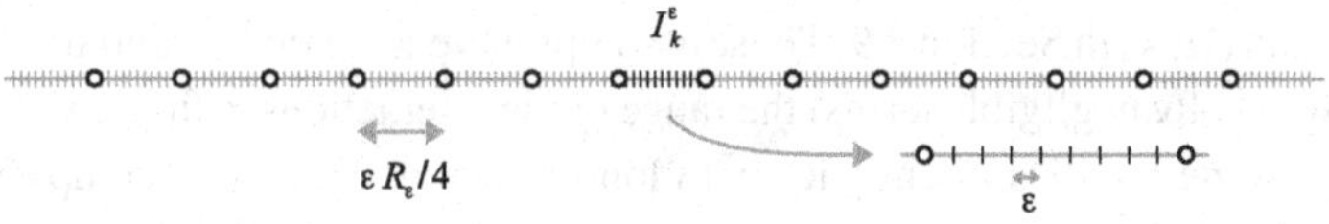

Figure 8.2 Coarse graining of the lattice $\varepsilon\mathbb{Z}$.

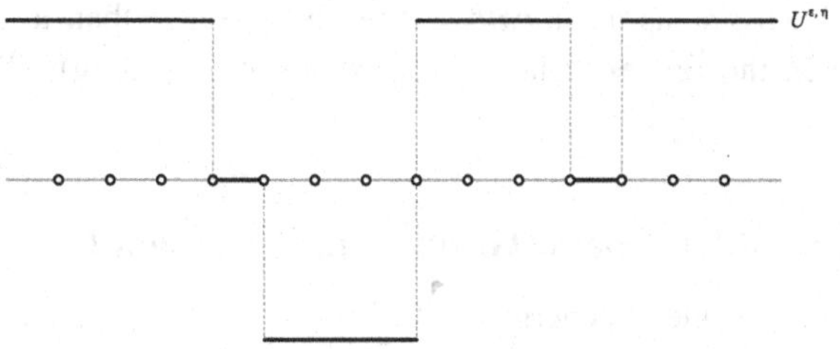

Figure 8.3 A coarse-grained function.

The Γ-limit is then finite on piecewise-constant functions, even though condition (H_f3) is not satisfied. In order to prove that the Γ-limit is finite exactly on this set of functions, we have to show a compactness result, and this can be done by a *coarse-graining* argument.

We subdivide the domain $\mathbb{R}$ in intervals I^ε_k, parametrized by $k \in \mathbb{Z}$, such that each interval contains a number of order R_ε of points of the lattice $\varepsilon\mathbb{Z}$. Since we want all points in each interval to interact with all points in the nearest intervals, we fix $\frac{\varepsilon R_\varepsilon}{4}$ as the size of I^ε_k (see Fig. 8.2).

Let u^ε be such that $E_\varepsilon(u^\varepsilon) \le c < +\infty$. We fix $\eta > 0$ and define a *coarse-grained function* $U^{\varepsilon,\eta} : \mathbb{Z} \to \{-1, 0, 1\}$ by setting

$$U^{\varepsilon,\eta}(k) = U^{\varepsilon,\eta}_k = \begin{cases} 1 & \text{if } \dfrac{4}{R_\varepsilon} \sum\limits_{\varepsilon i \in I^\varepsilon_k} u^\varepsilon_i > 1 - \eta, \\ -1 & \text{if } \dfrac{4}{R_\varepsilon} \sum\limits_{\varepsilon i \in I^\varepsilon_k} u^\varepsilon_i < -1 + \eta, \\ 0 & \text{otherwise} \end{cases} \tag{8.1}$$

(see Fig. 8.3).

Note that if $U^{\varepsilon,\eta}_k = 0$, then there exists a positive constant C_η such that

$$\sum_{\varepsilon i, \varepsilon j \in I^\varepsilon_k} \frac{1}{R^2_\varepsilon} (u^\varepsilon_i - u^\varepsilon_j)^2 \ge C_\eta;$$

hence, the equiboundedness of $E_\varepsilon(u^\varepsilon)$ gives the estimate

$$\#\{k \in \mathbb{Z}\colon U_k^{\varepsilon,\eta} = 0\} \le \frac{\sup_\varepsilon E_\varepsilon(u^\varepsilon)}{C_\eta} \le C'_\eta. \tag{8.2}$$

Moreover, if $\eta < \frac{1}{2}$, we have that if $U_k^{\varepsilon,\eta} = 1$ and $U_{k+1}^{\varepsilon,\eta} = -1$ (and conversely, by exchanging 1 and -1), then

$$\sum_{\varepsilon i,\varepsilon j \in I_k^\varepsilon \cup I_{k+1}^\varepsilon} \frac{1}{R_\varepsilon^2}(u_i^\varepsilon - u_j^\varepsilon)^2 \ge C$$

for some $C > 0$ independent of ε and η. Again using the uniform bound on $E_\varepsilon(u^\varepsilon)$, we get the estimate

$$\#\big\{k \in \mathbb{Z}\colon U_k^{\varepsilon,\eta}, U_{k+1}^{\varepsilon,\eta} \in \{-1,1\} \text{ and } U_k^{\varepsilon,\eta} \neq U_{k+1}^{\varepsilon,\eta}\big\} \le C'. \tag{8.3}$$

Estimates (8.2) and (8.3) ensure that, up to subsequences, $U^{\varepsilon,\eta} \to U^\eta$ as $\varepsilon \to 0$, where U^η is a piecewise-constant function assuming only the values -1 and $+1$, with a finite number of jump points. By using the definition of $U_k^{\varepsilon,\eta}$ as the average of u^ε in I_k^ε (up to a finite number of intervals), it follows that, for any $T > 0$,

$$\int_{[-T,T]} |U^{\varepsilon,\eta} - u^\varepsilon|\, dx \le cT\eta + r_\varepsilon, \tag{8.4}$$

where

$$\limsup_{\varepsilon\to 0} \frac{r_\varepsilon}{\varepsilon R_\varepsilon} < +\infty$$

since the total size of the union of the intervals where $U^{\varepsilon,\eta}$ vanishes is of order $\varepsilon R_\varepsilon$, and in each one of the other intervals the integral is less than $\frac{\varepsilon R_\varepsilon}{8}\eta$. In order to deduce the convergence of u^ε in $L^1_{\rm loc}(\mathbb{R})$, we have to prove that in fact the limit U^η does not depend on η. To check this, note that if we choose $\eta' < \eta$, by the definition of $U^{\varepsilon,\eta}$ we get the following monotonicity properties for the limit functions U^η and $U^{\eta'}$:

$$\{U^{\eta'} = 1\} \subseteq \{U^\eta = 1\} \text{ and } \{U^{\eta'} = -1\} \subseteq \{U^{\eta'} = -1\}. \tag{8.5}$$

Since both $\{U^{\eta'} = 1\} \cup \{U^{\eta'} = -1\}$ and $\{U^{\eta'} = 1\} \cup \{U^\eta = 1\}$ are equal to the whole $\mathbb{R}$, then the inclusions in (8.5) are in fact equalities; that is, $U^{\eta'} = U^\eta = U$. This allows us to deduce by (8.4) that, up to subsequences, $u^\varepsilon \to U$, concluding the proof of the compactness.

To compute the Γ-limit, let $u^\varepsilon \to u$ and consider, for any $r \in \mathbb{N}$, $r \le R_\varepsilon$ the restrictions of u^ε to the r disjoint sublattices $\mathcal{L}_\varepsilon^{k,r} = \varepsilon k + \varepsilon r\mathbb{Z}$, for $k \in \{1,\dots,r\}$. We localize the energies in a sufficiently small neighborhood $I(x)$ of a point $x \in S(u)$, and note that for $r \le R_\varepsilon$ (up to an arbitrary small fraction, less than

δR_ε for an arbitrarily small $\delta > 0$) the restrictions of u^ε to the lattices $\varepsilon k + \varepsilon r\mathbb{Z}$ have a change of sign. Hence

$$E_\varepsilon(u^\varepsilon; I(x)) \geq \frac{1}{R_\varepsilon^2} \sum_{r \leq R_\varepsilon} r \sum_{\varepsilon i, \varepsilon j \in \mathcal{L}_\varepsilon^{k,r} \cap I(x)} (u_i^\varepsilon - u_j^\varepsilon)^2 \geq \frac{1}{R_\varepsilon^2} \sum_{r \leq R_\varepsilon} (8r) - c\delta \geq 4 - c\delta.$$

Then

$$\liminf_{\varepsilon \to 0} E_\varepsilon(u^\varepsilon) \geq 4\#S(u).$$

The upper estimate can be proven by using the restrictions to $\varepsilon\mathbb{Z}$ of the function $u = 2\chi_{(x,+\infty)} - 1$ as a recovery sequence.

8.1.2 Coarse Graining in Dimension d

The model described in Section 8.1.1 can be extended to dimension d and to more general coefficients. We give only the statement of the result.

Let $a \colon \mathbb{R}^d \to [0, +\infty)$ be a continuous function and $R_\varepsilon > 0$ be such that

$$R_\varepsilon \to +\infty, \quad \varepsilon R_\varepsilon \to 0 \text{ as } \varepsilon \to 0.$$

Noting that the scaling of the energies in dimension d corresponding to the argument in the previous section is $\frac{1}{R_\varepsilon^{d+1}}$, we define

$$E_\varepsilon(u) = \sum_{\|i-j\| \leq R_\varepsilon} \varepsilon^{d-1} a_{ij}^\varepsilon (u_i - u_j)^2, \text{ where } a_{ij}^\varepsilon = \frac{1}{R_\varepsilon^{d+1}} a\Big(\frac{i-j}{R_\varepsilon}\Big). \tag{8.6}$$

In this expression the sum is taken over pairs (i, j) with $i, j \in \mathbb{Z}^d \cap \Omega$. The Γ-limit is an integral functional given by

$$4 \int_{\Omega \cap \partial^* A} \int_{\mathbb{R}^d} a(\xi) |\langle \xi, \nu \rangle| \, d\xi \, d\mathcal{H}^{d-1}.$$

The limit density

$$\varphi(\nu) = 4 \int_{\mathbb{R}^d} a(\xi) |\langle \xi, \nu \rangle| \, d\xi$$

generalizes the formula we have seen in the homogeneous case, where $\varphi(\nu) = 4 \sum_{k \in \mathbb{Z}^d} \alpha_k |\langle k, \nu \rangle|$; in this case, the effect of long-range interactions gives in the limit an integral term instead of a sum. Note that the corresponding Wulff shape may not be crystalline. In particular, it is not crystalline in the case when $a = \chi_{B_1}$, for which φ equals the constant

$$\tau = 4 \int_{B_1} |\xi_1| \, d\xi$$

and the limit is simply $\tau \mathcal{H}^{d-1}(\Omega \cap \partial^* A)$.

8.1.3 A Sparse Graph Sequence with Diffuse Interfaces

We now show with an example that ferromagnetic energies on sparse graph sequences may not be described by a sharp-interface model.

We consider the set of nodes given by $\mathcal{N}_\varepsilon = [0,1] \cap \varepsilon\mathbb{Z}$ and the set of connections $\mathcal{E}_\varepsilon$ given by the nearest neighbors and by pairs of points at distance $\lfloor \frac{1}{\sqrt{\varepsilon}} \rfloor$; that is,

$$\mathcal{E}_\varepsilon = \left\{ (\varepsilon i, \varepsilon j) \in \mathcal{N}_\varepsilon \times \mathcal{N}_\varepsilon : \|i-j\| = 1 \text{ or } \|i-j\| = \left\lfloor \frac{1}{\sqrt{\varepsilon}} \right\rfloor \right\},$$

as pictured in Fig. 8.4. Without loss of generality, we can suppose $\frac{1}{\sqrt{\varepsilon}} \in \mathbb{N}$.

We consider the energies

$$E_\varepsilon(u) = \frac{1}{8} \sum_{(\varepsilon i, \varepsilon j) \in \mathcal{E}_\varepsilon} \alpha_\varepsilon (u_i - u_j)^2$$

defined for $u\colon \mathcal{N}_\varepsilon \to \{-1,1\}$, where the scaling factor α_ε is to be determined in such a way that interfaces have finite (and nonvanishing) energy; that is, the limit is finite on piecewise-constant functions. We test the energies on $u = 2\chi_{[0,+\infty)} - 1$. The number of interacting pairs giving a contribution to the energy is $\frac{1}{\sqrt{\varepsilon}} + 1$; hence, if u^ε denotes the discretization of u, we have

$$E_\varepsilon(u^\varepsilon) = \alpha_\varepsilon \left(\frac{1}{\sqrt{\varepsilon}} + 1 \right),$$

which is equibounded if α_ε is of order $\sqrt{\varepsilon}$.

In the light of this computation, the energies we consider are scaled as

$$E_\varepsilon(u) = \frac{1}{8} \sum_{(\varepsilon i, \varepsilon j) \in \mathcal{E}_\varepsilon} \sqrt{\varepsilon} (u_i - u_j)^2. \tag{8.7}$$

We are going to "lift" the energies to $\mathbb{Z}^2$. Let $u\colon [0,1] \cap \varepsilon\mathbb{Z} \to \{-1,1\}$. In Fig. 8.5 we picture the interpolation of such a function and in Fig. 8.6 we represent it in terms of spins. We define a corresponding function

$$v_u\colon [0,1]^2 \cap \sqrt{\varepsilon}\mathbb{Z}^2 \to \{-1,1\}$$

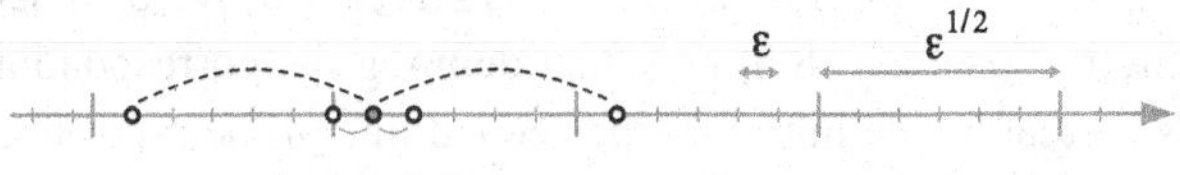

Figure 8.4 Connections of nearest neighbors and points at distance $\frac{1}{\sqrt{\varepsilon}}$.

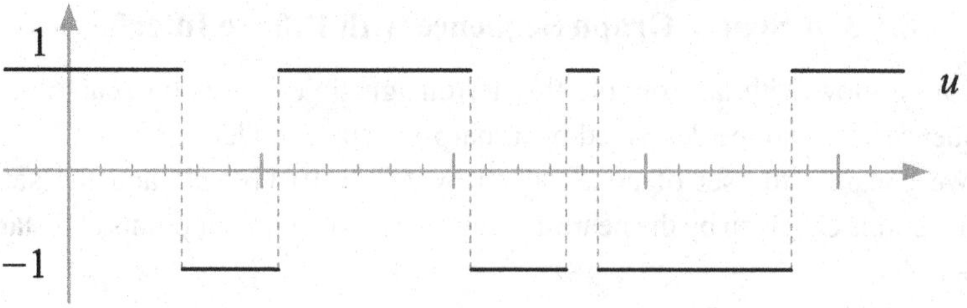

Figure 8.5 A one-dimensional spin function u.

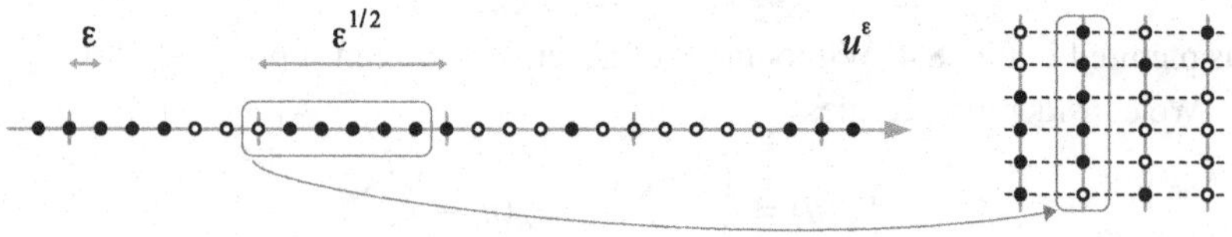

Figure 8.6 "Lifting" of u as in Fig. 8.5.

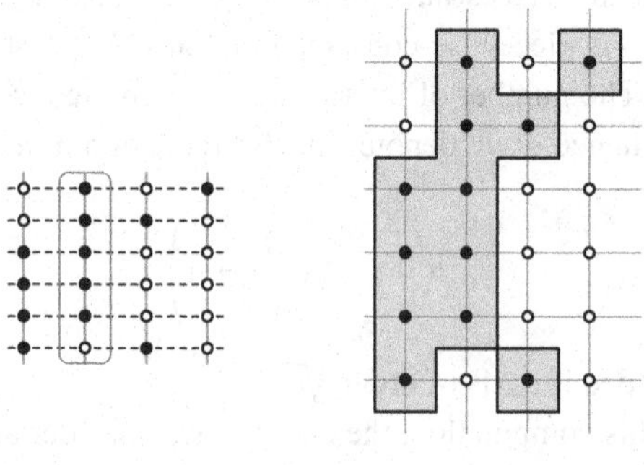

Figure 8.7 The function v_u corresponding to the lifting of u.

given by

$$v_u(\sqrt{\varepsilon}\, i_1, \sqrt{\varepsilon}\, i_2) = \begin{cases} u(\sqrt{\varepsilon}\, i_1 + \varepsilon i_2) & \text{if } \sqrt{\varepsilon}\, i_2 < 1, \\ v_u(\sqrt{\varepsilon}\, i_1, 0) = u(\sqrt{\varepsilon}\, i_1) & \text{if } \sqrt{\varepsilon}\, i_2 = 1. \end{cases}$$

This is obtained by identifying arrays of length $\sqrt{\varepsilon}$ with vertical segments in $\sqrt{\varepsilon}\mathbb{Z}^2$ of length 1 (see Fig. 8.6) and then defining the corresponding v_u (see Fig. 8.7). Note that the definition in the second line of the preceding function allows us to identify the upper and lower sides of the square $[0,1]^2$. In this way the "lifted energy" can be thought of as a nearest-neighbor energy taking into account an extension periodic in the vertical direction. In terms of v_u, the functional $E_\varepsilon(u)$ reads as

$$E_\varepsilon(u) = G_\varepsilon(v_u) = \frac{1}{8}\sum_{\langle \mathbf{i},\mathbf{j}\rangle} \sqrt{\varepsilon}\big(v_u(\sqrt{\varepsilon}\,\mathbf{i}) - v_u(\sqrt{\varepsilon}\,\mathbf{j})\big)^2,$$

where

$$\langle \mathbf{i},\mathbf{j}\rangle = \big\{\sqrt{\varepsilon}\,(\mathbf{i},\mathbf{j}) \in \sqrt{\varepsilon}\,\mathbb{Z}^2 \cap [0,1]^2\big\}.$$

For notational convenience, in this section we use boldface letters to indicate points $\mathbf{i} \in \mathbb{Z}^2$ in order to distinguish them from $i \in \mathbb{Z}$.

Given $A \subset [0,1] \times [0,1)$, $A^{\#}$ will denote the periodic extension of A to $[0,1]\times\mathbb{R}$. The sequence $\{G_\varepsilon\}$ Γ-converges as $\varepsilon \to 0$ to the crystalline perimeter given by

$$G(A) = \int_{[0,1]^2\cap\partial^*(A^{\#})} \|\nu\|_1 \, d\mathcal{H}^1$$

with respect to the convergence in $L^1_{\rm loc}$ of the piecewise-constant extensions of the spin functions.

Let $\{u^\varepsilon\}$ be a sequence with equibounded energy, and let u be the L^∞-weak* limit of u^ε (which we can assume exists up to subsequences). The corresponding functions $v^\varepsilon = v_{u^\varepsilon}$ converge, up to subsequences, to a set A of finite perimeter. In order to characterize u in terms of A we modify the sequence $\{u^\varepsilon\}$ so that the limit u is the same and the corresponding $\widehat{v}^\varepsilon$ converge to a set $\widehat{A}$ that is the subgraph of u. The function $\widehat{u}^\varepsilon$ is defined on each set $[\sqrt{\varepsilon}\,(k-1), \sqrt{\varepsilon}\,k) \cap \varepsilon\mathbb{Z}$ by permuting the values so that $i < j$ if $\widehat{u}^\varepsilon_i = 1$ and $\widehat{u}^\varepsilon_j = -1$ in this set. This operation does not modify the weak* limit. We define $\widehat{v}^\varepsilon = v_{\widehat{u}^\varepsilon}$. Then, the limit set $\widehat{A}$ is the subgraph of u, since we can write $u(x) = \int_0^1 v(x,y)\,dy$, where $v = 2\chi_{\widehat{A}} - 1$. Hence, $u \in BV(0,1)$.

Note that $G(\widehat{A}) \le G(A)$, so that it suffices to rewrite $G(\widehat{A})$ in terms of u in order to obtain a lower bound. The crystalline perimeter of $\widehat{A}$ can be written as

$$E(u) = 2\mathcal{H}^1(\{-1 < u(x) < 1\}) + |Du|(0,1).$$

In order to check this, it suffices to consider the case in which $\widehat{A}$ is a polyrectangle, in which case the horizontal segments of the boundary give the first term and the vertical segments correspond to the second one.

As for the construction of a recovery sequence, we only consider a constant function $u(x) = c$ with $c \in (-1,1)$. A general u can then be approximated by piecewise-constant functions so that the total variations converge. We discretize

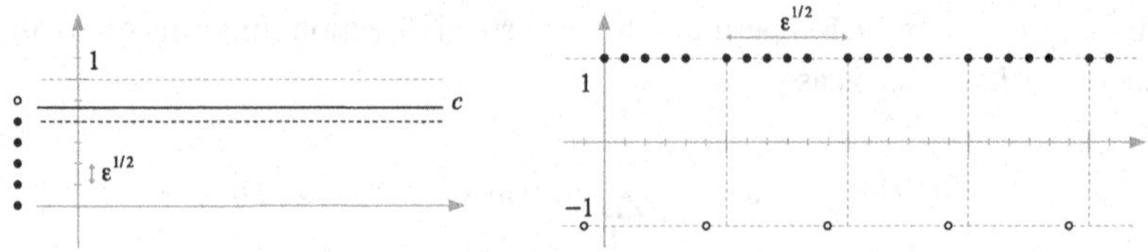

Figure 8.8 Diffuse interface.

the set $[0,1] \times [0,c]$ in $\sqrt{\varepsilon}\,\mathbb{Z}^2$, setting $v^\varepsilon(\sqrt{\varepsilon}\,\mathbf{i}) = 1$ if $\sqrt{\varepsilon}\,\mathbf{i} \in [0,1] \times [0,c]$ and $v^\varepsilon = -1$ otherwise. The ε-discretization of

$$u^\varepsilon(x) = \int_0^1 v^\varepsilon(x,y)\,dy,$$

where v^ε denotes the piecewise-constant extension of the discrete function v^ε, gives a recovery sequence for the Γ-limit. Note that for any interval $[\sqrt{\varepsilon}\,(k-1), \sqrt{\varepsilon}\,k)$ we have an interface in $\varepsilon\mathbb{Z}$ (see Fig. 8.8), which can be viewed as a *diffuse interface.*

8.2 Graphons

In this section, we consider sequences of general graphs $G_k = (\mathcal{N}_k, \mathcal{E}_k)$, where $\mathcal{N}_k$ denotes the set of nodes and $\mathcal{E}_k$ is the set of the connections of G_k; that is, a symmetric subset of $\mathcal{N}_k \times \mathcal{N}_k$.

8.2.1 Dense Graph Sequences

In order to quantify the density of a sequence of graphs, as a parameter we use the number of connections with respect to $(\#\mathcal{N}_k)^2$, which is of the order of the total number of pairs. We are interested in the case $\#\mathcal{N}_k \to +\infty$ as $k \to +\infty$.

Definition 8.1 (Dense and sparse [sequences of] graphs) Let $\{G_k\}$ be a sequence of graphs with $G_k = (\mathcal{N}_k, \mathcal{E}_k)$ and let $\#\mathcal{N}_k \to +\infty$ as $N_k \to +\infty$. The sequence is *dense* (or the graphs are dense) if

$$\liminf_{k\to+\infty} \frac{\#\mathcal{E}_k}{(\#\mathcal{N}_k)^2} \geq c > 0.$$

The sequence is *sparse* (or the graphs are sparse) if

$$\limsup_{k\to+\infty} \frac{\#\mathcal{E}_k}{(\#\mathcal{N}_k)^2} = 0.$$

Given a graph $G = (\mathcal{N}, \mathcal{E})$, we define the *energy of G* by setting

$$E^G(u) = \frac{1}{\#\mathcal{N}} \sum_{i,j \in \mathcal{N}} a_{ij}(u_i - u_j)^2 \tag{8.8}$$

for $u \colon \mathcal{N} \to \{-1, 1\}$, where

$$a_{ij} = \begin{cases} 1 & \text{if } (i,j) \in \mathcal{E}, \\ 0 & \text{if } (i,j) \in \mathcal{E}. \end{cases}$$

The matrix $A = (a_{ij})$ is called the *adjacency matrix* of the graph.

Let $n = \#\mathcal{N}$; then, we can use the set $\mathcal{N}_n = \{1, \ldots, n\}$ as the set of nodes, with a slight abuse of notation. Correspondingly, we will write the graph as $G_n = (\mathcal{N}_n, \mathcal{E}_n)$ and the adjacency matrix as $A^n = (a_{ij}^n)$.

Now, in view of analyzing the asymptotic behavior of the energies E^{G_n} as $n \to +\infty$, we introduce a *pixel representation* of the adjacency matrices on a common set of parameters $[0, 1] \times [0, 1]$. To that end, we consider the piecewise-constant *adjacency function* of the matrix A^n; that is, the function $a^n \colon [0, 1] \times [0, 1] \to \{0, 1\}$ defined as

$$a^n(x, y) = a_{ij}^n \text{ if } (x, y) \in I_i^n \times I_j^n, \tag{8.9}$$

where $I_1^n = [0, \frac{1}{n}]$ and $I_k^n = (\frac{k-1}{n}, \frac{k}{n}]$ for $k = 2, \ldots, n$.

Note that for a sparse sequence of graphs the adjacency functions a^n defined in (8.9) always converge to 0 strongly in L^1. The following examples show that the representation of the adjacency matrices and the corresponding weak limits is highly dependent on the choice of the labeling.

Example 8.2 (Parametrization of a graph and adjacency matrix) Let G be the graph pictured in Fig. 8.9(a). The adjacency matrix $A = (a_{ij})$ corresponding to the labeling is represented in Fig. 8.9(b), and the corresponding adjacency

$$\begin{pmatrix} 0 & 1 & 1 & 1 & 1 & 0 & 0 & 0 \\ 1 & 0 & 1 & 1 & 0 & 1 & 0 & 0 \\ 1 & 1 & 0 & 1 & 0 & 0 & 1 & 0 \\ 1 & 1 & 1 & 0 & 0 & 0 & 0 & 1 \\ 1 & 0 & 0 & 0 & 0 & 1 & 0 & 0 \\ 0 & 1 & 0 & 0 & 1 & 0 & 1 & 0 \\ 0 & 0 & 1 & 0 & 0 & 1 & 0 & 1 \\ 0 & 0 & 0 & 1 & 0 & 0 & 1 & 0 \end{pmatrix}$$

(a) (b) (c)

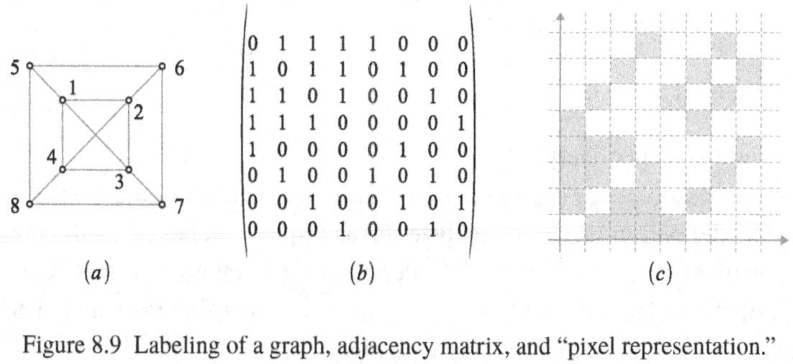

Figure 8.9 Labeling of a graph, adjacency matrix, and "pixel representation."

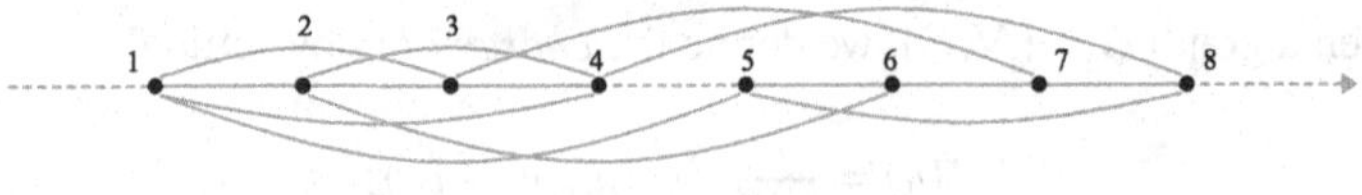

Figure 8.10 One-dimensional representation of the graph in Fig. 8.9.

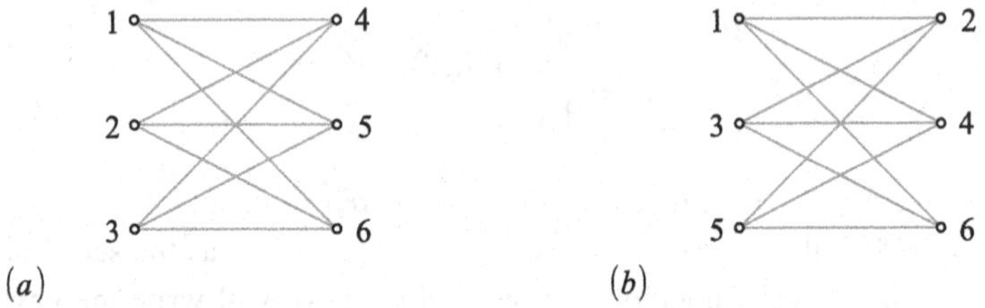

Figure 8.11 Different labelings of the complete bipartite graph.

function a given by (8.9) is pictured in Fig. 8.9(c). Note the different position of the origin in the representation of a and in the corresponding matrix.

The same graph can be represented in dimension 1. In Fig. 8.10 the graph is pictured as a set of eight points on the line, together with the corresponding connections.

Example 8.3 (Different labelings of the complete bipartite graph) We consider the *complete bipartite graph* with $2N$ nodes divided into two families of N nodes each, highlighting the dependence of the adjacency matrix on the choice of the labeling. In Fig. 8.11 two different parametrizations for $N = 3$ are pictured. The corresponding adjacency matrices are

$$\tilde{A} = \begin{pmatrix} 0 & 0 & 0 & 1 & 1 & 1 \\ 0 & 0 & 0 & 1 & 1 & 1 \\ 0 & 0 & 0 & 1 & 1 & 1 \\ 1 & 1 & 1 & 0 & 0 & 0 \\ 1 & 1 & 1 & 0 & 0 & 0 \\ 1 & 1 & 1 & 0 & 0 & 0 \end{pmatrix} \quad \text{and} \quad \hat{A} = \begin{pmatrix} 0 & 1 & 0 & 1 & 0 & 1 \\ 1 & 0 & 1 & 0 & 1 & 0 \\ 0 & 1 & 0 & 1 & 0 & 1 \\ 1 & 0 & 1 & 0 & 1 & 0 \\ 0 & 1 & 0 & 1 & 0 & 1 \\ 1 & 0 & 1 & 0 & 1 & 0 \end{pmatrix}$$

for (a) and (b) respectively.

Now, we associate to each parametrization the corresponding "pixel representation" defined in (8.9) and analyze the asymptotic behavior as the number of nodes diverges. Let $\widetilde{a}^{2N}$ and $\widehat{a}^{2N}$ denote the piecewise-constant representations of the adjacency matrices $\widetilde{A}^{2N}$ and $\widehat{A}^{2N}$ generalizing $\tilde{A}$ and $\hat{A}$ to the graphs with $2N$ nodes, respectively (see Fig. 8.12).

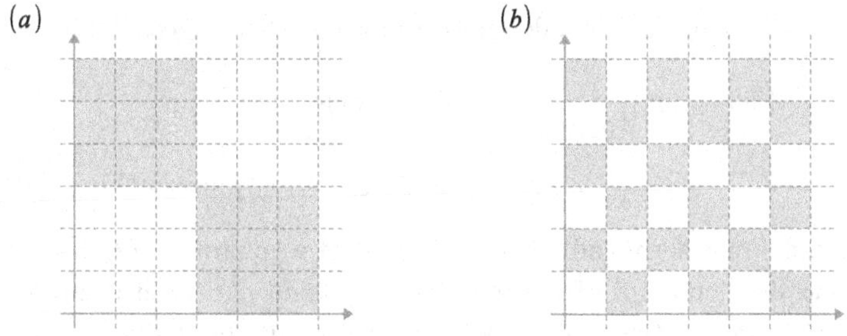

Figure 8.12 Different pixel representations of the complete bipartite graph.

We have that $\widetilde{a}^{2N}$ is actually constant, and equal to

$$\widetilde{a}(x,y) = \begin{cases} 0 & \text{in } \left[0, \frac{1}{2}\right]^2 \cup \left[\frac{1}{2}, 1\right]^2, \\ 1 & \text{otherwise,} \end{cases}$$

while $\widehat{a}^{2N} \overset{*}{\rightharpoonup} \widehat{a}(x,y) = \frac{1}{2}$ in L^∞.

Contrary to the case studied until now, where the limit is represented as a local functional in $\mathbb{R}^d$ starting from a lattice graph in $\varepsilon\mathbb{Z}^d$ with essentially finite range connections, here we embed arbitrary graphs in one-dimensional lattices in $\frac{1}{n}\mathbb{Z}$, and in order to analyze the asymptotic behavior of the energies defined on dense sequences of graphs, we give the definition of *graphon*. This concept extends the notion of (parametrized) graph, identified with its adjacency function a, to a "continuous set of nodes" given by $[0, 1]$, and a corresponding notion of norm.

Definition 8.4 (Graphons and cut norm) A *graphon* is a bounded measurable function $W\colon [0,1] \times [0,1] \to \mathbb{R}$ that is symmetric. The *cut norm* of a graphon W is defined as

$$\|W\|_\square = \sup_{T,S\subset(0,1)} \left| \int_S \int_T W(x,y)\,dx\,dy \right| \tag{8.10}$$

where the sup is taken over all measurable subsets of $(0, 1)$.

Note that in the definition of graphon we consider $\mathbb{R}$-valued functions, and not only taking values in $[0, 1]$, since we need to use the linear structure of the space.

Remark 8.5 An equivalent definition of the cut norm is given by

$$\|W\|_{\square} = \sup\left\{\left|\int_{(0,1)}\int_{(0,1)} W(x,y)f(x)g(y)\,dx\,dy\right| :\right.$$
$$\left. f,g\colon [0,1]\to[0,1] \text{ measurable}\right\}. \tag{8.11}$$

Remark 8.6 (cut norm and parametrizations of a graph) Let G be a graph. By using the definition of the adjacency function $a(x,y)$ in (8.9), to each parameterization of G we can associate a graphon W_G by setting

$$W_G(x,y) = a(x,y). \tag{8.12}$$

Note that if W_G and W'_G are associated to different parameterizations of the same G, then in general $\|W_G - W'_G\|_{\square} \neq 0$. If we consider the bipartite graph described in Example 8.3, and $\widetilde{W}_{2N}$ and $\widehat{W}_{2N}$ denote the graphons associated to parametrizations (a) and (b) respectively, we get $\|\widetilde{W}_{2N} - \widehat{W}_{2N}\|_{\square} = \frac{1}{2} > 0$.

Remark 8.7 (Cut-norm convergence, L^1-strong and L^1-weak convergence)

(i) Let $\{W_n\}$ be a sequence of graphons such that $\|W_n\|_{\square} \to 0$ as $n \to +\infty$. Then, if $\{f_n\}$ and $\{g_n\}$ are equibounded sequences of measurable functions, it follows that

$$\lim_{n\to+\infty}\int_{(0,1)}\int_{(0,1)} W_n(x,y)f_n(x)g_n(y)\,dx\,dy = 0. \tag{8.13}$$

If f_n and g_n are positive, then the result follows by the (equivalent) definition of cut norm in (8.11) since the sequences are equibounded. Otherwise, we can write $f_n g_n$ as the difference of products of positive functions

$$f_n g_n = \frac{1}{2}(\|f_n\|_\infty + f_n)(\|g_n\|_\infty + g_n) + \frac{1}{2}(\|f_n\|_\infty - f_n)(\|g_n\|_\infty - g_n) - \|f_n\|_\infty\|g_n\|_\infty$$

and conclude.

(ii) The convergence in the cut norm is stronger than the weak-L^1 convergence. Indeed, if $\|W_n - W\|_{\square} \to 0$ as $n \to +\infty$, then $W_n \rightharpoonup W$ in $L^1((0,1)\times(0,1))$ by using (8.11), which is uniform. Note that the weak convergence of W_n is not sufficient to have (8.13), which will be a key property to analyze the convergence of energies on graphs. On the contrary, the L^1-strong convergence implies the convergence in the cut norm, but does not provide sufficient compactness properties.

We now introduce a notion of metric on the set of graphons that is independent of the parametrizations. A map $\phi\colon [0,1] \to [0,1]$ is *measure preserving* if $\phi^{-1}(A)$ is measurable for all measurable $A \subset [0,1]$, and $|\phi^{-1}(A)| = |A|$, where $|\cdot|$

denotes the one-dimensional Lebesgue measure. Note that measure-preserving maps are a generalization of relabeling of nodes of a graph, in the sense that a permutation in the set of nodes (and hence in the set of the intervals I_k^n) corresponds to a measure-preserving transformation of $[0,1]$.

Let W be a graphon. For each measure-preserving map ϕ we define a corresponding graphon W_ϕ by setting

$$W_\phi(x,y) = W(\phi(x),\phi(y)). \tag{8.14}$$

Definition 8.8 (Cut metric) Let W, W' be graphons. The *cut distance* between W and W' is defined as

$$\delta_\square(W,W') = \inf_{\phi,\psi} \|W_\phi - W'_\psi\|_\square,$$

where the inf is taken over all measure-preserving maps of $[0,1]$, and W_ϕ, W'_ψ are defined as in (8.14).

Note that, by definition, the cut distance is independent of the parametrization.

We introduce the set $\mathcal{W}_0$ of the graphons with image in $[0,1]$; that is,

$$\mathcal{W}_0 = \{W\colon [0,1]^2 \to [0,1] \text{ measurable and symmetric}\}.$$

Note that $\mathcal{W}_0$ is the set of limits of graphs; that is, if $\{G_n\}$ is a sequence of graphs such that $\delta_\square(W_n, W) \to 0$ for some $W \in \mathcal{W}$, where W_n is the graphon corresponding to G_n, then $W \in \mathcal{W}_0$. Conversely, for all $W \in \mathcal{W}_0$ there exists a sequence $\{G_n\}$ of graphs such that $\delta_\square(W_n, W) \to 0$, where W_n is the graphon corresponding to G_n as in (8.12).

The following fundamental theorem describes topological properties of the cut metric and the cut norm.

Theorem 8.9 (Convergence of graphons) (i) *(completeness) Let $\{G_n\}$ be a sequence of graphs such that the corresponding sequence of graphons $\{W_n\}$ is a Cauchy sequence with respect to the metric $\delta_\square$; then, there exists $W \in \mathcal{W}_0$ such that $\delta_\square(W, W_n) \to 0$. The limit W is unique; that is, if $\delta_\square(W', W_n) \to 0$ for some $W' \in \mathcal{W}_0$, then $\delta_\square(W', W) = 0$.*

(ii) *Given a sequence $\{G_n\}$ of parametrized graphs and $W \in \mathcal{W}_0$ such that $\delta_\square(W, W_n) \to 0$, for all n there exists a measure-preserving map ϕ_n such that, setting $\widetilde{W}_n = (W_n)_{\phi_n}$,*

$$\|\widetilde{W}_n - W\|_\square \to 0.$$

(iii) *The convergence induced by the cut metric is compact.*

Remark 8.10 (A notion of topological convergence for graphs) The cut norm translates in analytical terms a notion of topological convergence of graphs. Given $F = (\mathcal{N}(F), \mathcal{E}(F))$ and $G = (\mathcal{N}(G), \mathcal{E}(G))$ graphs, we consider the set $\hom(F, G)$ of the homomorphisms of F on G, that is, the set of functions $\Phi \colon \mathcal{N}(F) \to \mathcal{N}(G)$ that are adjacency-preserving.

The structure of a graph G can be examined by considering the density of the homomorphisms of simple graphs F on G; we introduce the quantity

$$t(F, G) = \frac{\#\hom(F, G)}{(\#\mathcal{N}(G))^{\#\mathcal{N}(F)}}, \tag{8.15}$$

noting that $(\#\mathcal{N}(G))^{\#\mathcal{N}(F)}$ is the number of all possible functions between the set of nodes of F and the set of nodes of G.

A sequence of graphs G_n is said to be *left convergent* if there exists

$$\lim_{n\to+\infty} t(F, G_n)$$

for any F simple graph. Note that this definition is meaningful only for dense sequences of graphs, otherwise the limit is always 0. The relation with the convergence in the cut metric is given by the following result.

Proposition 8.11 *Let $\{G_n\}$ be a sequence of parametrized graphs with cardinality $\#\mathcal{N}(G_n) = n$, and let $\{W_n\}$ be the sequence of corresponding graphons given by* (8.12). *Then:*

(i) *the sequence $\{W_n\}$ is a Cauchy sequence with respect to the metric $\delta_\square$ if and only if the sequence $\{G_n\}$ is left convergent;*
(ii) *if $W \in \mathcal{W}_0$ is such that $\delta_\square(W, W_n) \to 0$, then, for any simple graph $F = (\mathcal{N}(F), \mathcal{E}(F))$,*

$$t(F, G_n) \to \int_{[0,1]^k} \prod_{(i,j)\in\mathcal{E}(F)} W(x_i, x_j)\, dx_1 \dots dx_k,$$

where $\mathcal{N}(F) = \{1, \dots, k\}$.

8.2.2 Convergence to Graphon Energies

Let $G_n = (\mathcal{N}_n, \mathcal{E}_n)$ be a parametrized graph such that $\#\mathcal{N}_n = n$, with the set of nodes parametrized as $\mathcal{N}_n = \{1, \dots, n\}$. For each $u \colon \frac{1}{n}\mathbb{Z} \to \{-1, 1\}$, we define the energy by

$$E_n(u) = \frac{1}{n^2} \sum_{i,j=1}^{n} a_{ij}^n (u_i - u_j)^2,$$

where $u_i = u(\frac{i}{n})$ and $(a^n_{ij})_{ij}$ is the adjacency matrix of the graph G_n. If we extend the spin function u to a piecewise-constant function by setting

$$u(x) = u_i \quad \text{if} \quad x \in I^n_i,$$

the energies can be written as

$$E_n(u) = \int_{(0,1)} \int_{(0,1)} a^n(x,y)(u(x) - u(y))^2 \, dx \, dy,$$

where a^n is defined in (8.9) and $u \in X_n$ is defined

$$X_n = \{u \colon [0,1] \to \{-1,1\} \colon u \text{ constant in each } I^n_i\}.$$

We now consider the Γ-limit of the sequence $\{E_n\}$, the convergence of the corresponding graphons in the cut norm describing the asymptotic behavior of the underlying environment.

Theorem 8.12 (Γ-convergence and volume constraint) *Let $\{G_n\}$ be a sequence of parametrized graphs such that the corresponding graphons W_n converge to a graphon W with respect to the cut norm. Then, the sequence $\{E_n\}$ Γ-converges, with respect to the L^∞-weak* convergence, to the functional F defined by*

$$F(u) = \begin{cases} 2\displaystyle\int_{(0,1)} \int_{(0,1)} W(x,y)(1 - u(x)u(y)) \, dx \, dy & \text{if } \|u\|_\infty \le 1, \\ +\infty & \text{otherwise in } L^\infty(0,1). \end{cases}$$

Moreover, if $\lambda_n \to \lambda$ with $\lambda_n \in [-1,1] \cap \frac{2}{n}\mathbb{Z}$, and

$$E_n^{\lambda_n}(u) = \begin{cases} E_n(u) & \text{if } \displaystyle\int_{(0,1)} u \, dx = \lambda_n, \\ +\infty & \text{otherwise,} \end{cases}$$

then the sequence $\{E_n^{\lambda_n}\}$ Γ-converges to the functional F^λ given by

$$F^\lambda(u) = \begin{cases} F(u) & \text{if } \displaystyle\int_{(0,1)} u \, dx = \lambda, \\ +\infty & \text{otherwise.} \end{cases}$$

Proof *Lower bound.* Let $u_n \overset{*}{\rightharpoonup} u$. Since $\|W_n - W\|_\square \to 0$, we apply Remark 8.7(a), obtaining

$$\lim_{n \to +\infty} \int_{(0,1)} \int_{(0,1)} (W_n(x,y) - W(x,y)) u_n(x) u_n(y) \, dx \, dy = 0. \tag{8.16}$$

Since $u_n(x) \in \{-1, 1\}$, it follows that $(u_n(x) - u_n(y))^2 = 2 - 2u_n(x)u_n(y)$, so that

$$\liminf_{n\to+\infty} E_n(u_n) = 2 \liminf_{n\to+\infty} \int_{(0,1)}\int_{(0,1)} W_n(x,y)(1 - u_n(x)u_n(y))\,dx\,dy$$
$$= 2 \int_{(0,1)}\int_{(0,1)} W(x,y)(1 - u_n(x)u_n(y))\,dx\,dy,$$

where we used (8.16) and the convergence

$$\int_{(0,1)}\int_{(0,1)} W_n(x,y)\,dx\,dy \to \int_{(0,1)}\int_{(0,1)} W(x,y)\,dx\,dy,$$

obtained by using as test functions $f(x) = g(x) \equiv 1$. Since

$$v_n(y) = \int_{(0,1)} W(x,y)u_n(x)\,dx \to v(y) = \int_{(0,1)} W(x,y)u(x)\,dx$$

strongly in $L^1(0,1)$ and $u_n \overset{*}{\rightharpoonup} u$ in $L^\infty(0,1)$, it follows that

$$\liminf_{n\to+\infty} \int_{(0,1)}\int_{(0,1)} W(x,y)u_n(x)u_n(y)\,dx\,dy = \liminf_{n\to+\infty} \int_{(0,1)} v_n(y)u_n(y)\,dy$$
$$= \int_{(0,1)}\int_{(0,1)} W(x,y)u(x)u(y)\,dx\,dy,$$

concluding the proof of the lower estimate.

Upper bound. Let $u\colon (0,1) \to [-1,1]$ and $n_h = h^2$ for any $h \in \mathbb{N}$. We consider a partition of $(0,1)$ in intervals with length $\frac{1}{h}$, so that the scale is larger than $\frac{1}{n_h} = \frac{1}{h^2}$. We set $J_k^h = (\frac{k-1}{h}, \frac{k}{h})$ and define

$$\tilde{u}_h = c_k^h = h \int_{J_k^h} u(x)\,dx \quad \text{in} \quad J_k^h.$$

We define the recovery sequence $\{u_{n_h}\}$ by setting, in each interval J_k^h,

$$u_{n_h}(x) = \begin{cases} 1 & \text{if } x \in \left(\dfrac{k-1}{h}, \dfrac{k-1}{h} + \dfrac{t_k^h}{h^2}\right), \\ -1 & \text{otherwise in } J_k^h, \end{cases}$$

where $t_k^h = \frac{h}{2}(c_k^h + 1)$. This is a good definition since $u(x) \in [-1,1]$ implies $1 \le t_k^h \le h$, and, up to a small error, we can assume that t_k^h is an integer.

The sequence $\{u_{n_h}\}$ weak* converges to u in $L^\infty(0,1)$. By using (8.16), we have

$$\lim_{h\to+\infty} E_{n_h}(u_{n_h}) = 2 \lim_{h\to+\infty} \int_{(0,1)}\int_{(0,1)} W(x,y)(1 - u_{n_h}(x)u_{n_h}(y))\,dx\,dy,$$

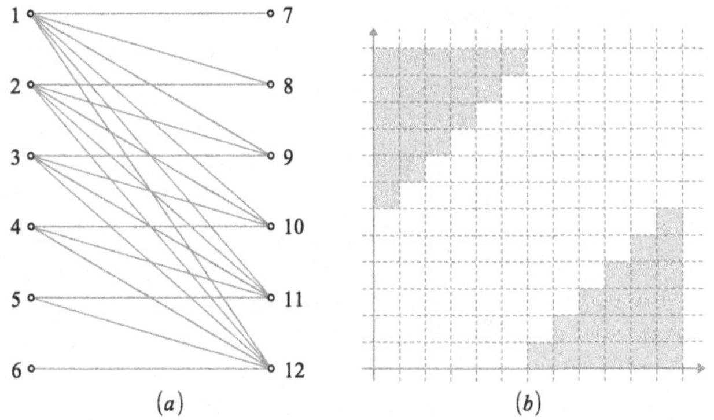

Figure 8.13 The half-graph with 12 nodes and its pixel representation.

and again by an argument of strong-weak convergence as in the proof of the lower bound we obtain

$$\lim_{h\to+\infty} E_{n_h}(u_{n_h}) = 2\int_{(0,1)}\int_{(0,1)} W(x,y)(1-u(x)u(y))\,dx\,dy.$$

Finally, we note that a slight variation of the construction of the recovery sequence is compatible with the volume constraint. □

Example 8.13 (Minimal-cut problem for the half-graph) Let G_{2k} be the half graph with $2k$ nodes, with the labeling of the nodes described in Fig. 8.13(a). The corresponding sequence of graphons W_{2k} is pictured in Fig. 8.13(b). Since W_{2k} strongly converges in $L^1((0,1)\times(0,1))$ to the function

$$W(x,y) = \begin{cases} 0 & \text{if } |x-y| \le \dfrac{1}{2}, \\ 1 & \text{otherwise in } (0,1)^2, \end{cases}$$

it follows that $\|W_{2k} - W\|_\square \to 0$ as $k \to +\infty$. Fixing $\lambda_{2k} = 0$ for all k, we can apply Theorem 8.12 and obtain that the sequence of the energies converges to the functional

$$F^0(u) = 2\int_0^{\frac{1}{2}}\int_{\frac{1}{2}+x}^{1}(1-u(x)u(y))\,dx\,dy$$
$$+2\int_{\frac{1}{2}}^{1}\int_0^{x-\frac{1}{2}}(1-u(x)u(y))\,dx\,dy.$$

We consider the minimal-cut problem for the limit functional F; that is,

$$\min\left\{F^0(u)\colon u\colon (0,1)\to[-1,1] \text{ such that } \int_0^1 u(x)\,dx = 0\right\}.$$

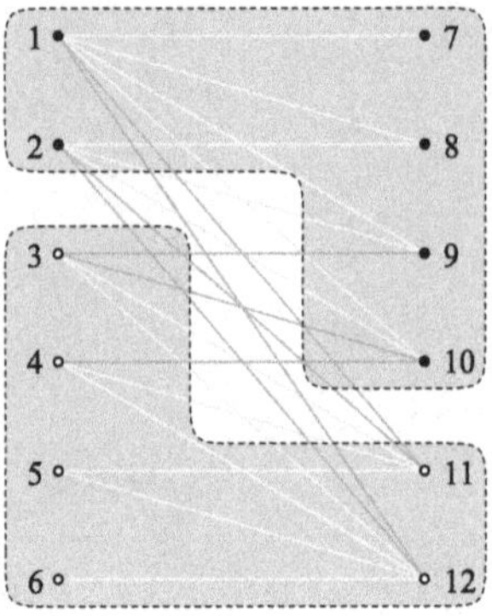

Figure 8.14 Minimal cut of the half-graph.

It can be proven that the minimum is attained at a function u such that $u(x) \in \{-1, 1\}$; then, it follows that a minimizer is given by

$$u(x) = \begin{cases} 1 & \text{in } \left[0, \frac{1}{6}\right] \cup \left[\frac{1}{2}, \frac{5}{6}\right], \\ -1 & \text{otherwise.} \end{cases}$$

In Fig. 8.14 we give a pictorial representation of the corresponding discrete solution of the minimal-cut problem in the case $k = 6$.

Bibliographical Notes to Chapter 8

Coarse-graining techniques are common in the treatment of multiscale problems. The variational version used here is due to Braides and Solci (2020). The example of diffuse interface is due to Braides et al. (2018a) and can be generalized to one-dimensional ferromagnetic energies obtained as projection of ferromagnetic energies also in dimension $d > 2$.

General references for graphons are the books by Lovász (2012) and by Janson (2013). We also mention the papers by Lovász and Szegedy (2006, 2007) and by Borgs et al. (2006, 2008, 2010, 2012). The connection with Γ-convergence highlighted here is due to Braides et al. (2020).

Appendix A
Multiscale Analysis

Energies depending on a small parameter ε can be studied at various scales as ε tends to 0. For spin systems, beside the surface scale, typically another meaningful scale at which the analysis of such energies can give useful information is the "bulk scale," that is, at which energies scale as the Lebesgue measure in $\mathbb{R}^d$. The preliminary study at the bulk scale and the subsequent study at the surface scale are a typical argument of *multiscale analysis*, for which the surface-scale analysis can be seen as a more detailed description of the behavior of minimizers of bulk-scale problems. This can be formalized by using the notation of developments by Γ-convergence. In the following sections we give the main ideas about developments and expansions, and briefly introduce some energies at the bulk scaling from which some spin energies studied in this book are derived.

A.1 Development by Γ-Convergence

The Fundamental Theorem of Γ-convergence can be iteratively applied to get a better description of the behavior of minimum problems for energies F_ε whenever the first Γ-limit $F^{(0)}$ possesses a nontrivial set of minimizers. The iterative process goes as follows.

(i) We conjecture a second *scale*; that is, an infinitesimal function of ε, say ε^α even though it is not necessarily a power of ε. This means that we find a converging sequence $\{x^\alpha_\varepsilon\}_\varepsilon$ such that $F_\varepsilon(x^\alpha_\varepsilon) = \min F^{(0)} + O(\varepsilon^\alpha)$.

(ii) We consider the scaled functionals

$$F^{(\alpha)}_\varepsilon(x) = \frac{F_\varepsilon(x) - \inf F}{\varepsilon^\alpha}. \tag{A.1}$$

Note that $F^{(\alpha)}_\varepsilon$ has the same minimizers as F_ε.

(iii) We compute the Γ-limit (*Γ-limit of* $\{F_\varepsilon\}$ *at scale* ε^α)

$$F^{(\alpha)}(x) = \Gamma\text{-}\lim_{\varepsilon\to 0} F_\varepsilon^{(\alpha)}(x). \tag{A.2}$$

If this limit is not trivial, then we have the following result, immediately obtained from the Fundamental Theorem of Γ-convergence, in the form elaborated by Anzellotti et al. (1994).

We set

$$m_\varepsilon = \inf F_\varepsilon, \qquad m^{(0)} = \inf F^{(0)}, \qquad m_\varepsilon^{(\alpha)} = \inf F_\varepsilon^{(\alpha)} \tag{A.3}$$

so that

$$m_\varepsilon^{(\alpha)} = \frac{m_\varepsilon - m^{(0)}}{\varepsilon^\alpha}. \tag{A.4}$$

Theorem A.1 (Development of minimum problems) *Under the coerciveness hypotheses of the Fundamental Theorem of* Γ*-convergence there exist the minima* $\min F^{(0)}$ *and* $m^{(\alpha)} = \min F^{(\alpha)}$, *and* $m_\varepsilon^{(\alpha)} \to m^{(\alpha)}$, *so that*

$$m_\varepsilon = m^{(0)} + \varepsilon^\alpha m^{(\alpha)} + o(\varepsilon^\alpha). \tag{A.5}$$

Moreover, if $\{x_\varepsilon\}$ *is a* ε^α*-minimizing sequence of* (F_ε)*, that is, such that* $F_\varepsilon(x_\varepsilon) = \inf F_\varepsilon + o(\varepsilon^\alpha)$*, and* x *is one of its limit points, then* x *is a minimizer for both* $F^{(0)}$ *and* $F^{(\alpha)}$.

Remark A.2 (i) From the definition of $F^{(\alpha)}$ its domain is contained in the set of minimizers of $F^{(0)}$; that is, $F^{(\alpha)}(x) = +\infty$ if $F^{(0)}(x) \neq \min F^{(0)}$.

(ii) In the hypotheses of the theorem there can still exist an intermediate scale, say still a polynomial ε^β with $\varepsilon^\alpha \ll \varepsilon^\beta \ll 1$, such that

$$F^{(\beta)}(x) = \Gamma\text{-}\lim_{\varepsilon\to 0} \frac{F_\varepsilon(x) - \inf F}{\varepsilon^\beta} \tag{A.6}$$

exists and is not trivial, but in that case we must have $\min F^{(\beta)} = 0$, otherwise development (A.5) would be violated.

(iii) If for all intermediate scale $\varepsilon^\alpha \ll s_\varepsilon \ll 1$ we have

$$\Gamma\text{-}\lim_{\varepsilon\to 0} \frac{F_\varepsilon(x) - \inf F}{s_\varepsilon} = \begin{cases} 0 & \text{if } F^{(0)}(x) = \min F^{(0)}, \\ +\infty & \text{otherwise,} \end{cases} \tag{A.7}$$

then we say that the development is *complete at scale* ε^α.

(iv) If the domain of $F^{(\alpha)}$ is dense in the set of minimizers of $F^{(0)}$, then the development is complete at scale ε^α.

(v) The process can be iterated, introducing a second scale, say $\varepsilon^\gamma \ll \varepsilon^\alpha$ and applying the process to $F_\varepsilon^{(\alpha)}$, and so on.

The definition of development by Γ-convergence just given allows us to extend the equivalence relation by Γ-convergence.

Definition A.3 (Equivalence by Γ-convergence at scale ε^α) We say that two sequences $\{F_\varepsilon\}$ and $\{G_\varepsilon\}$ are *equivalent by Γ-convergence at scale* ε^α if, from every sequence $\{\varepsilon_j\}$, we can extract a subsequence (still denoted by $\{\varepsilon_j\}$) and there exists a sequence $\{m_j\}$ of real numbers such that there exist the Γ-limits

$$\Gamma\text{-}\lim_j \frac{F_{\varepsilon_j} - m_j}{\varepsilon_j^\alpha}, \qquad \Gamma\text{-}\lim_j \frac{G_{\varepsilon_j} - m_j}{\varepsilon_j^\alpha},$$

they coincide, and are not trivial. If $\alpha = 0$, we simply say that $\{F_\varepsilon\}$ and $\{G_\varepsilon\}$ are *equivalent by Γ-convergence.*

In particular, in the hypotheses of the preceding theorem and taking $m_j = m_{\varepsilon_j}$, we have that F_ε is equivalent to $m_\varepsilon + \varepsilon^\alpha F^{(\alpha)}$ at scale ε^α. The introduction of the translations m_j has been done in order to distinguish between sequences F_ε diverging with different speed. This definition is at the core of the notion of expansion by Γ-convergence by Braides and Truskinovsky (2008), which allows one to study the convergence of minimum problems depending on a parameter.

A.2 Nearest-Neighbor Spin Energies

We now see how ferromagnetic spin energies can be seen as part of a development from pairwise nearest-neighbor energies at the bulk scaling.

We consider one-dimensional energies defined on functions $u\colon \varepsilon\mathbb{Z} \to \{-1, 1\}$ of the general form

$$E_\varepsilon(u) = \sum_{i=1}^N \varepsilon f(u_i, u_{i-1}),$$

with $N = \frac{1}{\varepsilon}$ and f symmetric.

We may rewrite

$$E_\varepsilon(u) = \sum_{i=1}^N \varepsilon(f(u_i, u_{i-1}) + g(u_i, u_{i-1})) - \sum_{i=1}^N \varepsilon g(u_i, u_{i-1}), \tag{A.8}$$

where $g(u,u) = f(-u,-u)$ and $g(1,-1) = g(-1,1) = \frac{1}{2}(f(1,1) + f(-1,-1))$. By adding such g, the integrand in the first sum has the same value in $(-1,-1)$ and $(1,1)$. Note that

$$g(u,v) = \frac{1}{2}g(u,u) + \frac{1}{2}g(v,v), \tag{A.9}$$

and hence we have

$$\sum_{i=1}^{N} \varepsilon g(u_i, u_{i-1}) = \sum_{i=1}^{N-1} \varepsilon\, g(u_i, u_i) + \frac{1}{2}\varepsilon\, g(u_0, u_0) + \frac{1}{2}\varepsilon\, g(u_N, u_N).$$

Since the last two terms can be neglected, the second sum in (A.8) can be rewritten as $\sum_{i=1}^{N-1} \varepsilon W_0(u_i)$, where W_0 is the affine function such that $W_0(1) = g(1,1)$ and $W_0(-1) = g(-1,-1)$. Note that this sum can also be written as $G(u) = \int_{(0,1)} W_0(u)\, dt$ after identification of the function u with its piecewise-constant interpolation. G is a continuous functional with respect to the weak L^1-convergence, and hence commutes with the Γ-limit. We may therefore consider only the first sum in (A.8). This is an example in which the addition of a continuous perturbation allows us to simplify the form of the functionals we consider.

Summing up, it is not restrictive to suppose that f be symmetric and $f(1,1) = f(-1,-1)$. Hence, excluding the trivial case f constant, there are the two cases: $f(1,1) < f(1,-1)$ and $f(1,1) > f(1,-1)$. Up to translations we may suppose that the two values taken by f are 0 and 1. We may rewrite the two cases as

$$f(u,v) = \frac{1}{4}(u-v)^2, \qquad f(u,v) = \frac{1}{4}(u+v)^2.$$

Note again that, up to multiplicative and additive constants, the two cases correspond, respectively, to

$$f(u,v) = -uv, \qquad f(u,v) = uv.$$

We only treat the first case since the second one can be reduced to the first by the change of variables $w_i = (-1)^i u_i$. We finally consider

$$E_\varepsilon(u) = \frac{1}{4}\sum_{i=1}^{N} \varepsilon(u_i - u_{i-1})^2 \qquad u_i \in \{-1, 1\}. \tag{A.10}$$

We may compare $E_\varepsilon(u)$ with an energy of the type $\widetilde{E}_\varepsilon(v) = \sum_i \varepsilon W(v_i)$ by setting

$$W(v) = \begin{cases} 0 & \text{if } v = -1 \text{ or } v = 1, \\ 1 & \text{if } v = 0, \\ +\infty & \text{otherwise,} \end{cases}$$

choosing

$$v_i = v(u)_i = \frac{u_i + u_{i-1}}{2}. \tag{A.11}$$

Note that if $u^\varepsilon \rightharpoonup u$, then also $v^\varepsilon = v(u^\varepsilon) \rightharpoonup u$, so that

$$\liminf_{\varepsilon\to 0} E_\varepsilon(u^\varepsilon) \geq \liminf_{\varepsilon\to 0} \widetilde{E}_\varepsilon(v^\varepsilon) \geq \int_{(0,1)} W^{**}(u)\,dt.$$

The Γ-limsup inequality for E_ε at u cannot be directly deduced from that for $\widetilde{E}_\varepsilon$ at the same u, since not all recovery sequences for the latter are derived from some u^ε by (A.11). In order to highlight the error due to the lack of this last correspondence, note that in order to weakly approximate a constant with functions oscillating we have again to introduce a mesoscopic scale $\varepsilon \ll \eta_\varepsilon \ll 1$, and take u^ε oscillating on this scale between 1 and −1. We need only to specify this construction when the limit function is a constant $0 < z < 1$. In that case we define the function $u_z : \mathbb{R} \to \{-1, 1\}$ as periodic of period 1 and such that

$$u_z(s) = \begin{cases} 1 & \text{if } 0 < s \leq \frac{z+1}{2} \\ -1 & \text{if } \frac{z+1}{2} < s \leq 1 \end{cases} \qquad \text{and} \qquad u^\varepsilon(i) = u_z\Big(\frac{\varepsilon i}{\eta_\varepsilon}\Big).$$

Since $u_z(\frac{t}{\eta_\varepsilon}) \rightharpoonup z$, also $u^\varepsilon \rightharpoonup z$ and it gives the desired value. Hence, we have

$$\Gamma\text{-}\lim_{\varepsilon\to 0} E_\varepsilon(u) = \begin{cases} 0 & \text{if } |u| \leq 1 \text{ almost everywhere,} \\ +\infty & \text{otherwise.} \end{cases}$$

We may now analyze the Γ-limit at the scale ε, which is the one-dimensional version of the ferromagnetic energies in Section 2.1. This scale is suggested from the fact that ε is the error that we make when we have a transition from a minimal state identically 1 or −1. In this case we simply have

$$E^{(1)}_\varepsilon(u) = \frac{1}{4}\sum_{i=1}^{N}(u_i - u_{i-1})^2 \qquad u_i \in \{-1, 1\}. \tag{A.12}$$

Hence, if $E^{(1)}_\varepsilon(u^\varepsilon) \leq c < +\infty$, then the piecewise-constant extension of u^ε has a set of discontinuity points $S(u^\varepsilon)$ of cardinality at most c. If

$$\liminf_{\varepsilon\to 0} \#S(u^\varepsilon) = M,$$

we may then suppose that, up to subsequences, $\#S(u^\varepsilon) = M$ for all ε. We can then let $S(u^\varepsilon) = \{t^j_\varepsilon\}_{j=1}^M$, with t^j_ε satisfying $0 < t^1_\varepsilon < t^2_\varepsilon < \cdots < t^M_\varepsilon < 1$ for all ε. Up to a further subsequence, we may also suppose that t^j_ε converge to points t^j with $0 \leq t^1 \leq \cdots \leq t^M \leq 1$ for all $j \in \{1, \ldots, M\}$. Hence, $\{u^\varepsilon\}$ converges (strongly) to a piecewise-constant function u, whose set of discontinuity points $S(u)$ is contained in $\{t^j : j = 1, \ldots, M\}$. We then have

$$\liminf_{\varepsilon\to 0} E^{(1)}_\varepsilon(u^\varepsilon) = \liminf_{\varepsilon\to 0} \#S(u^\varepsilon) \geq \#S(u).$$

The converse inequality is obtained by taking $u^\varepsilon(\varepsilon i) = u(\varepsilon i)$. Hence, the Γ-limit at scale ε is

$$F^{(1)}(u) = \begin{cases} \#S(u) & \text{if } u \text{ is piecewise constant,} \\ & \text{and } u \in \{-1,1\} \text{ almost everywhere,} \\ +\infty & \text{otherwise.} \end{cases}$$

We may hence conclude that E_ε is equivalent to $F_\varepsilon = \varepsilon F^{(1)}$ at scale ε.

Note that the functions on which $F^{(1)}$ is finite are dense for the weak topology in the set of the minimizers of $F^{(0)}$ and hence the development at scale ε is complete. Note also that we may also choose other functionals on the continuum equivalent to E_ε, for example, the functionals from the gradient theory of phase transitions

$$F_\varepsilon(u) = \int_0^1 (W_0(u) + \varepsilon^2|u'|^2)\,dt, \qquad u \in H^1(0,1),$$

with W_0 a double-well potential with minimum in -1 and 1 and such that

$$2\int_{-1}^1 \sqrt{W_0(s)}\,ds = 1.$$

A.3 The Blume–Emery–Griffiths Model

In its standard formulation the Blume–Emery–Griffiths model of mixtures of oil and water with a surfactant component in two dimensions can be described as follows. Given a bounded open set with Lipschitz boundary $\Omega \subset \mathbb{R}^2$, we define as usual $\mathcal{L}_\varepsilon = \mathcal{L}_\varepsilon(\Omega) = \varepsilon\mathbb{Z}^2 \cap \Omega$ and consider functions $u\colon \mathcal{L}_\varepsilon \to \{-1,0,1\}$. Here $u_i = 1$, $u_i = -1$, $u_i = 0$ parameterize the particle i of water, oil, and surfactant, respectively. We then introduce the family of nearest-neighbor energies

$$E_\varepsilon(u) = \sum_{\langle i,j\rangle} \varepsilon^2(-u_iu_j + k(u_iu_j)^2), \tag{A.13}$$

with $k \in \mathbb{R}$ a parameter that measures the strength of the quadratic interaction with respect to the biquadratic interactions. If we set $f(u,v) = -uv + k(uv)^2$, then we have $f(u,v) = f(v,u)$ and

$$\begin{aligned} &f(-1,-1) = f(1,1) = k-1, \\ &f(-1,1) = k+1, \\ &f(0,0) = f(0,1) = f(-1,0) = 0. \end{aligned}$$

If $k < 1$, then the ground states of E_ε are the constants $u = -1$ and $u = 1$, while if $k > 1$, then minimizers are all functions u with $u_i u_j = 0$ if $\|i - i\| = 1$. We may then compute the Γ-limit in the case $k < 1$ as in the case of the Ising system in the previous section, and obtain that it equals the trivial functional $F^{(0)}(u) = 4(k-1)|\Omega|$ with domain all u with $\|u\|_\infty \leq 1$.

As in the case of the previous section, the energy for a phase transition from a bulk -1 phase to a bulk 1 phase separated by an interface of finite length has an energy of order ε and suggests the correct scaling to track the energetic behavior of a phase-separation phenomenon. Observing that the absolute minimum value at scale ε is precisely given by

$$m_\varepsilon = \sum_{\langle i,j \rangle} \varepsilon^2(k-1),$$

we see that the term in the Γ-development at order ε is the Γ-limit of the family of discrete energies:

$$F_\varepsilon(u) = \frac{E_\varepsilon(u) - m_\varepsilon}{\varepsilon} = \sum_{\langle i,j \rangle} \varepsilon(1 - u_i u_j - k(1 - (u_i u_j)^2)). \qquad \text{(A.14)}$$

Energies as in (A.14) with $k \in (\frac{1}{3}, 1)$ have been analyzed in Section 6.1.3, with condition $k > \frac{1}{3}$ ensuring the presence of surfactant at interfaces.

A.4 Next-to-Nearest Antiferromagnetic Systems

We now consider bulk energies with next-to-nearest antiferromagnetic interactions. Their analysis justifies the treatment of energies in Example 7.1 as part of a development by Γ-convergence.

We define the one-dimensional spin energies with nearest and next-to-nearest interactions by setting

$$E_\varepsilon(u) = \sum_{i=0}^{N} \varepsilon(\alpha u_i u_{i-1} + u_{i-1} u_{i+1}).$$

Note that in the expression of the energies we prefer to use $u_i u_{i-1}$ rather than $-(u_i - u_{i-1})^2$. Since u is periodic, we may write

$$E_\varepsilon(u) = \sum_{i=0}^{N} \varepsilon\Big(\frac{1}{2}\alpha(u_i u_{i-1} + u_{i+1} u_i) + u_{i-1} u_{i+1}\Big).$$

Ground states of E_ε can be looked for among the functions (if they exist) that for all i minimize the corresponding term in the sum. Depending on α, we obtain the three cases:

(i) $\alpha < -2$. In this case the nearest-neighbor ferromagnetic term dominates and the minimizers are the constants -1 and 1.
(ii) $\alpha > 2$. In this case the oscillations between nearest neighbors dominate, and we have the two minimizers $(-1)^i$ and $(-1)^{i+1}$, of period 2.
(iii) $|\alpha| < 2$. In this case the interactions between second neighbors dominates; hence, we have four minimizers, corresponding to the four possible combinations of the antiferromagnetic oscillating minimizers for second neighbors. The ground states are 4-periodic functions of the form

$$(u^k)_i = v_{i+k}$$

for $k \in \{0, 1, 2, 3\}$, where $v = u^0$ is given by

$$v_j = \begin{cases} -1 & \text{if } j \in \{1, 2\}, \\ 1 & \text{if } j \in \{3, 4\}. \end{cases}$$

In the case $\alpha = -2$ both the set of minimizers for $\alpha < -2$ and for $|\alpha| < 2$ are ground states, so that we have six ground states, and analogously in the case $\alpha = 2$.

The interesting case is (iii). Note that in order to have absolute minimizers compatible with ground states we would have to suppose that N is a multiple of 4. In any case, we may scale our functional, noting that for ground states we have

$$\frac{1}{2}\alpha(v_i v_{i-1} + v_{i+1} v_i) + v_{i-1} v_{i+1} = -1$$

and consider the scaled energies

$$E_\varepsilon^{(1)}(u) = \frac{1}{\varepsilon}(E_\varepsilon(u) - \min E_\varepsilon) = \sum_{i=0}^{N} \Big(\frac{1}{2}\alpha(u_i u_{i-1} + u_{i+1} u_i) + u_{i-1} u_{i+1} + 1\Big).$$

In the case N is not a multiple of 4, the translation does not correspond to the value of the minimum of E_ε. These energies are studied in Example 7.1 with $\alpha = 1$.

Appendix B

Spin Systems as Limits of Elastic Interactions

In this section we note that spin systems can be interpreted as a limit case of nonlinear double-well interactions as the depth of the well becomes very steep, or, in the spirit of a multiscale approach, as part of a development starting from *soft spins*; that is, for real values of the parameters u_i. We present a study of soft spin systems with a nonconvex elastic potential by Braides and Yip (2012). A related result is by Van Gennip and Bertozzi (2012).

Let $W: \mathbb{R} \to [0, +\infty)$ be a locally Lipschitz double-well potential with wells in 1 and -1; that is, $W(u) = 0$ if and only if $u = 1$ or $u = -1$. Moreover, we suppose that W is coercive; that is,

$$\lim_{u \to +\infty} W(u) = \lim_{u \to -\infty} W(u) = +\infty,$$

and that W is convex close to -1 and to 1; that is, there exists $C_0 > 0$ such that the set $\{u: W(u) \leq C_0\}$ consists of two intervals on each of which W is convex. Standard examples include $W(u) = (1 - u^2)^2$ or $W(u) = (1 - |u|)^2$.

Let Ω be a bounded open subset of $\mathbb{R}^d$ with boundary of zero Lebesgue measure. As usual, for $\delta > 0$ we set $\mathcal{L}_\delta(\Omega) = \Omega \cap \delta\mathbb{Z}^d$.

We will analyze the asymptotic behavior of $E_{\varepsilon,\delta}$ defined by

$$E_{\varepsilon,\delta}(u) = \sum_{\delta i \in \mathcal{L}_\delta(\Omega)} \delta^d W(u_i) + \frac{\varepsilon^2}{2} \sum_{\langle i,j \rangle} \delta^d \Big| \frac{u_i - u_j}{\delta} \Big|^2 \tag{B.1}$$

on functions $u: \mathcal{L}_\delta(\Omega) \to \mathbb{R}$ by computing their Γ-developments which depend on the mutual behavior of δ and ε.

Whatever the dependence of δ on ε, the Γ-limit of $E_{\varepsilon,\delta}$ with respect to the weak L^1-convergence is given by

$$\int_\Omega W^{**}(u)\, dx,$$

where W^{**} is the convex envelope of W. The next-order Γ-limit is described by the following theorem.

Theorem B.1 *Let Ω, W, and $E_{\varepsilon,\delta}$ be just presented, and let $\delta = \delta(\varepsilon)$. We then have three different regimes, where we compute the Γ-limit with respect to the strong L^1-convergence. In all cases the domain of the Γ-limit is $BV(\Omega; \{-1, 1\})$ and is a surface term on the set $S(u)$.*

(i) (subcritical case $\delta \ll \varepsilon$) *If $\lim_{\varepsilon \to 0} \frac{\delta}{\varepsilon} = 0$, then we have*

$$\Gamma\text{-}\lim_{\varepsilon \to 0} \frac{1}{\varepsilon} E_{\varepsilon,\delta}(u) = C_W\, \mathcal{H}^{d-1}(\Omega \cap S(u)), \tag{B.2}$$

where $C_W = 2\int_{-1}^{1} \sqrt{W(s)}\, ds$, as in the continuous case.

(ii) (critical case $\delta \sim \varepsilon$) *If $\lim_{\varepsilon \to 0} \frac{\delta}{\varepsilon} = K$ with $0 < K < +\infty$, then*

$$\Gamma\text{-}\lim_{\varepsilon \to 0} \frac{1}{\varepsilon} E_{\varepsilon,\delta}(u) = \int_{\Omega \cap S(u)} \varphi_K(\nu)\, d\mathcal{H}^{d-1}, \tag{B.3}$$

where φ_K is given by the asymptotic formula

$$\varphi_K(\nu) = \lim_{N \to +\infty} \frac{1}{N^{d-1}} \inf\Big\{K \sum_i W(v_i) + \frac{1}{2K} \sum_{\langle i,j \rangle} |v_i - v_j|^2\Big\}, \tag{B.4}$$

where the indices i, j are restricted to the cube Q_N^ν and the infimum is taken on all v that are equal to $u^\nu(x) = \operatorname{sign}\langle x, \nu\rangle$ on a neighborhood of ∂Q_N^ν. Furthermore, φ_K is continuous.

(iii) (supercritical case $\varepsilon \ll \delta$) *If $\lim_{\varepsilon \to 0} \frac{\varepsilon}{\delta} = 0$, then we have*

$$\Gamma\text{-}\lim_{\varepsilon \to 0} \frac{\delta}{\varepsilon^2} E_{\varepsilon,\delta}(u) = 4 \int_{\Omega \cap S(u)} \|\nu\|_1\, d\mathcal{H}^{d-1}. \tag{B.5}$$

(iv) (interpolation) *For all $\nu \in S^{d-1}$ we have*

$$\lim_{K \to 0} \varphi_K(\nu) = c_W, \qquad \lim_{K \to +\infty} K\varphi_K(\nu) = 4\|\nu\|_1 . \tag{B.6}$$

Remark B.2 (i) In the terminology of Definition A.3, the family $\{\frac{\delta}{\varepsilon^2} E_{\varepsilon,\delta}\}_\varepsilon$ is equivalent to a family of nearest-neighbor interactions on a cubic lattice.

(ii) In the one-dimensional case, formula (B.4) reduces to the computation of an optimal-profile problem

$$C_K = \inf\Big\{K \sum_{i=-\infty}^{+\infty} W(v_i) + \frac{1}{K} \sum_{i=-\infty}^{+\infty} |v_i - v_{i-1}|^2\Big\}, \tag{B.7}$$

where the test functions υ satisfy the limit conditions $\lim_{t\to-\infty} \upsilon(t) = -1$ and $\lim_{t\to+\infty} \upsilon(t) = 1$. In particular, taking spin test functions with $\upsilon_i \in \{-1, 1\}$, we have

$$C_K \leq \frac{4}{K}. \tag{B.8}$$

(iii) For coordinate directions minimizers for $\varphi_K(e_k)$ are one-dimensional, so that

$$\varphi_K(e_k) = C_K,$$

which gives the estimate

$$\varphi_K(\nu) \leq C_K \|\nu\|_1, \tag{B.9}$$

since the right-hand side is the greatest positively one-homogeneous convex function satisfying $\varphi_K(e_k) = C_K$ for all k.

(iv) Note that in the super-critical case, the limit interfacial energy is *degenerate, or not uniformly convex.* This is understandable, as in this case the nonlinear term $W(u)$ dominates so that the energy concentrates on the spin function v, which takes on only values of 1 or -1. In this case, the energy is equivalent to bond-counting: the number of bonds between 1 and -1. It is likely that φ_K is uniformly convex for $0 < K < +\infty$ even though this is not immediately clear from its definition.

References

Achdou, Y., Camilli, F., Cutrì, A., and Tchou, N. 2013. Hamilton–Jacobi equations constrained on networks. *NoDEA Nonlinear Differential Equations Appl.*, **20**, 413–445.

Akcoglu, M. A., and Krengel, U. 1981. Ergodic theorems for superadditive processes. *J. Reine Angew. Math.*, **323**, 53–67.

Alicandro, R., and Cicalese, M. 2004. A general integral representation result for continuum limits of discrete energies with superlinear growth. *SIAM J. Math. Anal.*, **36**, 1–37.

Alicandro, R., and Gelli, M. S. 2016. Local and non local continuum limits of Ising type energies for spin systems. *SIAM J. Math. Anal.*, **48**, 895–931.

Alicandro, R., Braides, A., and Cicalese, M. 2006. Phase and anti-phase boundaries in binary discrete systems: a variational viewpoint. *Netw. Heterog. Media*, **1**, 85–107.

Alicandro, R., Cicalese, M., and Gloria, A. 2011a. Integral representation results for energies defined on stochastic lattices and application to nonlinear elasticity. *Arch. Ration. Mech. Anal.*, **200**, 881–943.

Alicandro, R., Cicalese, M., and Ponsiglione, M. 2011b. Variational equivalence between Ginzburg–Landau, XY spin systems and screw dislocations energies. *Indiana Univ. Math. J.*, **60**, 171–208.

Alicandro, R., Cicalese, M., and Sigalotti, L. 2012. Phase transitions in presence of surfactants: from discrete to continuum. *Interfaces Free Bound.*, **14**, 65–103.

Alicandro, R., Cicalese, M., and Ruf, M. 2015. Domain formation in magnetic polymer composites: an approach via stochastic homogenization. *Arch. Ration. Mech. Anal.*, **218**, 945–984.

Allaire, G. 2002. *Shape Optimization by the Homogenization Method.* Applied Mathematical Science, vol. 146. Springer, New York.

Ambrosio, L., and Braides, A. 1990a. Functionals defined on partitions in sets of finite perimeter. I: Integral representation and Γ-convergence. *J. Math. Pures Appl.*, **69**, 285–305.

Ambrosio, L., and Braides, A. 1990b. Functionals defined on partitions in sets of finite perimeter. II: semicontinuity, relaxation and homogenization. *J. Math. Pures Appl.*, **69**, 307–333.

Ambrosio, L., and Tortorelli, V. M. 1990. Approximation of functionals depending on jumps by elliptic functionals via Γ-convergence. *Comm. Pure Appl. Math.*, **43**, 999–1036.

Ambrosio, L., Fusco, N., and Pallara, D. 2000. *Functions of Bounded Variation and Free Discontinuity Problems*. Clarendon Press, Oxford.

Ambrosio, L., Gigli, N., and Savaré, G. 2008. *Gradient Flows in Metric Spaces and in the Space of Probability Measures*. Lectures in Mathematics ETH Zürich. Birkhäuser, Basel.

Anzellotti, G., Baldo, S., and Percivale, D. 1994. Dimension reduction in variational problems, asymptotic development in Γ-convergence and thin structures in elasticity. *Asymptot. Anal.*, **9**, 61–100.

Arbogast, T., Douglas, J. Jr., and Hornung, U. 1990. Derivation of the double porosity model of single phase flow via homogenization theory. *SIAM J. Math. Anal.*, **21**, 823–836.

Bach, A., Braides, A., and Cicalese, M. 2020. Discrete-to-continuum limits of multibody systems with bulk and surface long-range interactions. *SIAM J. Math. Anal.*, **52**, 3600–3665.

Bach, A., Cicalese, M., Kreutz, L., and Orlando, G. 2021. The antiferromagnetic XY model on the triangular lattice: chirality transitions at the surface scaling. *Calc. Var. Partial Differ. Equ.*, **60**, 149.

Bellettini, G., and Coscia, A. 1994. Discrete approximation of a free discontinuity problem. *Numer. Funct. Anal. Optim.*, **15**, 201–224.

Bellido, J. C., Mora-Corral, C., and Pedregal, P. 2015. Hyperelasticity as a Γ-limit of peridynamics when the horizon goes to zero. *Calc. Var. Partial Differ. Equ.*, **54**, 1643–1670.

Besicovitch, A. S. 1954. *Almost Periodic Functions*. Vol. 4. Dover, New York.

Bhargava, M., and Shankar, A. 2015. Ternary cubic forms having bounded invariants, and the existence of a positive proportion of elliptic curves having rank 0. *Ann. Math.*, **181**, 587–621.

Blake, A., and Zisserman, A. 1987. *Visual Reconstruction*. MIT Press Series in Artificial Intelligence. Massachusetts Institute of Technology Press, Cambridge, MA.

Blanc, X., and Lewin, M. 2015. The crystallization conjecture: a review. *EMS Surv. Math. Sci.*, **2**, 255–306.

Blanc, X., Le Bris, C., and Lions, P.-L. 2002. From molecular models to continuum mechanics. *Arch. Ration. Mech. Anal.*, **164**, 341–381.

Blanc, X., Le Bris, C., and Legoll, F. 2005. Analysis of a prototypical multiscale method coupling atomistic and continuum mechanics. *M2AN Math. Model. Numer. Anal.*, **39**, 797–826.

Blanc, X., Le Bris, C., and Lions, P.-L. 2006. Du discret au continu pour des modèles de réseaux aléatoires d'atomes. *C. R. Acad. Sci. Paris Ser. I*, **342**, 627–633.

Blanc, X., Le Bris, C., and Lions, P.-L. 2007. Atomistic to continuum limits for computational materials science. *ESAIM Math. Model. Numer. Anal.*, **41**, 391–426.

Blume, M., Emery, V. J., and Griffiths, R. B. 1971. Ising model for the λ transition and phase separation in He3-He4 mixtures. *Phys. Rev. A*, **4**, 1071–1077.

Boivin, D. 1990. First passage percolation: the stationary case. *Probab. Th. Rel. Fields*, **86**, 491–499.

Borgs, C., Chayes, J. T., Lovász, L., Sós, V. T., and Vesztergombi, K. 2006. Counting graph homomorphisms. Pages 315–371 in Klazar, M., Kratochvíl, J., Loebl, M., Matoušek, J., Valtr, P., Thomas, R. (eds.), *Topics in Discrete Mathematics*. Algorithms and Combinatorics, vol. 26. Springer, Berlin.

Borgs, C., Chayes, J. T., Lovász, L., Sós, V. T., and Vesztergombi, K. 2008. Convergent sequences of dense graphs I: Subgraph frequencies, metric properties and testing. *Adv. Math.*, **219**, 1801–1851.

Borgs, C., Chayes, J. T., and Lovász, L. 2010. Moments of two-variable functions and the uniqueness of graph limits. *Geom. Funct. Anal.*, **19**, 1597–1619.

Borgs, C., Chayes, J. T., Lovász, L., and Vesztergombi, K. 2012. Convergent sequences of dense graphs II: Multiway cuts and statistical physics. *Ann. Math.*, **176**, 151–219.

Bouchitté, G., Fonseca, I., Leoni, G., and Mascarenhas, L. 2002. A global method for relaxation in $W^{1,p}$ and in SBV^p. *Arch. Ration. Mech. Anal.*, **165**, 187–242.

Bowden, N. 1997. Self-assembly of mesoscale objects into ordered two-dimensional arrays. *Science*, **276**, 233–235.

Braides, A. 1998. *Approximation of Free-Discontinuity Problems*. Lecture Notes in Mathematics, vol. 1694. Springer, Berlin.

Braides, A. 2002. *Γ-convergence for Beginners*. Oxford Lecture Series in Mathematics and Its Applications, vol. 22. Oxford University Press, Oxford.

Braides, A. 2006. A handbook of Γ-convergence. Pages 101–213 in Chipot, M., and Quittner, P. (eds.), *Handbook of Differential Equations. Stationary Partial Differential Equations*. Vol. 3. Elsevier, Amsterdam.

Braides, A. 2015. An example of non-existence of plane-like minimizers for an almost-periodic Ising system. *J. Stat. Phys.*, **157**, 295–302.

Braides, A., and Caroccia, M. 2022. Asymptotic behavior of the Dirichlet energy on Poisson point clouds. *J. Nonlinear Science*, to appear.

Braides, A., and Chiadò Piat, V. 1995. A derivation formula for convex integral functionals on $BV(\Omega)$. *J. Convex Anal.*, **2**, 69–85.

Braides, A., and Chambolle, A. 2023. Ising systems, measures on the sphere, and zonoids. *Tunisian J. Math*, to appear.

Braides, A., and Cicalese, M. 2017. Interfaces, modulated phases and textures in lattice systems. *Arch. Ration. Mech. Anal.*, **223**, 977–1017.

Braides, A., and Defranceschi, A. 1998. *Homogenization of Multiple Integrals*. Oxford Lecture Series in Mathematics and Its Applications, vol. 12. Oxford University Press, Oxford.

Braides, A., and Francfort, G. A. 2004. Bounds on the effective behaviour of a square conducting lattice. *Proc. R. Soc. Lond. Ser. A Math. Phys. Eng. Sci.*, **460**, 1755–1769.

Braides, A., and Gelli, M. S. 2002. Limits of discrete systems with long-range interactions. *J. Convex Anal.*, **9**, 363–399.

Braides, A., and Kreutz, L. 2018a. Design of lattice surface energies. *Calc. Var. Partial Differ. Equ.*, **57**, 1–43.

Braides, A., and Kreutz, L. 2018b. An integral-representation result for continuum limits of discrete energies with multibody interactions. *SIAM J. Math. Anal.*, **50**, 1485–1520.

Braides, A., and Piatnitski, A. 2008. Overall properties of a discrete membrane with randomly distributed defects. *Arch. Ration. Mech. Anal.*, **189**, 301–323.

Braides, A., and Piatnitski, A. 2012. Variational problems with percolation: dilute spin systems at zero temperature. *J. Stat. Phys.*, **149**, 846–864.

Braides, A., and Piatnitski, A. 2013. Homogenization of surface and length energies for spin systems. *J. Funct. Anal.*, **264**, 1296–1328.

Braides, A., and Piatnitski, A. 2022. Homogenization of ferromagnetic energies on Poisson random sets in the plane. *Arch. Ration. Mech. Anal.*, **243**, 433–458.

Braides, A., and Solci, M. 2011. Interfacial energies on Penrose lattices. *Math. Models Methods Appl. Sci.*, **21**, 1193–1210.

Braides, A., and Solci, M. 2020. Compactness by coarse-graining in long-range lattice systems. *Adv. Nonlin. Studies*, **20**, 783–794.

Braides, A., and Solci, M. 2021 *Geometric Flows on Planar Lattices*. Birkhäuser, Cham.

Braides, A., and Truskinovsky, L. 2008. Asymptotic expansions by Γ-convergence. *Contin. Mech. Thermodyn.*, **20**, 21–62.

Braides, A., and Yip, N. K. 2012. A quantitative description of mesh dependence for the discretization of singularly perturbed non-convex problems. *SIAM J. Numer. Anal.*, **50**, 1883–1898.

Braides, A., Fonseca, I., and Francfort, G. A. 2000. 3D-2D asymptotic analysis for inhomogeneous thin films. *Indiana Univ. Math. J.*, **49**, 1367–1404.

Braides, A., Lew, G. A., and Ortiz, M. 2006. Effective cohesive behavior of layers of interatomic planes. *Arch. Ration. Mech. Anal.*, **180**, 151–182.

Braides, A., Maslennikov, M., and Sigalotti, L. 2008. Homogenization by blow-up. *Appl. Anal.*, **87**, 1341–1356.

Braides, A., Causin, A., and Solci, M. 2012. Interfacial energies on quasicrystals. *IMA J. Appl. Math.*, **77**, 816–836.

Braides, A., Chiadò Piat, V., and Solci, M. 2016a. Discrete double-porosity models for spin systems. *Math. Mech. Complex Syst.*, **4**, 79–102.

Braides, A., Garroni, A., and Palombaro, M. 2016b. Interfacial energies of systems of chiral molecules. *Multiscale Model. Simul.*, **14**, 1037–1062.

Braides, A., Conti, S., and Garroni, A. 2017. Density of polyhedral partitions. *Calc. Var. Partial Differ. Equ.*, **56**, 28.

Braides, A., Causin, A., and Solci, M. 2018a. Asymptotic analysis of a ferromagnetic Ising system with "diffuse" interfacial energy. *Ann. Mat. Pura Appl.*, **197**, 583–604.

Braides, A., Causin, A., Piatnitski, A., and Solci, M. 2018b. Asymptotic behaviour of ground states for mixtures of ferromagnetic and antiferromagnetic interactions in a dilute regime. *J. Stat. Phys.*, **171**, 1096–1111.

Braides, A., Cicalese, M., and Ruf, M. 2018c. Continuum limit and stochastic homogenization of discrete ferromagnetic thin films. *Anal. PDE*, **11**, 499–553.

Braides, A., Cermelli, P., and Dovetta, S. 2020. Γ-limit of the cut functional on dense graph sequences. *ESAIM Control Optim. Calc. Var.*, **26**, 26.

Burago, D., Burago, Y., and Ivanov, S. 2022. *A Course in Metric Geometry*. Vol. 33. American Mathematical Society.

Caffarelli, L. A., and de la Llave, R. 2001. Planelike minimizers in periodic media. *Comm. Pure Appl. Math.*, **54**, 1403–1441.

Caffarelli, L. A., and de la Llave, R. 2005. Interfaces of ground states in Ising models with periodic coefficients. *J. Stat. Phys.*, **118**, 687–719.

Carrillo, J. A., Chipot, M., and Huang, Y. 2014. On global minimizers of repulsive-attractive power-law interaction energies. *Philos. Trans. R. Soc. Lond. Ser. A Math. Phys. Eng. Sci.*, **372**, 20130399.

Cerf, R. 2006. *The Wulff Crystal in Ising and Percolation Models*. Lecture Notes in Mathematics, vol. 1878. Springer, Berlin.

Cerf, R., and Théret, M. 2011. Law of large numbers for the maximal flow through a domain of $\mathbb{R}^d$ in first passage percolation. *Trans. Amer. Math. Soc.*, **363**, 3665–3702.

Chambolle, A. 1995. Image segmentation by variational methods: Mumford and Shah functional and the discrete approximations. *SIAM J. Appl. Math.*, **55**, 827–863.

Chambolle, A., and Kreutz, L. 2023. Crystallinity of the homogenized energy density of periodic lattice systems. *Multiscale Model. Simul.*, **21**, 34–79.

Cicalese, M., and Solombrino, F. 2015. Frustrated ferromagnetic spin chains: a variational approach to chirality transitions. *J. Nonlinear Sci.*, **25**, 291–313.

Cicalese, M., Forster, M., and Orlando, G. 2019. Variational analysis of a two-dimensional frustrated spin system: emergence and rigidity of chirality transitions. *SIAM J. Math. Anal.*, **51**, 4848–4893.

Cicalese, M., Orlando, G., and Ruf, M. 2022. Emergence of concentration effects in the variational analysis of the N-clock model. *Comm. Pure Appl. Math.* **75**, 2279–2342

Dal Maso, G. 1993. *An Introduction to* Γ*-Convergence*. Birkhäuser, Boston.

Dal Maso, G., and Modica, L. 1986. Nonlinear stochastic homogenization and ergodic theory. *J. Reine Angew. Math.*, **368**, 28–42.

Daley, D. J., and Vere-Jones, D. 1988. *An Introduction to the Theory of Point Processes*. Springer, New York.

Daneri, S., and Runa, E. 2019. Exact periodic stripes for minimizers of a local/nonlocal interaction functional in general dimension. *Arch. Ration. Mech. Anal.*, **231**, 519–589.

De Giorgi, E. 1975. Sulla convergenza di alcune successioni di integrali del tipo dell'area. *Rend. Mat.*, **8**, 277–294.

De Giorgi, E. 2006. *Selecta*. Springer, Berlin.

De Giorgi, E., and Franzoni, T. 1975. Su un tipo di convergenza variazionale. *Atti Accad. Naz. Lincei Rend. Cl. Sci. Mat.*, **58**, 842–850.

De Giorgi, E., and Letta, G. 1977. Une notion générale de convergence faible pour des fonctions croissantes d'ensemble. *Ann. Scuola Norm. Sup. Pisa Cl. Sci.*, **4**, 61–99.

De Luca, L., and Friesecke, G. 2017. Crystallization in two dimensions and a discrete Gauss–Bonnet Theorem. *J. Nonlinear Sci.*, **28**, 69–90.

Deep, H. 2013. *Frustrated Spin Systems*. World Scientific.

Dembo, A., and Zeitouni, O. 1998. *Large Deviations Techniques and Applications*. Applications of Mathematics, vol. 38. Springer, New York.

E, W., and Li, D. 2009. On the crystallization of 2D hexagonal lattices. *Comm. Math. Phys.*, **286**, 1099–1140.

E, W., and Ming, P. 2007. Cauchy–Born rule and the stability of crystalline solids: static problems. *Arch. Ration. Mech. Anal.*, **183**, 241–297.

Edwards, S. F., and Anderson, P. W. 1975. Theory of spin glasses. *J. Phys. F: Metal Physics*, **5**, 965–974.

Elsey, M., and Esedoglu, S. 2018. Threshold dynamics for anisotropic surface energies. *Math. Comp.*, **87**, 1721–1756.

Figalli, A., and Zhang, Y. R. 2022. Strong stability for the Wulff inequality with a crystalline norm. *Comm. Pure Appl. Math.*, **75**, 422–446.

Fonseca, I., and Müller, S. 1992. Quasiconvex integrands and lower semicontinuity in L^1. *SIAM J. Math. Anal.*, **23**, 1081–1098.

Fonseca, I., Morini, M., and Slastikov, V. 2007. Surfactants in foam stability: a phase field model. *Arch. Ration. Mech. Anal.*, **183**, 411–456.

Friedrich, M., and Schmidt, B. 2014. An atomistic-to-continuum analysis of crystal cleavage in a two-dimensional model problem. *J. Nonlinear Sci.*, **24**, 145–183.

Friesecke, G., and Theil, F. 2002. Validity and failure of the Cauchy–Born hypothesis in a two-dimensional mass-spring lattice. *J. Nonlinear Sci.*, **12**, 445–478.

Friesecke, G., James, R. D., and Müller, S. 2006. A hierarchy of plate models derived from nonlinear elasticity by Gamma-convergence. *Arch. Ration. Mech. Anal.*, **180**, 183–236.

Frohlich, J., Israel, R., Lieb, E., and Simon, B. 1978. Phase transitions and reflection positivity. I. General theory and long range lattice models. *Comm. Math. Phys.*, **62**, 213–246.

Garavello, M., and Piccoli, B. 2006. *Traffic Flow on Networks*. AIMS Series on Applied Mathematics, vol. 1. American Institute of Mathematical Sciences, Springfield.

García Trillos, N., and Slepčev, D. 2015. On the rate of convergence of empirical measures in ∞-transportation distance. *Canad. J. Math.*, **67**, 1358–1383.

García Trillos, N., and Slepčev, D. 2016. Continuum limit of total variation on point clouds. *Arch. Ration. Mech. Anal.*, **220**, 193–241.

Garet, O., and Marchand, R. 2007. Large deviations for the chemical distance in supercritical Bernoulli percolation. *Ann. Probab.*, **35**, 833–866.

Giaquinta, M., Modica, G., and Souček, J. 1998. *Cartesian Currents in the Calculus of Variations II: Variational Integrals*. Springer, Berlin.

Giuliani, A., Lebowitz, J. L., and Lieb, E. H. 2006. Ising models with long-range dipolar and short range ferromagnetic interactions. *Phys. Rev. B*, **74**, 064420.

Giuliani, A., Lebowitz, J. L., and Lieb, E. H. 2007. Striped phases in two-dimensional dipole systems. *Phys. Rev. B*, **76**, 184426.

Giuliani, A., Lieb, E. H., and Seiringer, R. 2014. Formation of stripes and slabs near the ferromagnetic transition. *Comm. Math. Phys.*, **331**, 333–350.

Gompper, G., and Schick, M. 1994. *Self-Assembling Amphiphilic Systems*. Academic Press, London.

Grimmett, G. 1999. *Percolation*. Springer, Berlin.

Guerra, F., and Toninelli, F. L. 2002. The thermodynamic limit in mean field spin glass models. *Comm. Math. Phys.*, **230**, 71–79.

Heida, M., Kornhuber, R., and Podlesny, J. 2020. Fractal homogenization of multiscale interface problems. *Multiscale Model. Simul.*, **18**, 294–314.

Heitmann, R., and Radin, C. 1980. The ground state for sticky disks. *J. Stat. Phys.*, **22**, 281–287.

Imbert, C., Monneau, R., and Zidani, H. 2013. A Hamilton–Jacobi approach to junction problems and application to traffic flows. *ESAIM Control Optim. Calc. Var.*, **19**, 129–166.

Ishii, H., and Kumagai, T. 2021. Averaging of Hamilton–Jacobi equations along divergence-free vector fields. *Discrete Contin. Dyn. Syst.*, **41**, 1519–1542.

Janson, S. 2013. *Graphons, Cut Norm and Distance, Couplings and Rearrangements*. New York Journal of Mathematics Monographs, vol. 4. https://nyjm.albany.edu/m/.

Kesten, H. 1982. *Percolation Theory for Mathematicians*. Birkhäuser, Boston.

Kumar, P., and Mittal, K. L. 1999. *Handbook of Microemulsion Science and Technology*. Marcel Dekker, New York.

Le Bris, C., and Lions, P.-L. 2005. From atoms to crystals: a mathematical journey. *Bull. Amer. Math. Soc.*, **42**, 291–363.

Le Dret, H., and Raoult, A. 1995. The nonlinear membrane model as variational limit of nonlinear three-dimensional elasticity. *J. Math. Pures Appl.*, **74**, 549–578.

Lebowitz, J. L., and Mazel, A. E. 1998. Improved Peierls argument for high-dimensional Ising models. *J. Stat. Phys.*, **90**, 1051–1059.

Levitan, B. M., and Zhikov, V. V. 1982. *Almost Periodic Functions and Differential Equations*. Cambridge University Press Archive.

Lions, P.-L., and Souganidis, P. E. 2020. Effective transmission conditions for second-order elliptic equations on networks in the limit of thin domains. *C. R. Math. Acad. Sci. Paris*, **358**, 797–809.

Llaradji, M., Guo, H., and Zuckerman, M. J. 1991. Phase diagram of a lattice model for ternary mixtures of water, oil, and surfactants. *J. Phys. A*, **24**, L629.

Lovász, L. 2012. *Large Networks and Graph Limits*. Vol. 60. American Mathematical Society, Providence.

Lovász, L., and Szegedy, B. 2006. Limits of dense graph sequences. *J. Combin. Theory Ser. B*, **96**, 933–957.

Lovász, L., and Szegedy, B. 2007. Szémeredi's lemma for the analyst. *Geom. Funct. Anal.*, **17**, 252–270.

Ludwig, M. 2014. Anisotropic fractional perimeters. *J. Differ. Geom.*, **96**, 77–93.

Macek, R. W., and Silling, S. A. 2007. Peridynamics via finite element analysis. *Finite Elem. Anal. Des.*, **43**, 1169–1178.

Maggi, F. 2012. *Sets of Finite Perimeter and Geometric Variational Problems: an Introduction to Geometric Measure Theory*. Cambridge University Press, Cambridge.

Mezard, M., Parisi, G., and Virasoro, M. A. 1987. *Spin Glass Theory and Beyond*. World Scientific, Singapore.

Morgan, F. 1998. *Riemannian Geometry: a Beginner's Guide*. AK Peters, CRC Press.

Mumford, D., and Shah, J. 1989. Optimal approximations by piecewise smooth functions and associated variational problems. *Comm. Pure Appl. Math.*, **42**, 577–685.

Ohta, T., and Kawasaki, K. 1986. Equilibrium morphology of block copolymer melts. *Macromolecules*, **19**, 2621–2632.

Ortner, C., and Süli, E. 2008. Analysis of a quasicontinuum method in one dimension. *M2AN Math. Model. Numer. Anal.*, **42**, 57–91.

Parisi, G. 1980. The order parameter for spin glasses: a function on the interval 0–1. *J. Phys. A: Math. Gen.*, **13**, 1101–1112.

Peierls, R. 1936. On Ising's model of ferromagnetism. *Proc. Camb. Phil. Soc.*, **32**, 477–481.

Pimentel, L. P. R. 2013. On some fundamental aspects of polyominoes on random Voronoi tilings. *Braz. J. Probab. Stat.*, **27**, 54–69.

Presutti, E. 2009. *Scaling Limits in Statistical Mechanics and Microstructures in Continuum Mechanics*. Springer, Berlin.

Radin, C. 1981. The ground state for soft disks. *J. Stat. Phys.*, **26**, 365–373.

Raitums, U. 2001. On the local representation of G-closure. *Arch. Ration. Mech. Anal.*, **158**, 213–234.

Rosakis, P. 2014. Continuum surface energy from a lattice model. *Netw. Heterog. Media*, **9**, 453–476.

Schmidt, B. 2008. On the passage from atomic to continuum theory for thin films. *Arch. Ration. Mech. Anal.*, **190**, 1–55.

Schneider, R. 1988. *Convex Bodies: the Brunn–Minkowski Theory*. Cambridge University Press, Cambridge.

Scilla, G. 2013. Variational problems with percolation: rigid spin systems. *Adv. Math. Sci. Appl.*, **23**, 187–207.

Siconolfi, A., and Sorrentino, A. 2021. *Aubry–Mather theory on graphs*. arXiv:2203.16877.

Silling, S. A., and Lehoucq, R. B. 2010. Peridynamic theory of solid mechanics. Pages 73–168 in Aref, H., and van der Giessen, E. (eds.), *Advances in Applied Mechanics*, Advances in Applied Mechanics, vol. 44. Elsevier, Amsterdam.

Tadmor, E. B., Ortiz, M., and Phillips, R. 1996. Quasicontinuum analysis of defects in solids. *Philos. Mag. A*, **73**, 1529–1563.

Tartar, L. 2000. An introduction to the homogenization method in optimal design. Pages 47–156 of *Optimal Shape Design*. Springer, Berlin.

Theil, F. 2006. A proof of crystallization in two dimensions. *Comm. Math. Phys.*, **262**, 209–236.

Theil, F. 2011. Surface energies in a two-dimensional mass-spring model for crystals. *ESAIM Math. Model. Numer. Anal.*, **45**, 873–899.

Van Gennip, Y., and Bertozzi, A. L. 2012. Γ-convergence of graph Ginzburg–Landau functionals. *Adv. Differ. Equ.*, **17**, 1115–1180.

Whitesides, G. M. 2002. Self-assembly at all scales. *Science*, **295**, 2418–2421.

Wouts, M. 2009. Surface tension in the dilute Ising model: the Wulff construction. *Comm. Math. Phys.*, **289**, 157–204.

Index

Printed in the United States
by Baker & Taylor Publisher Services